Automotive Manufacturing Processes

Automotive Manufacturing Processes discusses basic principles and operational procedures of automotive manufacturing processes, issues in the automotive industry like material selection, and troubleshooting. Every chapter includes specific learning objectives, multiple-choice questions to test conceptual understanding of the subject and put theory into practice, review questions, solved problems, and unsolved exercises. It covers important topics including material decision-making processes, surface hardening processes, heat treatment processes, effects of friction and velocity distribution, the metallurgical spectrum of forging, and surface finishing processes.

Features:

- Discusses automotive manufacturing processes in a comprehensive manner with the help of applications.
- Provides case studies addressing issues in the automotive industry and manufacturing operations in the production of vehicles.
- Discussion on material properties while laying emphasis on the materials and processing parameters.
- Covers applications and case studies of the automotive industry.

The text will be useful for senior undergraduates, graduate students and academic researchers in areas including automobile engineering, industrial and manufacturing engineering and mechanical engineering.

Automotive Manufacturing Processes

A Case Study Approach

G. K. Awari, V. S. Kumbhar, R. B. Tirpude and
S. W. Rajurkar

CRC Press
Taylor & Francis Group
Boca Raton London New York

CRC Press is an imprint of the
Taylor & Francis Group, an **informa** business

First edition published 2023
by CRC Press
6000 Broken Sound Parkway NW, Suite 300, Boca Raton, FL 33487-2742

and by CRC Press
4 Park Square, Milton Park, Abingdon, Oxon, OX14 4RN

CRC Press is an imprint of Taylor & Francis Group, LLC

ISBN: 978-1-032-31238-5 (hbk)
ISBN: 978-1-032-43439-1 (pbk)
ISBN: 978-1-003-36732-1 (ebk)

DOI: 10.1201/9781003367321

Typeset in Times
by Newgen Publishing UK

Dedication

The book is dedicated to our parents
Late Smt Shashikala Awari
For her divine guidance and blessing.

Contents

Preface

The automotive manufacturing industry across the globe is one of the largest markets in the world. The term automotive or automobile manufacturing processes is used for automotive applications which can be transformed into the vehicle by using different types of materials and innovative processes. Modern automobiles are produced by using complex machinery components and therefore, it requires careful attention for enabling it to perform the task in a safe, economical and efficient way. Thus, for any mechanical, automobile, industrial, production or design engineer, it is essentially required to have a thorough knowledge of the various components, its manufacturing processes and its functions.

Several subjects on this area have been already been offered across the world and it was introduced in the model curriculum of AICTE UG/Diploma level from 2018 with different dimensions. Also, in the past many new technologies were introduced to the manufacturing sector. In view of this, useful resources for readers in the subject concerned are limited in number and therefore there is an imperative need for this book "*Automotive Manufacturing Processes: A Case Study Approach*". The book is organised into 12 chapters, and all the chapters of the book are synchronised with modern trends: (1) Automotive Materials; (2) Nonferrous Materials; (3) Heat Treatment; (4) Moulding and Casting; (5) Forging Process; (6) Welding Process; (7) Material Removal Processes; (8) Plastic Processing in the Automotive Industry; (9) Powder Metallurgy; (10) Surface Treatment; (11) Press Shop Process; and (12) Case Studies of Automotive Manufacturing Units.

An exclusive chapter on the case studies and applications of automotive manufacturing has been added to address the various issues and challenges of modern manufacturing industry. Many of the new diagrams (2D and 3D) and exploded views prepared with CAD software have been included to make the matter more understandable to focus on all processes found in contemporary automobiles and the emphasis is on those things that students need to know about in regard to the vehicles and the manufacturing processes of yesterday, today and tomorrow.

This book is presented to the student and teacher community, and industry professionals, containing comprehensive treatment of the subject matter in simple, lucid and enveloping a large number of systems that are properly graded, including typical examples, from an examination point of view. The book mainly is aimed at post-graduate, graduate and diploma engineering courses of most of the universities/board across the world. This book serves as an ideal resource for students of Mechanical Engineering, Automobile Engineering, Mechatronics Engineering, Production Engineering, Metallurgical Engineering, Industrial Engineering, Design Engineering and Technicians. It will be also a useful resource for amateur readers of the automotive manufacturing industry.

Evolving questions in the form of objective types and review questions are provided in this book and we hope all reader will immensely benefits from reading this textbook. This book is also useful to prepare the students for competitive examinations such as GATE, IES, Indian Engineering Services, MPSC, UPSC and other public sector undertakings. The authors' classroom and hands-on experiences fused with their research experiences in the automobile manufacturing industry are incorporated for greater insight into the automotive manufacturing processes.

The objective is to present a review of the technologically updated automotive manufacturing processes including advanced automotive materials currently in use as well as to identify the material and manufacturing challenges faced by the automotive industry.

SALIENT FEATURES OF THE BOOK

The main characteristics of this book are:

(1) The book is written in luculent language which can be understood by students of diploma in engineering courses (junior level) and technicians in the field. It progresses from beginner to more advanced material at an easy-to-follow pace, utilising examples throughout to aid readers interested in automobile manufacturing processes.
(2) It presents a selection of various manufacturing techniques and materials suitable for near-term application, with sufficient technical background to understand their domain of applicability and to consider variations to suit technical and organisational constraints.
(3) The modern trends in automotive manufacturing processes and innovative materials developed in the last decade are incorporated in the book.
(4) It promotes a vision of complete manufacturing processes and applications as integral to modern manufacturing engineering practices, that are equally as important and technically demanding as other aspects of development. This vision is consistent with current thinking on the subject.
(5) Several applications and case studies from the automotive industry have been provided for greater insight into the processes. The book includes 2D/3D drawings prepared with CAD software, self-explanatory figures and photographs to illustrate the concepts and techniques which will be inspiring for budding manufacturing entrepreneurs.
(6) Each chapter has learning objectives, and multiple choice questions at the end of each chapter to test the conceptual understanding of the subject and put theory into practice, with review questions included.

Acknowledgements

A mission's success is never the result of a single person's efforts; it is the result of the active participation of many people, either directly or indirectly involved in the project. We would like to express our gratitude and acknowledge the guiding lights which have imbued us with the right elements and assisted us in completing this mission. The authors' classroom and laboratory experiences have culminated in the book *Automotive Manufacturing Processes: A Case Study Approach*.

We would like to express our gratitude to Hon. Dr. Abhay Wagh, Director of Technical Education (DTE), Mumbai, Hon. Dr. Vinod Mohitkar, Director of MSBTE, Mumbai, Dr. M. B. Daigavane, Joint Director, DTE, RO, Nagpur and Principal, Government Polytechnic, Nagpur, Dr. D. R. Nandanwar, Ex. Joint Director, DTE, RO, Pune and Principal, Government Polytechnic, Awasari khurd, Dr. M. R. Chitlange, Secretary, Maharashtra State Board of Technical Education (MSBTE), Mumbai, Mr. Lilhare and Mrs. Mankar, Deputy Secretary, RBTE, Nagpur and Dr. R. V. Yenkar, Vice-Principal, Government Polytechnic, Nagpur, for their constant inspiration and motivation to develop the manuscript as per the aspirations of degree/diploma in engineering students.

We are grateful to Dr. Mohan Gaikwad-Patil, Chairman, Gaikwad-Patil Group of Institutions in Nagpur, Prof. Sandeep Gaikwad, Treasurer, Gaikwad-Patil Group of Institutions for enabling the use of laboratories facilities of the institute.

We are also grateful to Prof. Vidyadhar Kshirsagar, Prof. Yogesh Ramteke and Prof. Niteen Kakde from Tulsiramji Gaikwad-Patil College of Engineering Technology, Nagpur, for assisting us in compiling the case studies and preparation of CAD drawings using 3D CAD software.

We learned a lot while writing this book by consulting several articles, journals, online sources and open source material. We thank all of these authors, publications and publishers, many of which are listed in the bibliography. If someone has been unintentionally left out, we seek their forgiveness.

The authors also would like to express their gratitude to Prof. Dr. Jaji Varghese, Delhi Skill and Entrepreneurship University, Government of Delhi, Prof. Dr. S. Velumani, Velarar College of Engineering, Erode (TN), Prof. Dr. Abhijeet Digalwar, BITS Pilani (Rajasthan), Prof. Dr. Yousuf Ali, Lords Institute of Engineering, Hyderabad, Prof. Dr. D. K. Parbat, Government Polytechnic, Bramhapuri (MS), Prof. Dr. D. N. Kongre, Government Polytechnic, Nagpur, Prof. Dr. Vijay Mankar, Government Polytechnic, Nagpur, Prof. Dr. Shekhar Gajjal, Prof. K. M. Pawar, Prof. P. G. Gavade, Government Polytechnic, Awasari Khurd.

The authors are immensely impressed by the untiring help rendered by their friends, research scholars and colleagues, particularly Prof. Vishwajeet Ambade, from TGPCET, Nagpur, and from Government Polytechnic, Nagpur, such as Prof. Kisan Badole, Prof. Sandesh Goswami, Prof. Darshan Bapat and others in accomplishing the mammoth task of editing and the secretarial assistance provided in preparing the manuscript. The authors put on record their sincere thanks to them.

We sincerely thank our mentors, Prof. Dr. D. G. Wakde, Prof. Dr. L. B. Bhuyar, Prof. Dr. G. V. Gotmare, Government Polytechnic, Gondia, Dr. K. S. Dixit, Dr. H. N. Warhatkar and Dr. Achal Shahare for their assistance and inspiration. We are grateful to all of the automotive manufacturers for their support in developing the manuscript. We thank CRC Press, Taylor and Francis Group, particularly Mr. Gauravjeet Singh Reen, Senior Commissioning Editor-Engineering and Ms. Isha Ahuja for keeping our morale high, assisting us in preparing and maintaining our schedules, facilitating the work, providing regular updates, and standing by us patiently throughout this entire process.

Also, we are grateful to our family members for their prompt assistance in all of this book's efforts, without which this book would not have seen the light of day. The efforts of Mrs. Jaya Awari, Master Vedant Awari, Mr. Keshavrav Awari, Mr. Shankar Kumbhar, Mrs. Ranjana Kumbhar, Mrs. Shubhangi Kumbhar, baby Vedashree Kumbhar, MAA Shila, mother Laxmibai, Mrs. Dr. Pranali Tirpude, Master Pushak Tirpude, Ms. Anushka Tirpude, Mrs Nilima Rajurkar and Ku Sonal Rajurkar in developing the book's manuscript are acknowledged. Last but not least, we bow our heads to the Almighty God's majesty and our parents for making our experience one of the most technologically satisfying of our lives.

We hope that this book will serve its readers' purposes and we would like to continue to receive their assistance and suggestions. Suggestions to improve the book's quality and style are always welcome, accepted and will be incorporated into future editions.

About the Authors

G. K. Awari has completed his Ph.D. in Mechanical Engineering and he has more than 30 years of teaching experience at UG/PG/Diploma and Research level and 10 years as an approved Principal of **A+** Grade NAAC Accredited Engineering Institute. He has been nominated by AICTE as "Margdarshak for NBA Accreditation" of mentee institutes and also works as "Chief Coordinator" of AICTE mentor institutes for achieving NBA Accreditation to MBI.

He has more than 30 Scopus indexed journal papers, 10 patents, 10 copyrights to his name. Recently he has been awarded with a patent titled as "HYBRID TOOL CONCEPT FOR BORING, REAMING & CHAMFERING IN A SINGLE TOOL" and a commercial product has been developed for Mahindra and Mahindra, Nagpur. He has been also granted an Australian Patent on "3D PRINTING of COST EFFECTIVE HUMAN SKULL MODELS AND SKULL INPLANTS". He has executed more than eight funded projects of AICTE as a Principal Investigator and organised 10 AICTE-ISTE-SRM University-funded National Workshops and 10 International Conferences as a convener. Total grants received for his projects are Rs. 1.14 Crore.

He has contributed to the development academics curricula as a Board of Study (BOS) Member at Goa University, Sant Gadgeba Amravati University, YCCE, Nagpur (An Autonomous Institute) and RTM Nagpur University. He is presently BOS member in GH Raisoni University, Saikheda, MP, Delhi Skill and Entrepreneurship University (DSEU), New Delhi and Chairman BOS, Automobile Engineering at Government Polytechnic (GP), Nagpur.

He has authored a total of 10 books in the engineering and technology domain and the following four books have been authored by him for CRC Press, Taylor and Francis Group.

1. *Quantitative Techniques in Business, Management and Finance: A Case Study Approach.*
2. *Automotive Systems: Principles and Practice.*
3. *Additive Manufacturing and 3D Printing Technology: Principles and Practice.*
4. *Ethics in Information Technology.*

Dr. Awari is recognised as a Research Supervisor at four Indian universities. A total of 18 research scholars have completed their PhD in Mechanical Engineering under his supervision from RTM Nagpur University and SG Amravati University. Currently six research scholars are pursuing doctoral research work under his supervision. He is also a reviewer of various renowned international journals such as *ASME*, *ASCE* and *Thermal Science*.

Dr. Awari belives in the co-learning process and participative management and currently he is Head of Automobile Engineering Department at Government Polytechnic, Nagpur.

V. S. Kumbhar received Bachelor of Engineering and Master of Engineering degree in Automobile engineering from Shivaji University Kolhapur, Maharashtra, India, in 2007 and 2013, respectively. He has a total of 14 years of experience (four years of industrial experience in automobile manufacturing engineering and 10 years of teaching experience at undergraduate level). He has taught various subjects such as Automobile Engines, Automobile Systems, Automobile Design, Machine Design, Strength of Materials, Advanced Engines and Automobile Engineering. His areas of interest are fuels and combustion, and emission control systems.

He has more than 11 international journal publications to his credit and authored a book on "*Automotive Systems*" for CRC Press, Taylor and Francis Group. He has contributed to the development of the Academics as Subject Expert for Maharashtra State Board of Technical Education, MSBTE, Mumbai, and at Government Polytechnic, Nagpur. He had contributed in academics as a BOS member of Automobile Engineering at Government Polytechnic (GP), Nagpur. Prof. Kumbhar is pursuing a Ph.D. in Mechanical Engineering at GH Raisoni University, Amravati, and currently he is working at the Automobile Engineering Department of Government Polytechnic, Awsari Khurd.

R. B. Tirpude received Bachelor of Engineering degree in Mechanical Engineering from RTM Nagpur University, Nagpur, Maharashtra, India, in 1998, M.E. degree in CAD/CAM from Sant Gadgebaba Amravati University, Amravati, Maharashtra, India, in 2008 and Ph.D. in Mechanical Engineering in 2022 from RTM Nagpur University, Nagpur. He has earnt his MBA (operations) from Delhi University, Delhi, India, in 2012.

He has a total of 21 years of teaching experience at undergraduate and research level. He has taught various subjects such as Computer Aided Drafting, Hydraulics and Pneumatics, Operation Research, Fluid Power, Machine Design and Automobile Engineering. His area of interest is emission control systems. He has total 19 international journal publications, 10 international conference publications, 11 national conference publications and 10 patents to his name. He has also authored a book on "**Automotive Systems**" for CRC Press, Taylor and Francis Group. He has been honoured with "ISTE Narsee Munji Best Project Award" and "Shastri Gaurav Purskar-2022". He is also a recipient of "Best Teacher Award" from Government Polytechnic Nagpur as well as Dr. Babasaheb Ambedkar National Association of Engineers, Nagpur chapter. He has contributed in the development Academics as Board of Study (BOS) member at MSBTE and various autonomous institutes of the Govt. of Maharashtra. He is presently a BOS member of Automobile Engineering at Government Polytechnic

(GP), Nagpur. He is currently working in the Automobile Engineering Department at Government Polytechnic, Nagpur, handling various collaborations of the department with leading automobile manufacturing industries.

S. W. Rajurkar received Bachelor of Engineering degree from Visvesvaraya Regional College of Engineering, Nagpur (VRCE) now (VNIT), Nagpur University, Nagpur, Maharashtra, India, in 1989 and M.E. degree from Thapar University, Patiala, Punjab, India, in 2001, all in Mechanical Engineering. He gained his Ph.D. from Malaviya National Institute of Technology (MNIT), Jaipur, Rajasthan, India, in 2010. He has more than 30 years of teaching experience at diploma, undergraduate, postgraduate and research levels. He has taught various subjects including Metrology & Quality Control, Manufacturing Process, Industrial Management, Refrigeration & Air Conditioning, Operation Research, Machine Design, Engineering Drawing and Automobile Engineering. His area of interest is Supply Chain Management & Industrial Engineering.

He has 50 citations including a total of seven international journal publications, 16 international conference publications, four national conference publications and four patents. Three research scholars have commenced their Ph.D. in Mechanical Engineering under his supervision in two Indian universities and one research scholar has completed a Ph.D. in Mechanical Engineering under his supervision. He is also a recipient of "ISTE Best Polytechnic Teacher Award for Maharashtra & Goa Section for the Year 2016".

He has contributed in the development of Academics as Board of Governance (BOG) Member at Government College of Engineering, Chandrapur, and also worked as a Board of Study (BOS) Member at Government Polytechnic, Nagpur. He is also recognised by AICTE as "Margadarshak" for NBA Accreditation of mentee institutes.

He has authored a book on "*Supply Chain Modelling for Perishable Food Products*" by Lambert Academic Publishing. He has also developed videos/e-content modules for the benefit of students/teachers. He has worked as a resource person for various Faculty Development Programs sponsored by DTE, Mumbai, MSBTE, Mumbai on "NBA", "Implementation of Outcome Based Education" held at GCOE, Chandrapur, Government Polytechnic, Nagpur, Sakoli and Gondia and at many other institutions. He has also delivered an invited expert lecture on "Logistics & Supply Chain Management" at Management Training Program for MCVC/ITI Teachers at G.P. Nagpur. He is presently working at the Government College of Engineering, Chandrapur, as Associate Professor and Head of Department, Mechanical Engineering Department and Dean (Administration).

Abbreviations

AC	Alternating current
AJM	Abrasive jet machining
Al_2O_3-H_2O	Hydrated aluminium oxide
ASTM	American Society for Testing of Materials
AW	Arc welding
AWS	American Welding Society
BCC	Body-centred cubic
CBN	Cubic boron nitride
CFRP	Carbon fibre-reinforced plastic
CNC	Computer numerical control
CuFeS	Chalcopyrite
CVD	Chemical vapour deposition
DC	Direct current
FCC	Face-centred cubic
Fe	Ferrous
Fe_3C	Cementite
GMAW	Gas metal-arc welding
GN	Gas nitriding
GNC	Gas nitro-carburizing
HCl	Hydrochloric acid
HDPE	High-density polyethylene
HIP	Hot isostatic pressing
HPC	High performance ceramics
HSS	High speed steel
LBW	Laser beam welding
LDPE	Low-density polyethylene
LTE	Local thermodynamic equilibrium
$Mg(OH)_2$	Magnesium hydroxide
MIG	Metal inert-gas
MRR	Material removal rate
MSE	Material science and engineering.
MT	Metric tons
NIMS	National Institute of Materials Science
OFW	Oxyfuel welding
PACVD	Plasma-assisted chemical vapor deposition
PAW	Plasma arc welding
PCD	Polycrystalline diamond
PEW	Percussion welding
PN	Plasma nitriding
PNC	Plasma nitro-carburizing
PVC	Polyvinyl chloride
PVD	Physical vapour deposition

REP	Rotating electrode process
RPW	Resistance projection welding
RSEW	Resistance seam welding
RSW	Resistance spot welding
RW	Resistance welding
SMAW	Shielded metal arc welding
TCHT	Thermochemical heat treatment
$TiCl_4$	Titanium tetrachloride
TIG	Tungsten inert gas
TTT	Time–temperature transformation
UHMWPE	Ultrahigh-molecular-weight polyethylene
UW	Upset welding
ZTA	Zirconia toughened alumina

PREFIXES, SUFFIXES AND SYMBOLS

R	Chip thickness ratio
K	Chip reduction ratio
Lc	Length of chip
V	Cutting speed
Vc	Chip velocity
α	Tool rake angle
t	Depth of cut
ϕ	Shear angle
V_c	Velocity of chip flow
Fs	Shear force or metal resistance to shear during chip formation
Fc	Backing up or compressive normal force
N	Force exerted by tool on the chip
β	Angle of friction
Fa	Axial feed
FV	Vertical component of the resultant force
Fs	Shear force on shear plane
A	Cross-sectional area of uncut chip
t	Depth of cut or uncut chip thickness
b	Width of cut
r	Mean shear stress
σ	Mean normal stress
Fr	Radial force acting in horizontal plane
Ft	Cutting force or tangential force
H	Heat Produced
D	Diameter of workpiece
f	Feed
V	Volume of metal
I	Flowing current in ampere
t	Time for which current passes
A_i	Initial cross-sectional area of the workpiece

$\mathbf{A_f}$	Final cross-sectional area of the workpiece
σ_{Y_0}	Yield stress
M	Bending moment
L	Neutral plane
R	Radius of curvature
I	Second moment of the beam cross-section
σ_x	Pulling stress
h	Throat of the butt weld
P	Tensile force on the plates
l	Length of the weld
σ_t	Permissible tensile stress

1 Automotive Materials

LEARNING OBJECTIVES

- To understand the ideal material requirements for various automotive components based on applications.
- To comprehend the classification of materials used in vehicular applications in detail.
- To get details of various types of ferrous materials, their properties, alloys and their vehicular applications.
- To know the process of material selection for automotive applications based on their properties.

1.1 INTRODUCTION

The selection and utilisation of suitable materials is the most important aspect of the automotive design, manufacturing, and production, because it is closely associated with the in-service behaviour, life, economy and effectiveness, legal requirements performance, environmental aspects, and involves material science and engineering. Materials science and engineering (MSE) is an interdisciplinary field which involves the invention of new materials and improvements to existing material already in use in all respects such as the enhancement of properties, range of applications, etc. through microstructure–composition–synthesis–processing and their relationships.

The composition of a material is associated with its detailed chemistry and their quantities, while the structure involves the layout and arrangement of the various atoms involved in it. Automotive research not only is associated with the development of new materials, synthesis, processing of materials, and manufacturing aspects related to the production of components. Synthesis is the way by which materials are made from naturally occurring or manmade chemicals, while processing involves the conversion of these materials into useful size and shapes as per requirements.

1.2 REQUIREMENTS OF IDEAL AUTOMOTIVE MATERIALS

The necessity for the use of advanced materials in automotive applications is based on the following different vehicular development trends:

DOI: 10.1201/9781003367321-1

- Increasing and forceful customer demand for clean, safe, affordable, energy-efficient, aesthetically pleasant, high comfort levels, with better handling and ride characteristics, with the best possible ergonomics and serviceability using intelligent vehicles.
- Continuous improvements to fuel economy and emission control strategies, as well as weight reduction, are important aspects in vehicle design.
- Functional requirements of the components, legislation requirements as per motor vehicle rules, and environmental concerns, and also economics and market aspects, in-service behaviour, and reliability of the components are some of the driving forces for the selection and use of the most suitable material.

The common requirements of the automotive material is as follows:

1. Mechanical, chemical, thermal, electrical, and physical properties
2. Strength to weight ratio
3. Strength to volume ratio
4. Availability
5. Aesthetics
6. Manufacturability
7. Cost
8. Surface finish
9. Biological effects
10. Storage
11. Life
12. Continuous supply
13. Pollution and recycling
14. Reliability

1.3 CLASSIFICATION OF MATERIALS USED IN AUTOMOBILES

There are different ways of classifying materials as explained below.

1.3.1 Metals and Alloys

Steels, aluminium, magnesium, zinc, cast iron, titanium, alloys, copper, and nickel come under this category. These metals have good electrical and thermal conductivity as well as relatively high strength, high stiffness, ductility, or formability, and shock resistance. There are rare applications of metals in their pure form; to enhance the properties of the materials used for the various components as per the requirements alloys can be generated and used. Common applications of this material are in taking the load or working as a structural member. As far as automotive applications are concerned, various chassis and body components such as the body structure, body panels, engine, and transmission elements are made from metals and alloys.

1.3.2 Ceramics

Ceramics are inorganic crystalline materials. Beach sand and rocks are examples of naturally occurring ceramics. Ceramics have better wear, heat, and chemical resistance in corrosive environments as compared to metals and polymers. There are special applications for advanced ceramics which can be prepared with the help of advanced and special processes by using naturally occurring ceramics. Advanced ceramics can be used in the biomedical, electrical electronics, communication, computer, and advanced automotive areas.

1.3.3 Glasses and Glass-Ceramic

Glass is a non-crystalline amorphous solid that is often transparent and has widespread practical, technological, and decorative uses and can be derived from molten silica. Glasses are also used in houses, cars, computers, mirrors, and television screens in the various areas of engineering and technology. Glasses can be thermally treated (tempered) to make them stronger. Forming glasses and nucleating (creating) small crystals within them by a special thermal process creates materials that are known as glass ceramics. Zerodur is a glass-ceramic material which is used for special applications such as mirror substrates for large telescopes (e.g., the Chandra and Hubble telescopes). The advantages of glass ceramics include that they can be produced by any glass-forming technique, flexibility in the design of micro- and nanostructure as per required applications, low porosity, and the possibility of the combination of a variety of desired properties. Glasses and glass-ceramics can be manufactured by melting and casting processes.

1.3.4 Polymers (Plastics)

Polymers can be prepared by using the process of polymerisation. Polymers are the macromolecules formed by the combinations of a large number of micromolecules called monomers. Rubbers as well as various types of adhesives come under the category of polymers. Most polymers have good electrical resistance. Good thermal insulation and a high strength-to-weight ratio are important features of polymers. With this toughness, resilience, resistance to corrosion, lack of conductivity (heat and electrical), colour, transparency, processing, and low cost are important properties associated with polymers as an effect of their long-chain molecular structure. As far as the automotive field is concerned, polymers have a variety of applications in various sectors such as the vehicle interior and electrical and electronic parts, tires, various hose pipes, and insulations.

1.3.5 Semiconductors

Silicon, germanium, and gallium arsenide-based semiconductors are some of the electronic materials widely used in various applications. The electrical conductivity of semiconductor materials is between that of ceramic insulators and metal conductors. Semiconductors possess specific electrical properties. A conductor is a material which carries electricity, and an insulator is a material which does not carry electricity. In semiconductors, the conductivity level is regulated so that it can be used in electronic devices such as transistors, diodes, and integrated circuits as per the required applications; for many applications there is a requirement for large single semiconductor crystals. These can be manufactured with the help of molten materials and films of semiconductors with advanced processes can be utilised. Semiconductors are a part of advanced electronic systems used in vehicles.

1.3.6 Composite Materials

Composites involve the combinations of various properties of different materials. Composites are made of at least two materials, with different properties which are not found in a single material, for example concrete, plywood, and fiberglass. Combinations of various thermal, mechanical, and electrical properties can be achieved with composite materials. By dispersing glass fibres in a polymer matrix, fiberglass can be manufactured. There are many applications of composite materials in the aircraft, aerospace, and automotive fields due to their light weight, shock resistance, reduced weight, etc.

1.3.7 Others

The components of a vehicle such as the seat covers and various sections can be made with leather, different types of fluids and wood, fabric, etc. as per aesthetic requirements and comfort-related factors.

Materials used in automotives can be classified according to their purpose, such as:

1) Materials used in automotive body manufacturing as per strength requirements.
2) Materials used for lubricants, coolants, and as a fuel.
3) Materials used for the electrical and electronic accessories.
4) Materials used for the engine, fuel line, and chassis components.
5) Materials used for the dashboard panels and associated parts.
6) Materials used for the interior and cushion seats, etc.
7) Materials used for the wheels and tires.

The most widely used material in the current vehicular industry is plain carbon steel, and there are now a great efforts being taken by the research community to replace it with advanced high-strength metals, light non-ferrous alloys, such as aluminium, magnesium, and titanium alloys, and a range of composites, including carbon fibre composites, metal matrix composites, and nanocomposites. Titanium alloy also has automotive applications. The material distribution of a typical vehicle is as described in Table 1.1.

In each vehicle close to 300–400 parts are used, and these parts have to work in the different operating conditions (Figure 1.1).

TABLE 1.1
Material Distributions in a Typical Vehicle

Materials	Percentage of Vehicle Weight	Major Areas of Application
Steel	55	Body structure, body panels, engine and transmission components, suspension components, driveline components
Cast iron	9	Engine components, brakes, suspension components
Aluminium	8.5	Engine block, wheel, radiator
Copper	1.5	Wiring, electrical components
Polymers and polymer matrix composites	9	Interior components, electrical and electronic components, under-the-hood components, fuel-line components
Elastomers	4	Tires, trims, gaskets
Glass	3	Glazing
Other	4	Carpets, fluids, lubricants, etc.

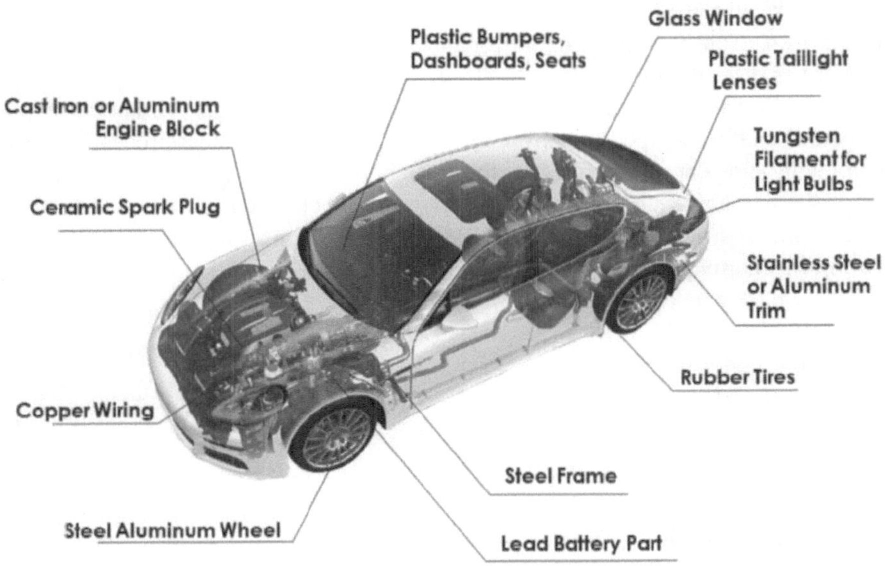

FIGURE 1.1 Various vehicle components and their materials.

1.4 PROPERTIES OF AUTOMOTIVE MATERIALS

Both the product designer and the manufacturer need to have a detailed understanding of the properties and qualities of materials so that they can be choose and use more efficiently. Each material has certain properties which make it more suitable for some applications than others. Vehicle construction materials, in general, must be able to withstand the action of forces without experiencing substantial distortion and must incorporate a high degree of operational protection. This is especially important when it comes to vehicle systems and other modes of transport. Mechanical properties are correlated with the behaviour of the material and the application of force. These are the features of which vehicle builders are initially concerned when evaluating a particular material, such as the chassis, bodywork, and suspension systems. Sometimes, additional features are also important for the components, such as electrical and electronic control systems, which may also require magnetic components. Optical, thermal, and chemical properties are critical for components such as windscreens, heat exchangers, and corrosion control systems.

1.5 FERROUS MATERIALS AND THEIR ALLOYS: CAST IRON AND STEELS—TYPES, PROPERTIES, AND APPLICATIONS

Ferrous materials/metals suggest that the main constituent is iron, such as pig iron, wrought iron, cast iron, steel, and their alloys. The ease of fabrication, greater strength, wider availability, and low cost of ferrous materials enable its use in various applications such as the manufacturing and building construction sectors. Pig iron is the main raw material used for ferrous metals which can be obtained by smelting iron ore, coke, and limestone in a blast furnace. Ferrous means iron. Iron is another name for pure ferrite, Fe. The various types of iron ores are as described in Table 1.2.

Ferrous materials are usually stronger and harder, and their characteristics and properties can be changed as per requirements with heat treatment processes as well as alloying elements. According to the carbon content, the structural and physical properties and nomenclature of iron materials can be changed.

The principal ferrous metals and alloys used in engineering are classified under the following groups:

A. Pig iron
B. Wrought iron
C. Cast iron
D. Steel

TABLE 1.2
Various Types of Iron Ore

Sr. No.	Iron Ore	Colour	Iron (%)
1	Hematites Fe_3O_4	Red	70
2	Magnetite Fe_2O_3	Black	72
3	Limonite	Brown	62.5
4	Siderite	Brown	48

a) Carbon steel
   (i) Low-carbon steels (0.008–0.30%C)
   (ii) Medium-carbon steels (0.30–0.60%C)
   (iii) High-carbon steels (0.60–2.00%C)
b) Alloy steel

A. Pig iron

Pig iron is an element that, through chemical reduction of iron ore in a blast furnace, all iron and steel products are derived. Pig iron can be obtained from the iron core, very high-quality hard coke, and the most commonly used flux material is limestone. The iron cores are in the form of hydrates or oxides of the metal, carbonates. The flux is mixed with ashes of the fuel and fusible products with the help of ore which separates from the metal as slag.

B. Wrought iron

The metal with sufficient ductility for hot and cold deformation processes is basically known as wrought iron. Wrought iron can be obtained by remelting pig iron in a puddling furnace. Wrought iron is the purest form of pig iron with the lowest amount of slag forged into fibres. The chemical analysis of this metal shows as much as 99% iron and small traces of carbon, phosphorus, manganese, silicon, sulphur, and slag. The fundamental properties of wrought iron include toughness, malleability, and ductility. The ultimate tensile strength is about 350 N/mm^2 and the melting point is 1530^0C.It has better corrosion resistance than mild steel but cannot be tempered or hardened. Its applications include form bars, plates, boiler tubes, bolts and nuts, chains, etc.

C. Cast iron

In cast iron, the carbon content varies between 2 to 6.67%; basically cast irons are alloys of iron and carbon. Commercial cast irons have dedicated compositions as per their application requirements and the range of carbon content varies from 2.3 to 3.75% and it also consists of elements such as silicon, phosphorus, sulphur, and manganese in substantial amounts. As an effect of the higher carbon content it possesses poor ductility and malleability, and cannot be forged, rolled, drawn, or pressed into the desired shape, but instead is formed by melting and casting to the required final shape and size, thus resulting in the name "cast irons." Cast irons have the following characteristics:

1. They are more cost efficient as compared to other commercial alloys.
2. They can be easily melted down due to their lower melting temperature ($1150–1250^0$C) as compared to steels ($1350–1500^0$C).
3. As an effect of the high fluidity of melting and low shrinkage during the solidification process, they can be easily cast.
4. Their corrosion resistance is fairly good.
5. In general, they are brittle, with lower tensile strength, and so cannot be used to make components that are subjected to shock and their mechanical properties are inferior to those of steels.

1.5.1 Cast Iron Classifications

Cast irons are classified according to the following criteria:

(a) On the basis of the furnace used in their manufacture:
  (1) Cupola cast iron
  (2) Electric furnace cast iron
  (3) Air furnace cast iron
  (4) Duplex cast iron
(b) On the basis of the composition and purity:
  (1) Low-carbon, low-silicon cast iron
  (2) High-carbon, low-sulphur cast iron
  (3) Nickel alloy cast iron
(c) On the basis of the microstructure and appearance of fracture:
  (1) Gray cast iron
  (2) White cast iron
  (3) Malleable cast iron
  (4) Nodular cast iron
  (5) Mottled cast iron
  (6) Chilled cast iron

1.5.1.1 Gray Cast Iron

- *Process*

Gray cast iron is obtained with the melting of pig iron, coke, and scrap in a cupola furnace and allowing it to cool for slower solidification. During the solidification process carbon is deposited in the form of graphite flakes. Basically, this type of iron when machined or fractured shows a greyish colour, and so it called grey cast iron. This greyish colour is a result of the free graphite content. It has a dull grey crystalline or granular structure and due to reflection of the free graphite flakes in strong light it provides a glistering effect. In tension, the ultimate tensile strength is 120–300 N/mm^2 and in compression it is 600–750 N/mm^2. The typical composition is C = 2.5–3.8%, Si = 1.1–2.8%, Mn = 0.4–1.0%, P = less than 0.15%, S = less than 0.1%, Fe = the remainder.

Characteristics

As an effect of high fluidity it can be easily cast into complex shapes and thin sections with lower cost. It can be easily identified by the presence of flakes of graphite in a matrix of ferrite and pearlite or austenite; a total graphite flake contributes about 10% of the metal volume. It can be easily machined and possesses machinability, high vibration damping capacity, high resistance to wear, high compressive strength, low tensile strength, low impact strength, and lower ductility.

- *Applications*

Gray cast irons have wider uses in the manufacturing of engine frames, machine bases, drainage pipes, elevator counter weights, pump housings, cylinders, and pistons for internal combustion engines, fly wheels, etc. (Table 1.3).

TABLE 1.3
Typical Compositions (%) of Irons and Cast Irons

Material	Carbon	Silicon	Manganese	Sulphur	Phosphorus
Pig iron	3.0–4.0	0.5–3.0	0.1–1.0	0.02–0.1	0.03–2.0
Wrought iron	0.02–0.08	0.1–0.2	0.02–0.1	0.02–0.04	0.05–0.2
Gray cast iron	2.50–3.75	1.00–2.50	0.40–1.00	0.06–0.12	0.10–1.00
White cast iron	1.75–2.30	0.85–1.20	0.10–0.40	0.12–0.35	0.05–0.20
Malleable cast iron	2.20–3.60	0.40–1.10	0.10–0.4	0.03–0.30	0.10–0.20

1.5.1.2 White Cast Iron

- *Process*

White cast iron can be produced by melting pig iron, coke, and steel scrap in a cupola furnace and allowing it to cool in such a way that there will be rapid solidification. During this solidification process the carbon is stored in the form of iron carbide (Cementite, Fe3C compound). After fracture it shows a whitish colour, and so it is called white cast iron. The typical composition of the white cast iron is C = 3.2–3.6%, Si = 0.4–1.1 %, Mg = 0.1–0.4%, P = less than 0.3%, S = less than 0.2%, Fe = remainder.

- *Characteristics*

The characteristics of the white cast iron are as follows:

a) White cast iron is very hard, having poor machinability, brittle, and wear resistant, it also possesses poor mechanical properties.
b) Its solidification range is 2650–2065°F, and shrinkage is about 1/8 inch per foot.
c) It possesses high tensile strength and low compressive strength.

- *Applications*

Its applications include wearing plates, road roller surface, grinding balls, dies, and extrusion nozzles. White cast irons have wider applications as a raw material for manufacturing malleable cast iron.

1.5.1.3 Malleable Cast Iron

- *Process*

This is produced from white cast iron by annealing heat treatment which consists of slower heating of the white cast iron. The process is established at around 900^{0}C and holding at this temperature for a long time before then cooling to room temperature.

- *Properties*

It has good mechanical properties like ductility and malleability as compared to white cast iron and grey cast iron. It can be used for the forging of intricate shapes.

- *Applications*

Its applications include connecting rods, transmission gears, differential cases, flanges, pipe fittings, valve parts, marine services, etc.

1.5.1.4 Nodular Cast Iron

- *Process*

This cast iron contains graphite in the form of nodules or spheroids. It is also known as spheroidal graphite iron, high-strength iron, or ductile iron. It is produced from grey cast iron by the addition of a small quantity of magnesium or cerium just before pouring. Additional magnesium converts the graphite into the spheroidal or nodular form. This conversion enhances the strength of the metal.

- *Properties*

Nodular cast iron has good ductility and malleability, which is a significant property.

- *Applications*

Its applications include valves, gears, pump bodies, crankshafts, and other automotive and machine components.

1.5.1.5 Mottled Cast Iron

In the microstructure of mottled cast iron there free cementite as well as graphite flakes. For some compositions, with carbon and silicon contents, such structures are notified under the existing conditions of cooling. As discussed earlier, faster cooling provides a white structure and slow cooling a grey structure. However, for intermediate cooling rates, a mottled structure is observed. It is observed that regions between the surface and centre of a chilled casting, do not possess good properties so this stage can be avoided.

1.5.1.6 Chilled Cast Iron

- *Process*

The quick cooling process is commonly known as chilling and the cast iron produced with this process is called chilled cast iron. This type of cast iron can be identified by its white structure at the surface and grey structure in the centre. Basically, the outer surface of the casting gets chilled, sometimes up to a depth of 1–2 mm. Sometimes, chills are used to get the required surface properties such as antifriction and antiwear characteristics. The composition of the melt is adjusted in such a way that usual cooling gives a grey structure, while rapid cooling gives white structure and the properties include low notch sensitivity, hardness, machinability, wear resistance, damping capacity, etc.

- *Applications*

Its applications include railway-freight-car wheels, hammers and dies, crushing rollers, grinding balls, road rollers, etc.

1.5.2 Plain Carbon Steel

Plain carbon steels are classified into three groups depending on the carbon content. These are:

(A) Low-carbon steels (0.008–0.30%C)
(B) Medium-carbon steels (0.30–0.60%C)
(C) High-carbon steels (0.60–2.00%C)

1.5.2.1 Low-Carbon Steels

Composition: 0.008% to 0.25% carbon and the remainder iron with impurities. The microstructure consists of ferrite and perlite constituents.

- *Properties*

They are soft and weak, tough, ductile, malleable, machinable, weldable, and nonhardenable by heat treatment, and the least expensive to produce.

- *Applications*

Steels with 0.008% to 0.15% carbon are widely used for fabrication-related work, for example, rivets, screws, wires, and nails. Steels with a carbon content in the range of 0.15% to 0.30% are widely used as structural steel (mild steel) and can be utilised for building bars, grills, angles, beams, channels, etc.

1.5.2.2 Medium-Carbon Steels

Medium-carbon steel consists of 0.30% to 0.60% carbon by weight with the remainder being iron with impurities. These types of alloys have a tempered martensite microstructure. For improving the properties it can be heat treated by austentising, quenching, and then tempering. A number of strength ductility combinations can be achieved with the addition of chromium, nickel, and molybdenum. Sometimes, with the cost of ductility and toughness, these types of steels are stronger than low-carbon steel.

- *Properties*

These are poorly ductile and malleable, medium hard, medium tough, and slightly difficult to machine, weld and harden. They are also called machinery steels.

- *Applications*

They are used for bolts, wheel spokes, axles, lock washers, large forging dies, springs, wires, cylinder liners, hammers, rods, turbine rotors, crank pins, railway rails, and railway wheels.

1.5.2.3 High-Carbon Steels

Composition: 0.60% to 1.4.0% carbon and the remainder iron with some impurities.

- *Properties*

These are hard, wear resistant, brittle, difficult to machine, difficult to weld, and can be hardened by heat treatment as per the application requirements. The hardness produced after hardening is high. They are also called tool or die steels and basically consist of chromium, vanadium, tungsten, and molybdenum.

- *Applications*

They are used for forging dies, punches, hammers, chisels, vice jaws, shear blades, drills, cutters, knives, balls, razor blades, and races for ball bearings, mandrels, files, metal cutting saws, wire drawing dies, and reamers.

1.5.3 Alloy Steels

Alloy steel may be defined as steel to which elements other than carbon with various compositions and heat treatments are used to enhance their properties. The important alloying elements in steel are nickel, chromium, molybdenum, cobalt, vanadium, manganese, silicon, tungsten, etc.

1.5.3.1 Stainless Steels

Stainless steels are a group of highly alloyed steels designed to provide high corrosion resistance generally for the ambient temperature with alloying elements like chromium, at least 15%. The chromium in the alloy forms a thin, impervious oxide film in an oxidising atmosphere, which protects the surface from corrosion. Nickel can be used as an alloying element in certain stainless steels to increase corrosion resistance from environmental conditions. Carbon is used to provide strength and hardenability to the metal; however, increasing the carbon content has an adverse effect on corrosion protection property because there is formation of chromium carbide which reduces the amount of free Cr available in the alloy. With corrosion resistance, it possesses creep properties, good thermal conductivity, machinability, high cold and hot deformation properties, and excellent surface finish; these properties are desirable in many applications. Stainless steels are more expensive than plain C or low-alloy steels.

All types of stainless steels are subdivided into three metallurgical classes:

(a) Austenitic
(b) Ferritic
(c) Martensitic

1.5.3.2 Austenitic Stainless Steels

These possess an austenitic microstructure at room temperature, and along with chromium, nickel is also used as an alloying element because nickel is a very strong stabiliser for austenitic steel. These steels have the highest corrosion resistance amongst all stainless steels and the greatest strength and scale resistance at higher temperature operations. They retain their ductility at temperatures close to absolute zero. They have the following composition: G 0.03–0.25%, Mn 2 to 10%, Si 1 to 2%, Cr 16 to 26%, Ni 3.5 to 22%, P and S normal, and Mo and Ti in some cases.

- *Applications*

Their applications include in the aircraft industry, chemical processing, food processing, and household items, and the dairy industry mainly uses austenitic stainless steel.

1.5.3.3 Ferritic Stainless Steels

The microstructure of these steel is generally ferritic. These steels have a lower carbon to chromium ratio, which eliminates the effect of thermal changes so they are not hardened by heat treatment. These steels have good ductility and are magnetic in nature. Ferritic steels are more corrosion resistant than martensite steels. The chemical composition of ferritic steel is C 0.08–0.20%, Si 1%, Mn 1–1.5%, and Cr 11–27%.

- *Applications*

Their applications include dairy components, kitchenware, automobile fittings, and in food processing plants.

1.5.3.4 Martensitic Stainless Steels

The martensitic microstructure in the hardened condition is the key feature of martensitic stainless steels. Because of the higher carbon–chromium ratio martensitic stainless steels are the only types hardenable by heat treatment, but the hardening reduces their corrosion resistance. Martensitic stainless steels are magnetic in all conditions and also have the best thermal conductivity of the stainless types. Ductility, hardness, and ability to hold on edge are important characteristics of martensitic steels. Martensitic stainless steels can be easily cold and hot worked with low carbon content, have better machining ability as well as good toughness, and good corrosion resistance to weather and some chemicals. Martensitic stainless steels have the following composition: C 0.15–1.2%, Si 1%, Mn 1%, and Cr 11.5–18%.

- *Applications*

Their applications include turbine blades, ball bearings, and table cutlery.

1.5.3.5 Tool Steels

Tool steels are a class of (usually) highly alloyed steels designed for specific application requirements such as high strength, hot hardness, wear resistance, hardness, and toughness under impact at room and elevated temperatures for industrial cutting tools, dies, and moulds. To obtain these properties especially hard tool steels are heat treated. Important aspects of the high levels of alloying elements are:

(1) Improved hardenability
(2) Less distortion during heat treatment
(3) Hot hardness
(4) Formation of hard metallic carbides for abrasion resistance
(5) Enhanced toughness under impact.

The tool steels can be divided into major types, according to the application and composition. The AISI uses a classification scheme that includes a prefix letter to identify the tool steel.

- **T, M High-speed steels**

High-speed steels are used as cutting tools in machining operations like drilling and tapping operations, and are formulated for high wear resistance and hot hardness. This tungsten molybdenum steel contains 6% tungsten, 6% molybdenum, 4% chromium and 2% vanadium. The two AISI designations indicate the principal alloying element: T for tungsten and M for molybdenum.

- **H Hot-working tool**

Hot-working tool steels are intended for hot-working dies in applications like extrusion, forging, die-casting, etc.

- **O, A, D Cold-work tool steels**

Cold-work tool steels are die steels used for cold working operations such as cold extrusion, sheet metal pressworking, and certain forging operations. The designation D stands for die and A, O stands for air- and oil-hardening. They all provide good wear resistance and low distortion.

- **W Water-hardening tool**

Water-hardening tool steels have a high carbon content with few to no other alloy components. They are hardened only by rapid quenching of water. They are commonly used due to their low cost, but are limited to low-temperature applications. They can be used for closed heating dies.

- **S Shock-resistant tool steels**

Shock-resistant tool steels are applicable in cases of high toughness requirements as in many sheet metal forming processes like punching, shearing, and bending operations.

- **P Mould steels**

Mould steels are used to make moulds used for moulding plastics and rubber components.

- **L Low-alloy tool steels**

Low-alloy tool steels are generally reserved for special applications. Plain carbon, low alloy, and stainless steels are used for various tool and die applications. Cast irons and certain nonferrous alloys are also suitable for some tooling applications to manufacture various components.

- **Effects of alloying elements on steel**

The alloying is done as per the specific application requirements or to fulfil the specific purposes or needs such as to increase wear resistance and corrosion resistance and to enhance electrical and magnetic properties, which is not possible using the elements like plain carbon steel. The key alloying elements used in steel are nickel, chromium, molybdenum, cobalt, vanadium, manganese, silicon, and tungsten. These elements may be used independently or in various combinations along with heat treatments to produce the required properties in the steel.

1. Nickel

Nickel enhances the toughness and strength of steel. The nickel content varies from 2 to 5% and carbon content from 0.1 to 0.5%. For this particular range, nickel contributes great strength and hardness with improved elastic limit, ductility, and resistance to corrosion. A 25% nickel content in the alloy gives maximum toughness and provides good resistance to corrosion, rusting, and burning at elevated temperatures. Alloys containing nickel can be used in the manufacture of spark plugs for petrol engines, boiler tubes, valves for use with superheated steam, valves for internal combustion engines, etc. Alloy steel with 36% nickel is called invar. It shows nearly zero coefficient of expansion, and so it can be widely used for measuring instruments and length standards for daily use.

2. Chromium

A combination of hardness with high strength and high elastic limit can be achieved with the help of this alloying element in steel. It enhances the anticorrosive properties of steel. Commonly, chrome steels contain chromium in the range of 0.5–2% and carbon in the range of 0.1–1.5%. Applications of chrome steel include balls, rollers, and races for bearings. A nickel chrome steel contains 3.25% nickel, 1.5% chromium, and 0.25% carbon. Applications include car crankshafts, axles, and gears, which require great strength and hardness.

3. Vanadium

vanadium aids in obtaining a fine grain structure in tool steel. A very small amount of vanadium (less than 0.2%) produces a noticeable increase in tensile strength as well as elastic limit in both low- and medium-carbon steels without affecting the ductility. In chrome-vanadium steel, chromium, vanadium, and carbon contents vary in the ranges of 0.5–1.5%, 0.15–0.3%, and 0.13–1.1%, respectively. It has extremely good tensile strength, endurance limit, ductility, and elastic limit. Applications of chrome vanadium steel include as springs, gears, pins shafts, and various forged parts.

4. Tungsten

Tungsten avoids grain growth, and also increases the case depth of quenched steel and maintains properties like hardness even when heated to a red colour or elevated temperature. It is commonly used in conjunction with another element. Steel with tungsten, carbon 3–18% and 0.2–1.5%, respectively, can be used for cutting tools. The applications of these tungsten steels include cutting tools, permanent magnets, dies, valves, taps, etc.

5. Cobalt

At high temperatures cobalt provides red hardness by retention of hard carbides. During heat-treatment it is subjected to decarburisation. For magnets it increases the hardness and strength with residual magnetism and a coercive magnetic force in steel.

6. Manganese

For both hot-rolled and heat-treated conditions manganese enhances the strength of steel. The manganese alloy steels contain manganese and carbon at about 1.5% and 0.40–0.55%, respectively. It is widely used for making gear shafts, axles, and other parts where fair ductility and high strength are required. The main use of manganese steel is in machinery parts which are subjected to severe wear.

7. Silicon

Silicon steels show similar behaviour to nickel steels. These types of steels have a high elastic limit as compared to another carbon steels. Silicon steels contain silicon and carbon in the range of 1–2% and 0.1–0.4%, respectively, along with other alloying elements. Basically, these types of steel utilised for electrical machinery, springs, and corrosion-resistant materials for valves in internal combustion engines.

8. Molybdenum

Molybdenum steel contains a very small quantity of molybdenum (0.15–0.30%) with chromium and manganese (0.5–0.8%). This type of steels provides extra tensile strength and is used for automobile parts and airplane fuselage.

1.6 MATERIAL SELECTION PROCESS IN THE AUTOMOTIVE INDUSTRY

The selection of the best material for a specific application is a very complex process that involves a variety of factors. The basic factors associated with selection of the most suitable material include operating parameters, manufacturing and production process, functional requirements, cost considerations, aesthetics, ergonomics, etc.

Manufacturing process

The following factors are important during material selection from a manufacturing process point of view: machinability, plasticity, malleability, ductility, casting properties, surface treatments, weldability, heat treatments, tooling requirements, surface finish, and forming process.

Functional requirements

Functional requirements considered during material selection are as follows: strength, hardness, rigidity, toughness, thermal conductivity, fatigue, electrical conductivity, creep, and aesthetics.

Cost considerations

The cost considerations affect the cost of a product and hence it is essential to consider this during material selection for a particular manufacturing process: raw material, raw material availability, transportation, processing, storage, manpower, special treatments, inspections, packaging properties, inventory, taxes, and related charges.

Operating parameters

Operating parameters influencing material selection include: pressure, temperature, flow, type of material, corrosion requirements, environment, fire protection, weathering, and biological effects.

1.7 SUMMARY

Material selection, as a part of material science, is an important aspect of the automotive manufacturing process, as the selection and use of proper materials directly affects various parameters of the vehicle such as service behaviour, life, economy, effectiveness, performance of legal requirements, environmental aspects, etc. To fulfil the various demands of customers it is essential to ensure clean, safe, cost affordable, energy efficient, aesthetically pleasant, and high comfort qualities. The selection of the best material for an automotive application depends on the application requirements such as material used in the dashboard panel, chassis components, seats, wheels and tires, etc. Each material has certain properties such as mechanical, thermal, electrical, optical, chemical, etc., which make it more suitable for some applications than others. Material selection in the automotive application involves the manufacturing process, operating parameters, cost considerations, functional requirements, etc.

1.8 REVIEW QUESTIONS

1) State any four types of material used in engineering applications.
2) State any four requirements of an ideal automotive material.
3) List the types of materials used in the vehicle body.
4) List the applications of cast iron in the automobile sector.
5) State any eight mechanical properties of a material.
6) Define these terms and state their applications:
   a) Plain carbon steel
   b) Mild steel
   c) Low-carbon steel
   d) Medium-carbon steel
   e) High-carbon steel
7) Describe the physical properties of a material.
8) Describe the mechanical properties of a material.
9) Describe the thermal properties of a material.
10) Classify engineering materials.
11) Classify cast iron.
12) Describe any two properties of a material used in the automobile with an example.
13) Describe the requirements of an automotive material.

2 Nonferrous Materials

LEARNING OBJECTIVES

- To understand the details of conventional and modern non ferrous materials their distinctive features and their vehicular applications.
- To cite distinctive physical and mechanical characteristics of nonferrous alloys.
- To understand the details of material decision making process
- To understand the business factors affecting material selection.

2.1 INTRODUCTION

Nonferrous metals are those which do not contain iron as the base material. The range of these nonferrous metals includes low-strength metals like aluminium and copper, and high-temperature resistant and high-strength alloys like titanium, molybdenum, etc. There are wider applications of nonferrous metals due to the following advantages:

1. Anticorrosive properties
2. Good magnetic and electrical properties
3. Ease of cold forming
4. Fusibility and castability
5. Lower density
6. Attractive colour
7. Good formability.

The nonferrous metals used in engineering are copper, aluminium, lead, zinc, tin, nickel, etc. and their alloys.

2.2 NONFERROUS MATERIALS AND THEIR ALLOYS

Nonferrous metals are alloys or metals that do not contain any appreciable amounts of iron. All pure metals are nonferrous elements, except for iron (Fe), which is also called ferrite from the Latin "ferrum", meaning "iron". Nonferrous

DOI: 10.1201/9781003367321-2

metals tend to be more expensive than ferrous metals but are used for their desirable properties, including light weight (aluminium), high conductivity (copper), and nonmagnetic properties or resistance to corrosion (zinc). Some nonferrous materials are used in the iron and steel industries, such as bauxite, which is used for flux in blast furnaces. Other nonferrous metals, including chromite, pyrolusite and wolframite, are used to make ferrous alloys. However, many nonferrous metals have low melting points, making them less suitable for applications at high temperatures. The details of copper, aluminium, lead, zinc, tin and nickel are presented in the following section.

2.2.1 Copper and Its Alloys

Copper is not available in pure form directly from the ground. It is extracted from its ores, through a series of processes. As far as India is concerned, the copper ores are found in Khetri in Rajasthan and Ghatsila in Bihar. The main ore used in copper extraction is copper pyrites. Copper is now extracted from ores which are generally extracted from open pit mines and having low grade, which are mostly sulphides, such as chalcopyrite ($CuFeS_2$). The ore is crushed and concentrated by flotation, and then smelted (melted or fused) and refined. This process is known as pyrometallurgy in which heat is used for refining the metal. This results in the production of a crude form of copper known as blister copper of 98–99% purity. In the process of electrolysis, pure copper is deposited on the cathode for the final refining. The copper can be processed through hydrometry with the help of chemical and electrolytic reactions. This produces a highly pure (99.9%) copper, which is again melted and cast into the required shapes.

2.2.1.1 Properties and Uses

1. Anticorrosive properties as a noble metal.
2. Good nonmagnetic properties.
3. It is soft, malleable, ductile and flexible, making casting, forging, rolling and drawing into wires easy.
4. Good thermal and electrical conductivity.
5. Tough and strong.
6. It can be easily polished, and possesses a pleasing appearance.
7. It can be brazed, welded, or soldered.
8. It possesses excellent machinability.

Copper is used in the following appliances:

1. Electrical accessories
2. Tubing for heat exchangers
3. Pure copper can be used as a solid lubricant in hot metal-forming processes
4. It is an important ingredient in copper alloys like brass and bronze
5. Copper is widely used in the form of tubes in mechanical engineering.
6. It is used with tin, zinc, nickel and aluminium for making useful alloys.

2.2.2 Copper Alloys

The required strength and hardness of copper alloys can be obtained with the help of heat treatment as per the application requirement.

The copper alloys are broadly classified into the following two groups:

1. Copper–zinc alloys (brasses); in brass zinc is the main alloying metal about with 90% Cu and 10% Sn.
2. Copper–tin alloys (bronzes), in bronze tin is the main alloying metal with about 65% Cu and 35% Zn. The important alloying elements like beryllium bronze or phosphor bronze are generally used for springs and bearings as they have sufficient hardness.

A) Brasses

The copper–zinc alloy is also known as brass. There is a wide use of brass in various applications. Based on the proportion of copper and zinc, the brass can be categorised into different types. The properties of brass may be greatly affected by small quantities of alloying elements. Brasses have very good anticorrosive properties and can be utilised for soldering purposes. As far as manufacturability is concerned, they can be easily fabricated by processes like spinning and electroplated with nickel and chromium-like metals.

All brasses are fundamentally alloys of copper and zinc. The types of brasses are as follows.

(1) α-Brasses

These consist of Cu (copper) up to 36%. They are soft, malleable and ductile, have an annealed condition, and fairly good corrosion resistance. All the α-brasses of these types are suitable for cold rolling, wire drawing, press work and other related operations. Some of the important brasses in this group are categorised as follows.

(i) Cap copper

This contains zinc between 2–5%. Zinc is used as a deoxidiser in the process of deoxidation of copper. In the absence of zinc, copper oxide present in the microstructure reduces the malleability and ductility. It is very ductile and is used for detonator caps, especially in ammunition factories.

(ii) Gilding metals

The zinc content in gilding metals varies from 5 to 15% and according to the zinc amount they have different shades of colour from reddish to yellowish. As far as their applications are concerned they can be used for bullet envelopes, drawn containers, condenser tubes, coins, needles, emblems and jewellery because of their golden colour.

(iii) Cartridge brass (70–30 brass)

Cartridge brass contains about 30% zinc. Amongst all the brasses, it has the maximum ductility and malleability. As far as manufacturability is concerned, these are used for forming by deep drawing, trimming, and stretching, spinning and related press work operations. It is also called 70–30 brass. There is wide use of cartridge brass such as cartridge cases, radiator fins, lamp fixtures, rivets and springs.

(2) α–β Brasses

Zinc in the commercial α–β brasses varies between 32–40%. These are hard and strong as compared to α-brasses and they can be fabricated by hot working processes. Further classification of this type of brass is as follows.

(i) Muntz metal

This contains about 40% zinc and the remainder is copper. Hot-worked Muntz metal or 60–40 brass has a tensile strength of 35–40 kg/mm^2 and a hardness of 100–120 VPN. Muntz metal can be used for shafts, utensils, pump parts, nuts and bolts, and condenser tubes used in refrigeration systems and similar applications where corrosion is not as severe.

(ii) Naval brass

For special applications like marine uses, the addition of about 1% tin to Muntz metal increases the corrosion resistance. This brass is also called naval brass or tobin bronze. Brass with 39% zinc and 1% tin is used for welding rods, marine components, propeller shafts, nuts and bolts, piston rods, etc.

(3) Brazing brass

Brass having a 50–50 composition of zinc and copper can be used for brazing purposes. The 50% zinc brass melts at lower temperature (~ 870°C) and can be utilised for joining brasses used for commercial purposes in various applications. Due to its brittleness, it can only be used for the brazing purposes rather than other engineering applications.

B) Bronze

Bronzes are the alloys of copper containing alloying elements other than zinc. In these alloys zinc may be present in very small amount. As engineering applications are concerned, important types of bronzes are as follows:

(i) Aluminium bronze

Composition: In this bronze, the content of aluminium varies between 4–11% with the remainder being copper. Other elements such as Fe, Ni, Mn and Si are also added to improve certain properties as per the application requirements.

Properties:

(a) Good ductility and toughness strength
(b) Bearing properties
(c) Anticorrosive properties
(d) Good fatigue resistance

Applications:
It can be used in jewellery, heavy-duty parts, heat exchangers, gear, bearings and bushes, marine equipment.

(ii) Tin bronze

Composition: This bronze contains about 88% Cu, 10% Sn and 2% Zn.

Properties:

(a) Good ductility and malleability
(b) Good anticorrosive properties

Applications: They are used in pumps, gears, coins, heavy load bearings and marine fittings.

(iii) Gun metal

Composition: This consists of zinc in the range of 2–5% and tin in the range of 5–10% with the remainder being copper.

Properties:

(a) Anticorrosive
(b) Higher tensile strength
(c) Zinc acts as a deoxidiser and also improves the fluidity of melt

Applications:

(a) It can be used for gun barrels and other ordnance parts
(b) Bearings, marine castings, gears and steam pipe fittings

(iv) Phosphor bronze

As per its constituents, phosphor bronze can be subdivided into two main categories:

(a) Cast phosphor bronze
(b) Wrought phosphor bronze

(a) Cast phosphor bronze

This contains phosphorus in the range of 5–13% with the remainder being copper. It can be used in bearings, slide valves, gear wheels and gudgeon pins. Phosphorus bronze having a composition of 12% tin and 0.3% phosphorus has a hardness of 100 BHN. It possesses good tensile strength with 5% elongation.

(b) Wrought phosphor bronze

In this bronze, the tin percentage varies in the range of 2.5–8.5%, phosphorus in the range of 0.1–0.35%, with the remainder being copper. It possesses good corrosion resistance and high strength and is mainly used for springs.

v) Bearing materials

These can be used in the construction of machines, engines or parts of equipment where there are rotary or reciprocating motions. A good bearing material should possess the following properties:

i. High compressive strength
ii. Sufficient hardness and high wear resistance
iii. Lower friction coefficient

The types of bearing materials include copper–lead alloy, white metal alloy and tin bronzes.

White metal alloy (Babbitt)

This is a tin-base white metal and consists of 88% tin, 8% antimony and the remaining 4% copper. It is a soft material, having low strength and a low friction coefficient. Babbitt metal can be utilised in manufacturing of fine and heavy-duty bearings and it does not affect the shaft greatly when there is a failure of lubricants.

2.2.3 Aluminium and Its Alloys

Aluminium is a chemical element with the symbol Al and atomic number 13. It has a density lower than those of other common metals and approximately one-third that of steel. It has a great affinity towards oxygen, and forms a protective layer of oxide on the surface when exposed to air. Aluminium visually resembles silver, both in its colour and in its great ability to reflect light. It is soft, nonmagnetic and ductile.

- **Aluminium Production**

The principal ore for aluminium production is bauxite, which consists largely of hydrated aluminium oxide (Al_2O_3-H_2O) with various other oxides. Extraction of the aluminium from bauxite can be summarised in three steps:

(1) Washing to separate dirt and clay and crushing the ore into fine powders;
(2) The Bayer process, in which the bauxite is converted to pure alumina (Al_2O3); in which the fine powder is treated with caustic soda; and
(3) It can be dissolved in a molten sodium fluoride and aluminium fluoride bath at 940–980^{0}C through the process of electrolysis, which is used so that the alumina is separated into aluminium at the cathode and oxygen gas (O_2) at the anode.

- **Properties and uses**

The density of aluminium is about a third of that of steel or brass. It is a silvery white and light metal. As a result of better thermal and electrical conductivity than copper, it is extensively used for heavy conductors and overhead cables. It possesses better anticorrosive properties than many other metals, and is used in the manufacture of containers for the chemical industry. It has good ductility, and so it can be easily forged, rolled and die cast. Light in weight, it is extensively used in the aircraft industry. The melting point of pure aluminium is about 650°C. It becomes hard after cold working operations. Also, air and water have no effect on it.

- **Aluminium alloys**

The small quantities of another alloying element convert this soft, weak material into a hard and strong metal, without affecting the weight. Alloys are classified as cast and wrought. Aluminium alloys are further classified according to the heat treatment used.

1. Duralumin

The copper content of duralumin varies between 3.5–4.5%, manganese from 0.4% to 0.7%, magnesium in the range of 0.4–0.7% and the remainder is aluminium. Duralumin possesses high tensile strength, high machinability, heat treatment, strength and forming properties. It is widely used for forging, stamping, sheets, bars and rivets, tubes, aircraft and automobile parts due to its low weight.

2. Y-alloy

This is also known as copper–aluminium alloy. To increase its strength and machinability copper can be added to pure aluminium. The copper content of Y-alloy varies from 3.5% to 4.5%, nickel percentage varies from 1.8% to 2.3% nickel, the magnesium content varies in the range of 1.2–1.7%, with 0.6% each of silicon, manganese and iron. Y-alloy is useful for cast uses such as cylinder heads, pistons and other components of aero-engines. It can be utilised in the form of sheets and strips.

3. Magnalium

This is manufactured by melting aluminium in a vacuum with 2–10% magnesium and cooling it in a vacuum or under pressure of about 100–200 atmospheres. It also contains about 1.75% copper. As a result of its light weight and good mechanical properties, its major application is in the aircraft and automotive sectors.

4. Hindalium

This consists of aluminium and magnesium with a small quantity of chromium. It is the tradename of the aluminium alloy produced by Hindustan Aluminum Corporation Ltd., Renukoot (UP). It is produced as a rolled product in 16 gauges which are generally used for anodised utensils.

2.2.4 Magnesium and Its Alloys

Magnesium alloys are mixtures of magnesium (the lightest structural metal) with other metals (called an alloy), often aluminium, zinc, manganese, silicon, copper, rare earths and zirconium. Magnesium alloys have a hexagonal lattice structure, which affects the fundamental properties of these alloys.

Most magnesium is commercially produced with the use of seawater, which contains about 0.13% $MgCl_2$. For extraction of Mg, either electrolysis or a thermal reduction method can be used. In the electrolysis process a batch of seawater is mixed with milk of lime—calcium hydroxide ($Ca(OH)_2$). The resulting reaction precipitates magnesium hydroxide ($Mg(OH)_2$), which settles at the bottom and is collected in the form of slurry. The slurry is then filtered to increase the $Mg(OH)_2$ content and then mixed with hydrochloric acid (HCl); the resulting product is a more concentrated $MgCl_2$ as compared with the original seawater. To decompose the salt into magnesium (Mg) and chlorine gas (Cl_2) an electrolysis process is used. The magnesium is then cast into ingots for further processing as per requirements in various shapes. In the thermal reduction method, ores of magnesium are broken down and mixed with reducing agents such as ferrosilicon. The mixture can be heat-treated in a vacuum which forms vapours of magnesium which can be condensed to form crystals of magnesium. These crystals are then melted, refined and poured into ingots and the shapes created as per the processing or application requirements. The chlorine is also recycled to form more $MgCl_2$.

- **Properties**

Magnesium has a high strength-to-weight ratio, and it is 35% and 75% lighter than aluminium and iron, respectively. For equal stiffness, magnesium alloys weigh ~25% less than aluminium. Magnesium also has better electromagnetic shielding characteristics, high thermal conductivity, excellent castability, good ductility, and good damping characteristics than aluminium, and it also can be easily recycled.

- **Magnesium alloys**

Magnesium alloys can be subdivided into two main categories: cast and wrought alloys. Magnesium alloys are typically used as cast alloys; however, research into wrought alloys has shown tremendous growth during the last decade. Cast magnesium alloys have wider applications in the automotive, electronics, and aerospace industries. The most common cast magnesium alloys are AZ63, ZE41, ZC63, HK31, HZ32, QE22, AZ91, AM50, AZ81, ZK51, ZK61, QH21, WE54, WE43 and Elektron 21. The most common wrought magnesium alloys are AZ31, M1A, HK31, HM21, AZ61, AZ80, Elektron 675, ZK60, ZE41 and ZC71. Here, the prefix letters designate two main alloying metals in magnesium alloys that were developed as per ASTM B275 where A: aluminium; E: rare earths; C: copper; O: silver; S: silicon; T: tin; H: thorium; K: zirconium; M: manganese; L: lithium; W: yttrium; B: bismuth; Z: zinc; R: chromium; D: cadmium; N: nickel; and F: iron.

- **Characteristics of the magnesium alloys**

This is the lightest structural metal in use (its density is 1.74 g/cm^3 as compared to 2.7 g/cm^3 for aluminium alloys). It has a higher strength-to-density ratio as compared to aluminium alloys. It has high damping capacity. As compared to aluminium alloys and steel, the room temperature formability of wrought magnesium alloys is also lower.

- **Applications**

Magnesium components are widely used by major automotive companies, including TATA Motors, General Motors (GM), Ford, Volkswagen, and Toyota, and other automotive industries. Magnesium alloys are currently being considered for several applications, including:

- Powertrain applications, transfer cases, transmission cases, engine blocks
- Steering components
- Chassis components: wheels, subframe, engine cradle, control arm, etc.
- Interior: trim plates, seat components, instrument panels, etc.
- Body: tailgate/liftgate inner, door inner, roof frame, sunroof panel, bumper beam, radiator support, shotgun, A and B pillars, hood outer/fender, decklid/hood inner, decklid/door outer, dash panel, frame rail, etc.

2.2.5 Titanium Alloy

Titanium is a low-density element having approximately 60% of the density of steel and super alloys that can be strengthened effectively by alloying and deformation processing. Important Ti ores are rutile, having 98–99% TiO_2 and limenite having a composition of FeO and TiO_2. By reaction with chlorine gas, the TiO_2 from the ores of titanium can be converted into titanium tetrachloride ($TiCl_4$). With the use of distillation processes, impurities can be removed from this and highly concentrated $TiCl_4$ is reduced to metallic titanium reacting with the magnesium; this is called the Kroll process.

Titanium has good heat-transfer properties and is nonmagnetic. Its coefficient of thermal expansion is somewhat less than half that of aluminium and lower than that of steel. As compared to steel, titanium and its alloys have higher melting points, but for structural applications maximum useful temperatures generally range from 427°C (800°F) to 538°C to 595°C (1000°F to 1100°F), based on composition.

Pure titanium melts at 1670°C and has a low density of 4.51 g/cc (60% heavier than aluminium and 40% lighter than steel). Titanium has high affinity to oxygen, it is a strong deoxidiser, and can catch fire and cause severe damage. Ti is stronger than Al. The high strength and low weight make it very useful as a structural metal for various applications. As there is a protective thin oxide surface film it has excellent corrosion resistance. It can be used as a biomaterial. Ti can be used in elevated temperature components. As a downside, pure Ti has lower strength, although alloying can be carried out to improve the strength. Care should be taken during processing because oxygen, nitrogen and hydrogen can cause titanium to become more brittle. Titanium can also be cast using a vacuum furnace. As a result of its excellent corrosion resistance and high strength-to-weight ratio, titanium can be used in a variety of applications, including the following:

- Aircraft – body structure, engine parts, sporting equipment, valve and pump parts, desalination, chemical processing, turbine engine parts, marine hardware, medical implants and prosthetic devices.
- The use of Ti in bikes and automotives is increasing constantly.

Pure Ti exhibits two phases – hexagonal α-phase at room temperature and BCC β-phase above 882°C. (α+β) alloys have moderate strength, low density, reasonable ductility and good creep resistance. Metastable β alloys are stronger, heavier and less ductile as compared to α alloys. Creep strength is reduced with increasing β content. (α+β) alloys shows a good strength–ductility combination.

Compositions, Properties and Applications of Some Ti Alloys

Alloy type	Comp. wt.% (UNS No.)	Condition	Ys (MPa)	UTS (MPa)	%El	Applications
Cp Ti	99.1 Ti (R50500)	Annealed	414	484	45	Airframe skins, marine and chemical processing equipment
α	Ti-5Al-2.5Sn(R54520)	Annealed	784	826	16	Gas turbine engine casting and rings, chemical processing equipment

Alloy type	Comp. wt.% (UNS No.)	Condition	Ys (MPa)	UTS (MPa)	%El	Applications
Near-α	Ti-8Sl-1Mo-1V(R5810)	Annealed	890	950	15	Forged jet engine components, compressor disc, plate, hubs
$\alpha-\beta$	Ti-6Al-4V(R56400)	Annealed	877	947	14	Prosthetic implants, airframe components
$\alpha-\beta$	Ti-6Al-6V-2Sn(R56620)	Annealed	985	1050	14	Rocket engine case, airframe structure
β	Ti-10V-2Fe-3Al	Heat-treated (aging)	1150	1223	10	High-strength airframe components, parts requiring uniform tensile stresses

2.3 CERAMICS – PROPERTIES OF CERAMIC MATERIALS AND APPLICATIONS

Ceramic materials are compounds of metallic and non-metallic elements, available in the form of oxides, carbides, borides and nitrides and having different compositions and forms. Most have crystalline structures, but unlike metals, the interatomic bonds are either ionic, predominately ionic or they have some covalent characteristics. As a result of the strength of the primary bonds most ceramics have high melting temperature, compressive strength and melting temperature. A ceramic is an inorganic non-metallic solid made up of either metal or non-metal compounds that have been shaped and then hardened by heating to high temperatures. The unavailability of the free electrons makes it a poor electrical conductor with high thermal resistance and results in the many transparent thin sections. Ceramics can be classified into two categories – traditional ceramics and industrial ceramics – used for various industrial applications such as automotives, aerospace, cutting tools used in manufacturing industries, heat exchangers, semiconductors, seals, etc. Common properties of the ceramics include high strength, brittleness, elastic modulus and hardness at elevated temperature.

2.4 POLYMER MATERIALS

2.4.1 Thermo-Setting Plastics, Rubber

A polymer is a long-chain-molecule compound, with each molecule having repeated units connected together which are formed by the polymerisation process. In a single polymer molecule there may be thousands, even millions, of units. The characteristics of the polymer depends on the structure of each molecule, its size and shape and its arrangement.

Characteristics/properties of polymers

- Resistance to corrosion and chemicals
- Noise reduction
- Wide choice of colours and transparencies

- Low electrical and thermal conductivity
- High strength-to-weight ratio
- Low density
- Relatively low cost
- Ease of manufacturing and complexity of design possibilities

Polymers are mainly classified as thermoplastics or thermoplastic polymers, thermosetting plastics or thermosets and elastomers.

2.4.2 Thermoplastics

Thermoplastics are composed of long chains of linear macromolecules produced by joining together monomers; they typically behave in a ductile manner after heating which converts them into a viscous liquid state and after cooling they again become solids; these cycles can be repeated multiple times without degradation of the polymers. The chains may be linear or branched, similar to a few trees that are tangled up together. Individual chains are intertwined and there are relatively weak van der Waals bonds between atoms of chains. This determines the strength of the bond. The chains in thermoplastics can be untangled through the application of a tensile stress. However, the behaviour of the thermoplastics depends on various parameters, including the structure, its composition, temperature and its rate of deformation. Common examples are acrylics, cellulosics, polyethylenes, nylons and polyvinyl chloride. Thermoplastics can be amorphous or crystalline. They can be processed into shapes by heating to elevated temperatures as per application requirements. Thermoplastics are easily recycled.

2.4.3 Thermosetting Polymers

Thermosetting polymers consist of long chains (linear or branched) of molecules that are strongly cross-linked to one another to form three-dimensional network structures. As an effect of the nature of the bonds the hardness and strength are not affected by the temperature and deformation rate. There are two stages to the polymerisation process in thermosetting plastic, in the first stage molecules are partially polymerised into linear chains and in the second there is completion of cross-linking under heat and pressure during the moulding and shaping process. Thermosetting polymers are similar to a bunch of strings that are knotted to one another in many ways rather than only being twisted together, and each string may have other side strings attached to it. Thermosets are generally more brittle and stronger as compared to thermoplastics. Thermosets do not melt upon heating but begin to decompose. It is difficult to reprocess or recycle these after cross-linking.

Elastomers

These are known as rubbers. They have an elastic deformation of >200%. These may be thermoplastics or thermosets with light cross-linking. The polymer chains are made up of coil-like molecules that can be extended reversibly by applying a force. Thermoplastic elastomers are a form of polymer. They have the manufacturing simplicity of thermoplastics as well as the elasticity of rubber.

2.4.4 Applications of Polymers

Polymers may sound like a very industry-specific thing, and to some degree they are, but they are actually a relatively large part of many widely recognizable industries. Polymer testing and consultancy for plastic has applications in such industries as aerospace, automotive, electronics, packaging and medical devices. Polymers are incredibly diverse elements that represent such fields of engineering as avionics through biomedical applications, drug-delivery systems, biosensor devices, tissue engineering, cosmetics, etc. The application of polymers and their subsequent composites is still advancing and increasing quickly due to their easy manufacture. When considering a polymer application, how the material behaves over time needs to be ascertained to enable its real value to be assessed. It is important to understand that polymeric materials may include raw materials, polymer compounds, foams, structural adhesives and composites, fillers, fibres, films, membranes, emulsions, coatings, rubbers, sealing materials, adhesive resins, solvents, inks and pigments. Keep in mind that the following are some of the industries in which you would see the use and application of various polymeric materials and polymers.

2.4.4.1 Properties and applications of thermoplastics

There are several commonly available types of thermoplastics used in industry. Each possesses distinct characteristics, and some thermoplastics have subtypes with even more unique properties. One of these types of plastics will be suitable for almost any project. The properties and applications of various types of thermoplastics are presented in following sections.

1. Acetals

Acetals (from acetic and alcohol) have good strength and stiffness, and good resistance to creep, abrasion, moisture, heat and chemicals. Typical applications include mechanical parts and components requiring high performance over a long period of time, such as bearings, cams, gears, bushings and rolls, and also impellers, wear surfaces, pipes, valves, shower heads and housings.

2. Acrylics

Acrylics (such as PMMA) possess moderate strength, good optical properties and weather resistance. They are transparent (but can be made opaque), are generally resistant to chemicals and have good electrical resistance. Typical applications include optical lenses, lighted signs, displays, window glazing, skylights, automotive headlight lenses, windshields, lighting fixtures and furniture.

3. Acrylonitrile-butadiene-styrene

Acrylonitrile-butadiene-styrene is rigid and dimensionally stable. It has good impact, abrasion and chemical resistance; good strength and toughness; good low-temperature properties; and high electrical resistance. Typical applications include pipes, fittings, chrome-plated plumbing supplies, helmets, tool handles, automotive components, boat hulls, telephones, luggage, housing, appliances, refrigerator liners and decorative panels.

4. Cellulosics

Cellulosics have a wide range of mechanical properties, depending on their composition. They can be rigid, strong and tough; however, they weather poorly and are affected by heat and chemicals. Typical applications include tool handles, pens, knobs, frames for eyeglasses, safety goggles, machine guards, helmets, tubing and pipes, lighting fixtures, rigid containers, steering wheels, packaging film, signs, billiard balls, toys and decorative parts.

5. Fluorocarbons

Fluorocarbons possess good resistance to high temperature (e.g., a melting point of 327°C for Teflon), chemicals, weather and electricity. They also have unique non-adhesive properties and low friction. Typical applications include linings for chemical-processing equipment, non-stick coatings for cookware, electrical insulation for high-temperature wire and cable, gaskets, low-friction surfaces, bearings and seals. Polyamides (from the words poly, amine, and carboxyl acid) are available in two main types: nylons and aramids.

- **Nylons** have good mechanical properties and abrasion resistance; they also are self-lubricating and resistant to most chemicals. All nylons are hygroscopic (absorb water); the moisture absorption reduces desirable mechanical properties and increases part dimensions. Typical applications include electrical parts, gears, bushings, bearings, rolls, zippers, fasteners, combs, wear-resistant surfaces, tubing, guides and surgical equipment.
- **Aramids** (aromatic polyamides) have very high tensile strength and stiffness. Typical applications include fibres for reinforced plastics, bulletproof vests, cables and radial tires.

6. Polycarbonates

Polycarbonates are versatile; they have good mechanical and electrical properties, high impact resistance, and they can be made resistant to chemicals. Typical applications include optical lenses, safety helmets, food-processing equipment, bullet-resistant window glazing, bottles, signs, windshields, electrical insulators, load-bearing electrical components, business machine components, medical apparatus, guards for machinery and parts requiring dimensional stability.

7. Polyesters

Polyesters have good mechanical, electrical and chemical properties; good abrasion resistance; and low friction. Typical applications include gears, cams, rolls, load-bearing members, pumps and electromechanical components.

8. Polyethylenes

Polyethylenes possess good electrical and chemical properties; their mechanical properties depend on the composition and structure. The three major polyethylene classes are:

(1) Low density (LDPE)
(2) High density (HDPE)
(3) Ultrahigh molecular weight (UHMWPE).

Typical applications for LDPE and HDPE are houseware, bottles, garbage cans, ducts, bumpers, luggage, toys, tubing, bottles and packaging materials. UHMWPE is used in parts requiring high-impact toughness and resistance to abrasive wear; examples include artificial knee and hip joints. Polyimides have the structure of a thermoplastic, but the non-melting characteristic of a thermoset.

9. Polypropylenes

Polypropylenes have good mechanical, electrical and chemical properties and good resistance to tearing. Typical applications include medical devices, appliance parts, automotive trim and components, wire insulation, TV cabinets, luggage, ropes, pipes, fittings, dairy-product drinking cups and juice containers, and weather stripping.

10. Polystyrenes

Polystyrenes generally have average properties and are somewhat brittle, but inexpensive. Typical applications include disposable containers, packaging, trays for meats, cookies and candy, foam insulation, appliances, automotive and radio/TV components; houseware; and toys and furniture parts (as a substitute for wood).

11. Polysulphones

Polysulphones have excellent resistance to heat, water and steam; they have dielectric properties that remain virtually unaffected by humidity, are highly resistant to some chemicals, but are attacked by organic solvents. Typical applications include steam irons, hot-water containers, coffeemakers, medical equipment that requires sterilisation, aircraft cabin interiors, power-tools, appliance housings and electrical insulators.

12. Polyvinyl chloride

Polyvinyl chloride has a wide range of properties, is inexpensive and water resistant, and can be made rigid or flexible; it is not suitable for applications requiring strength and heat resistance. Rigid PVC is tough and hard; it is used for signs and in the construction industry (e.g., in pipes and conduits). Flexible PVC is used in wire and cable coatings, in low-pressure flexible tubing and hoses, and in footwear, imitation leather, upholstery, records, gaskets, seals, trim, film, sheet and coatings.

2.4.4.2 Applications and use of thermosetting plastics

The applications and uses for thermosetting plastics have grown substantially over the years. The success and benefits of this type of plastic are due in large part to its favourable plastic properties. Thermoset plastic polymers cross-link together when curing to form an irreversible chemical bond. It is one of two organic polymer-based plastic materials, the other being thermoplastic. However, the difference between thermoplastic vs. thermoset plastics lies in their ability to be remelted or reshaped; thermoset plastics permanently form a rigid three-dimensional structural network that immobilises the molecules. Applications of various types of thermosetting plastics are as follows.

1. Alkyds

Alkyds (from alkyl, meaning alcohol, and acid) possess good electrical insulating properties, impact resistance, dimensional stability and low water absorption. Typical applications are in electrical and electronic components.

2. Aminos

Aminos have properties that depend on composition; generally, they are hard, rigid and resistant to abrasion, creep and electrical arcing. Typical applications include small-appliance housings, countertops, toilet seats, handles and distributor caps. Urea typically is used for electrical and electronic components, and melamine for dinnerware.

3. Epoxies

Epoxies have excellent mechanical and electrical properties, good dimensional stability, strong adhesive properties and good resistance to heat and chemicals. Typical applications include electrical components requiring mechanical strength and high insulation, tools and dies, and adhesives. Fibre-reinforced epoxies have excellent mechanical properties and are used in pressure vessels, rocket-motor casings, tanks and similar structural components.

4. Phenolics

Phenolics are rigid (though brittle) and dimensionally stable; they have high resistance to heat, water, electricity and chemicals. Typical applications include knobs, handles, laminated panels and telephones; as a bonding material to hold abrasive grains together in grinding wheels; and electrical components (such as wiring devices, connectors and insulators).

5. Polyesters

Polyesters have good mechanical, chemical and electrical properties; they generally are reinforced with glass (or other) fibres and also are available as casting resins. Typical applications include boats, luggage, chairs, automotive bodies, swimming pools and materials for impregnating cloth and paper.

6. Polyimides

Polyimides possess good mechanical, physical and electrical properties at elevated temperatures; they also have good creep resistance, low friction and low wear characteristics. Polyimides have the non-melting characteristic of a thermoset, but the structure of a thermoplastic. Typical applications include pump components (bearings, seals, valve seats, retainer rings and piston rings), electrical connectors for high-temperature use, aerospace parts, high-strength impact-resistant structures, sports equipment and safety vests.

7. Silicones

Silicones have properties that depend on their composition; generally, they weather well, possess excellent electrical properties over a wide range of humidity and temperature, and resist chemicals and heat. Typical applications include electrical components requiring strength at elevated temperatures, such as oven gaskets, heat seals and waterproof materials.

2.4.4.3 Rubber

Rubber is an elastic substance that is obtained by coagulating the milky juice of any of various tropical plants. It is essentially a polymer of isoprene, and is prepared as sheets and then dried. There are several types of rubbers as illustrated below.

1. Natural rubber

The base for natural rubber is latex, a milk-like sap obtained from the inner bark of a tropical tree. Natural rubber has good resistance to abrasion and fatigue, and high friction, but low resistance to oil, heat, ozone and sunlight. Typical applications are tires, seals, couplings and engine mounts.

2. Synthetic rubbers

Examples of synthetic rubbers are butyl, styrene butadiene, polybutadiene and ethylene propylene. Compared to natural rubber, they have better resistance to heat, gasoline and chemicals, and have a higher range of useful temperatures. Synthetic rubbers that are resistant to oil are neoprene, nitrile, urethane and silicone. Typical applications of synthetic rubbers are tyres, shock absorbers, seals and belts.

3. Silicones

Silicones have the highest useful temperature range of elastomers (up to 315°C), but other properties, such as strength and resistance to wear and oils, generally are inferior to those of other elastomers. Typical applications of silicones are seals, gaskets, thermal insulation, high-temperature electrical switches and electronic apparatus.

4. Polyurethane

This elastomer has very good overall properties of high strength, stiffness and hardness, and also has exceptional resistance to abrasion, cutting and tearing. Typical applications of polyurethane are seals, gaskets, cushioning, diaphragms for the rubber forming of sheet metals and automotive body parts.

2.5 GLASS FIBRE AND CARBON FIBRE: PROPERTIES AND APPLICATIONS

Glass fibre is a material made up of several fine fibres of glass. The product is one of the most versatile industrial materials known today. It has comparable mechanical properties to other fibres such as carbon fibre and polymers. Glass fibre is used as a reinforcing agent for many polymer products in order to form a very durable and lightweight material, known as fibreglass. Fibreglass offers some unique advantages over other materials due to its thickness, weight and strength. With such a wide range of properties, the material can satisfy design and project objectives in many industrial applications.

2.5.1 Glass Fibre

Glass fibres are the most widely used and least expensive of all fibres. The composite material is called glass-fibre reinforced plastic (GFRP), and may contain between 30% and 60% glass fibres. The fibres are made by drawing molten glass through small openings in a platinum die, which are then elongated, cooled and wound on a roll. Glass fibre is manufactured from various raw materials, namely, silica sand, alumina, limestone, clay and boric acid. Types of glass fibres include glass fibres classified as A, E, S, etc., which have particular fields of applications, these are:

1. A – glass fibre for acid resistance
2. C – glass fibre for improved acid resistance
3. D – glass fibre for electronic applications
4. E – glass fibre for electrical insulation
5. S – glass fibre for HS

Properties of glass fibre include:

1. High ratio of surface area to weight
2. Susceptible to chemical attack.

Applications of glass fibre include:

- The automobile industry is one of the largest users of glass fibre.
- Polymer matrix composites containing glass fibres are used to manufacture exterior body panels, pultruded body panels and air ducts, engine components, bumper beams, etc.
- There is wider use of this material in automotive applications since it possesses less weight so there is an enhancement in the fuel efficiency.

2.5.2 Carbon Fibre

Carbon fibre is a material consisting of extremely thin fibres of about 0.005–0.010 mm in diameter consisting of carbon atoms. Carbon fibre is a polymer, and is also referred to as graphite fibre. It is a very strong material as well as being lightweight. Carbon fibre is five-times stronger than steel and about twice as stiff.

Properties of carbon fibre include:

- High strength-to-weight ratio
- Good rigidity
- Resistant to corrosion
- Conducts electricity
- Resistant to fatigue
- Good tensile strength but brittle
- Fire resistant/not flammable
- High thermal conductivity
- Low coefficient of thermal expansion and low abrasion
- Non-poisonous
- Biologically inert and permeable to X-rays
- Self-lubricating
- Excellent shield against electromagnetic interference
- Relatively expensive
- Requires specialised experience and equipment for use
- High damping
- Electromagnetic properties

Applications of carbon fibre include:

- Racing car chassis
- Hoods
- Car emblems
- Mufflers
- Interior panels of a car
- Steering wheels

2.6 MODERN MATERIALS USED IN THE AUTOMOTIVE INDUSTRY

The economic performance of automotive manufacturing units is concerned with the development of new materials with long-term durability, economic viability, high component quality, and system reliability. Because of these multifunctional aspects of system integration, environmental factors such as energy performance, CO_2 reduction with system reliability and sustainability, lightweight engineering, and high fuel efficiency may influence the future of automotive materials. Carbon fibres are also gaining popularity as a new light-weight material, with 53% of major manufacturers currently using these composites. Carbon fibres for automotive use are projected to exceed 11,000 metric tons (MT) per year by 2025, up from about 7,000 MT in 2017. To minimise vehicle weight, Ford is using natural fibre-reinforced composites. Cellulose tree fibres are utilised to reinforce plastics and storage bins in the armrests of the Lincoln MKX mid-size SUV.

3M has developed hollow glass microspheres known as "glass bubbles", which are made of water-resistant and chemically stable soda-lime borosilicate glass. This can help reduce the composite weight by up to 40%. It is suitable for use as a filler in sheet or bulk-moulded composites. Tata Steel has introduced a replacement for the existing flooring and sidewall material used in buses and trailers. Coretinium is a polypropylene honeycomb core that has been optimised to minimise the weight of floors and sidewalls. It is currently used for commercial vehicles in the EU region.

SABIC, a chemical manufacturing firm, unveiled a fibre-reinforced thermoplastic composite bulkhead to replace conventional metal and thermoset plates. It is projected to have a 35% mass reduction as compared to current metals. Setex woven UDMAX tape was used to make the bulkhead. EcoPaXX PA 410, a bio-based polyamide produced by DSM Engineering Plastics, is used in the crankshaft cover of Volkswagen's MDB-4 TDI diesel engines.

The Cadillac CT6's body is made of carbon fibre-reinforced plastic (CFRP). CFRP components, such as the hood, splitter and wheel spats, are extremely lightweight, resulting in improved performance, fuel economy and traction at high speeds. BASF collaborated with Magna and Ford to create a carbon-fibre composite grill opening reinforcement. It was featured in the Ford Shelby GT350 Mustang in 2016. This material is 24% lighter than standard metal. Acrodur by BASF is a water-based, low-emission binder that strengthens natural fibres to produce a long-lasting, stable and lightweight solution for car roof frames. It was employed in the Mercedes-Benz E-Class.

2.7 MATERIAL DECISION-MAKING PROCESS

It is estimated that there are between 40,000 and 80,000 materials and at least 1000 different ways to handle them. When choosing materials for engineering designs, a thorough understanding of the functional specifications for each particular product is needed, as are a number of essential criteria or attributes to take into account. A material selection characteristic is one that affects the selection of a material for a specific application. These include chemical properties, physical properties, magnetic properties, electrical properties, mechanical properties, manufacturing properties (formability, machinability, castability, weldability, heat treatability, etc.), product shape, material cost, material effect on climate, performance characteristics, availability, fashion, consumer trends, cultural aspects, aesthetics, and recycling targets, etc. There are four simple steps for achieving a match with design specifications:

(1) A system for converting design criteria into material and process specifications.
(2) A tool for sorting out those that do not meet the requirements, leaving only a subset of the original menu.
(3) A framework for rating the surviving materials and processes and selecting those with the most potential.
(4) A method of looking for supporting information about the top-ranked candidates, presenting as much context information as possible about their strengths, shortcomings, history of usage and future potential.

2.8 BUSINESS FACTORS AFFECTING MATERIAL SELECTION

The business factors affecting material selection are as follows:

- Uninterrupted availability of the material
  It is essential to confirm the availability of the material required so as to maintain the supply and demand of the manufactured products.
- Political issues
  If there is any political interference it will directly affect the supply, purchase cost of the raw material, etc.
- Cost of manufacturing
  The cost of manufacturing consists of raw material cost, cost of processing, tooling cost, testing and certification cost, labour cost, etc.
- Environmental and social costs
  Since environmental regulations are becoming increasingly strict, it is essential to consider any environmental burden during the processing of the raw material as well as recurring costs.
- Governmental laws and regulations related to imports
  Government policies and rules and regulations directly affect the availability of materials.
- Foreign trade regulations related to limits of exports from other countries

2.9 SUMMARY

Nonferrous metals are those which do not contain iron as the base material. Copper, aluminium and magnesium and their alloys, ceramics, and polymers have wide automotive applications. The future of automotive materials may be influenced by environmental factors such as energy performance, CO_2 reduction with system reliability and sustainability, lightweight engineering, and high fuel efficiency. A material selection characteristic is one that affects the selection of a material for a specific application.

These include the physical properties, electrical properties, magnetic properties, mechanical properties, chemical properties, manufacturing properties, material cost, product shape, material effect on the climate, performance characteristics, availability, fashion, consumer trends, cultural aspects, aesthetics and recycling targets, etc. It is estimated that there are between 40,000 and 80,000 materials and at least 1000 different ways to handle them, so the decision-making process based on various criteria is important. Interrupted supply and availability, manufacturing cost, environmental and social issues, government laws and regulations, etc. are business factors related to material selection.

2.10 REVIEW QUESTIONS

1) List various properties of nonferrous materials.
2) List various nonferrous materials with their applications.
3) List various advanced automotive materials with their applications.
4) Describe the various factors affecting material selection.
5) Describe various polymers used in automotive applications with their properties.

3 Heat Treatment

LEARNING OBJECTIVES

- To comprehend the nuances of the phase concept and the phase transformation process.
- To instil a fundamental understanding of modern heat treatment processes, their requirements, and reheating principles.
- To comprehend the complexities of the heat treatment process used in automotive applications.

3.1 INTRODUCTION: CONCEPT OF PHASE AND PHASE TRANSFORMATION

The performance, functioning and working of any element/component during its service life depend on its composition, structure, ingredients and manufacturing and surface processing history and applied heat treatments. Strength, toughness, hardness, ductility, malleability and wear resistance are important properties of the material that have major impact through alloying elements and the heat-treatment processes employed. The properties of alloys which are not heat treatable are improved by mechanical working, such as rolling, forging and extrusion. Heat treatments are an important process through which modifications in the microstructure are carried out to improve the properties as per the application requirements such as crankshaft, camshafts, dies and moulds. Therefore, it is essential to study the alloy making process and the special heat treatments to be applied in materials science.
The objectives/aims of these processes are as follows:

- **To soften** – to improve plasticity (or slip band formation) by modifying the size, shape and distribution of micro-constituents.
- **To stress-relieve** – to allow relaxation of residual stresses by raising the temperature in order to lower the yield strength and enhance recovery.
- **To homogenise** – to obtain compositional uniformity in a phase by the diffusion of alloying elements at higher temperature, for example, austenitising and solutionising.

DOI: 10.1201/9781003367321-3

- **To toughen** – to develop the ability to absorb energy or withstand occasional stresses in the plastic range without fracturing (to increase the total area under the stress–strain curve).
- **To harden** – to provide slip interference by altering the size, shape and distribution of micro-constituents by grain refinement, quench-hardening or age hardening.
- **To add chemical elements through the surface** – to improve wear and fatigue resistance caused by the development of surface-compressive residual stresses arising from the absorption of interstitial solute atoms under a suitable thermal cycle (carburising, nitriding, etc.).
- **To accomplish special surface processing operating** – such as to develop magnetic properties by producing a coarse-grained structure by high-temperature treatment, to remove chemical elements from the surface (by heat expulsion or chemical method at elevated temperature) or to improve the electrical properties and cold, subzero, deep cryogenic treatment.

3.1.1 Concept of Phase

A physically distinct and homogeneous portion or part in a material is called the phase of the material. Each phase is a homogeneous part of the total mass, and has its own characteristics and properties. For example, water has three phases – liquid water, solid ice, and steam. A phase has specific characteristics are as follows:

- The same structure or atomic arrangement throughout;
- Roughly the same composition and properties throughout;
- A definite interface between the phase and any surrounding or adjoining phases;
- Consider a mixture of sand and water as an example of a two-phase system. These two different components have their own distinct structures, characteristics and properties;
- There is a clear boundary in this mixture between the water (one phase) and the sand particles (the second phase).
- Another example is ice in water: the two phases are the same chemical compound of exactly the same chemical elements (hydrogen and oxygen), even though their properties are very different.
- It is not necessary that one phase be a liquid; for example, sand suspended in ice is also a two-phase system.
- A typical example of a two-phase system in metals occurs when lead is added to copper in the molten state.

Phase diagram

A graphical representation of the relationships between environmental constraints (e.g. temperature and sometimes pressure), composition and regions of phase stability, ordinarily under conditions of equilibrium is called a phase diagram. Most phase diagrams are prepared by using slow cooling conditions whereby phases are in equilibrium. One can get following important information from the phase diagrams, such as:

- Phases at different composition and temperature;
- Equilibrium solubility of one element or compound in another element;
- Melting points of different phases in an alloy;
- Temperature of solidification or range of solidification of an alloy.

Phase equilibrium

The state of a system where the phase characteristics remain constant over indefinite time periods is called phase equilibrium. At equilibrium the free energy is a minimum. In an equilibrium diagram, liquid is one phase and solid solution is another phase.

3.1.2 Phase Transformation

The microstructure of an alloy is constituted as an effect of a change in the number or/and characteristics of the phases. In single- and two-phase alloys ordinarily the microstructure development involves some type of phase transformation. The transformation of a metal from one phase to another is a part of the metallurgical process. Every transformation depends on the two factors: a thermodynamic factor that determines whether the transformation rate is possible and a kinetic factor that determines whether transformation is possible at a practical rate.
These transformations are usually divided into three groups:

- Simple diffusion-dependent transformations: In these transformations there are no changes in either the composition or number of phases present. These transformations consist of the solidification of pure metals, allotropic transformations and recrystallisation and grain growth.
- Diffusion-dependent transformations: In these transformations, there is some alteration in phase compositions and also in the number of phases present. The final microstructure generally consists of two phases.
- Diffusion less transformation: In these transformations there is a development of the metastable phase.

The time required for the transformation to go to completion is quite important in the control of the microstructure of a material.

3.1.2.1 Applications of the phase transformations

Phase transformations are usually observed in microstructural changes in cooling or freezing (dendrite formation), in castings, in amorphous structures (solidification phenomenon in glassy structures), in heat treatment or in the binary phase diagram of Fe–Fe_3C system and TTT diagrams for eutectoid steels.

3.1.2.2 Phase transformation in Fe–C systems

Iron and carbon make a series of alloys, including a number of steels and cast iron, and are the most important subjects in the study of steels and cast irons. Iron and carbon are polymorphic elements. Steels are alloys containing up to 1.2% of carbon with cast irons containing 2.3–4.2% of carbon. Carbon alloys greater than 4.6% have poor properties and are not preferred. Fe–C alloys are very important for modern industry due to the wide range of applications of ferrous metallurgy in manufacturing, casting and associated manufacturing processes. Theoretically speaking, it is based on the transformation of phases in metals, e.g. in the manufacture of iron, the transformation of liquid iron into a solid state, as in the casting of metals; heat treatment processes where different mechanical properties can be achieved by changing the structure of the crystal. The Fe–C system provides vital information on heat treatment, based on polymorphic transformation as well as eutectoid decomposition. The phases that can be found in Fe–C alloys are mostly liquid solution, ferrite, austenite, cementite and carbon-free (as graphite)

3.1.3 Cooling Curve for Pure Iron

Iron can be generated in more than one crystalline form. Figure 3.1 shows the cooling curve for pure iron. The melting point for pure iron is 1539°C. If it is cooled from its molten state, it solidifies the bcc phase first. This form of iron is known as ferrite. As the temperature drops to 1400°C, it changes from the phase to the FCC phase. This is commonly referred to as iron or austenite. Later, at 910°C, it is again transformed into a BCC crystal known as ferrite (δ). Each of these transformations is linked to a change in volume and enthalpy. As a result, the cooling curve is expected to show steps or discontinuities at these three different temperatures.

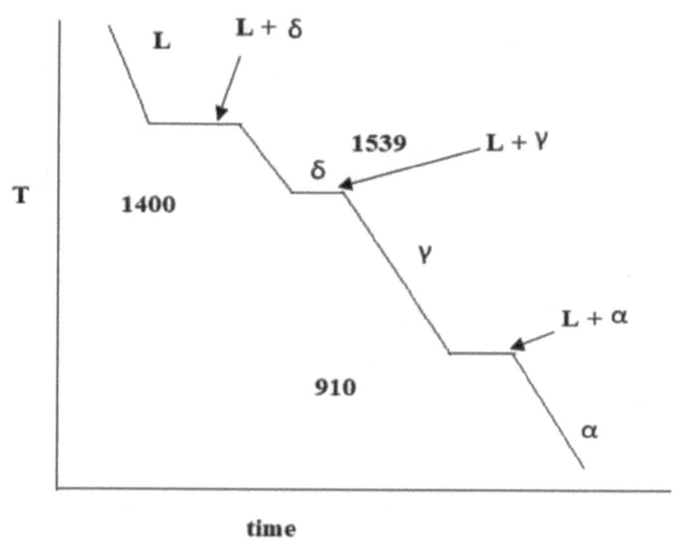

FIGURE 3.1 Cooling curve for pure iron.

3.2 IRON–IRON CARBIDE PHASE EQUILIBRIUM DIAGRAM

1. Ferrite is also known as a stable solution. It is an interstitial solid solution which contains a small amount of carbon dissolved in bcc iron. Figure 3.2 shows an iron–iron carbide equilibrium diagram. It is a type of iron which is stable below 912°C. The maximum solubility of ferrite is 0.025% C and 0.008% C at 723°C and room temperature, respectively. The soft and ductile structure is shown in the diagram.
2. Interstitial solid carbon solution dissolved in fcc iron is nothing but austenite. The maximum solubility of austenite is 2.0% C at 1130°C. It is highly formable and most heat treatments begin with this single phase. Austenite is not generally stable at room temperature. However, it is possible to produce austenite at room temperature under certain conditions.
3. A supersaturated solid carbon solution in ferrite is nothing but martensite. The formation of martensite can be formed when the steel is cooled so quickly that the transition from austenite to pearlite is suppressed. Interstitial carbon atoms distort bcc ferrite into the BC-tetragonal structure (BCT) responsible for the hardness of the quenched steel.
4. The eutectic mixture of austenite and cementite is referred to as ledeburite. Ledeburite contains 4.3% C and is produced at 1147°C.

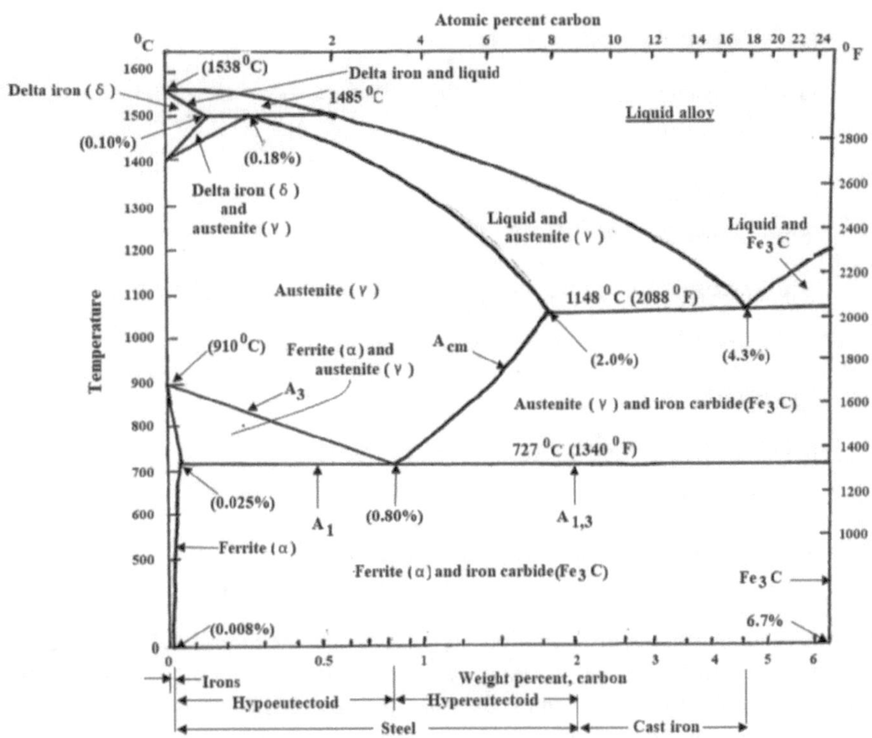

FIGURE 3.2 Fe–Fe_3C diagram.

5. Cementite or iron carbide is a brittle and very strong, intermetallic alloy of iron and carbon, as Fe_3C, which contains 6.67% C. Cementite or iron carbide is the hardest structure that appears in the diagram but its exact melting point is unknown. Ledeburite crystal structure is orthombic. Ledeburite has low tensile strength (approximately 5,000 psi) but high compressive strength.
6. The 0.80% C-containing eutectoid mixture is known as pearlite and is formed at 723°C with very slow cooling. It is a very fine plate-like or lamellar mixture of ferrite and cementite. The white ferritic backdrop or matrix comprises thin cementite plates.

Transformations

- **Peritectic reaction**

  Liquid+Solid1↔Solid2

  L(0.53wt%C)+δ(0.09wt%C)↔γ(0.17wt%C) at 1495°C

  Liquid-18.18wt% +δ-ferrite 81.82 wt%→100wt% γ

- **Eutectic reaction**

  Liquid↔Solid1+Solid2

  Liquid (4.3wt%C) ↔ γ(2.11wt%C) + Fe3C (6.67wt%C) at 1147°C

  Liquid-100 wt% →51.97wt% γ +Fe3C (48.11wt%)

  The phase mixture of austenite and cementite formed at eutectic temperature is called ledeburite.

- **Eutectoid reaction**

  Solid1↔Solid2+Solid3

  γ(0.77wt%C) ↔ α(0.0218wt%C) + Fe3C(6.67wt%C) at 727°C

  γ (100 wt%) →α(89 wt%) +Fe3C(11wt%)

  Typical density

  α ferrite=7.87 gcm-3

  Fe3C=7.7 gcm-3

  Volume ratio of α- ferrite: Fe3C=7.9:1

The time–temperature transformation curve is shown in Figure 3.3. The log time scale given on the X axis and the temperature scale is provided on the Y axis. There are two horizontal lines shown in the diagram, the upward line represents the M_s temperature line, i.e. the start of the martensite temperature transformation line, while the lower line represents the M_f line, i.e. the finish of the martensite transformation line.

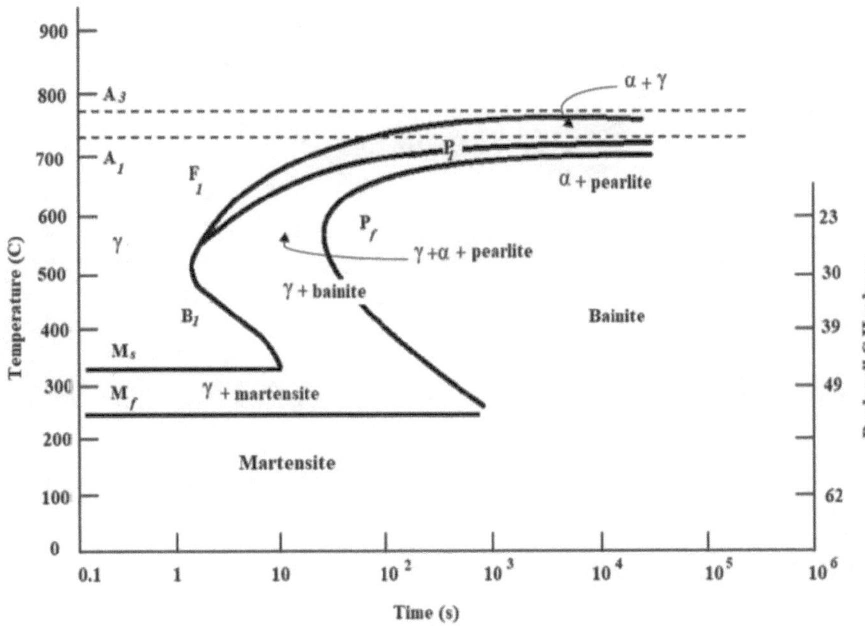

FIGURE 3.3 Time–temperature transformation diagram.

At the extreme top horizontal line shown in the diagram is the Ac1 line at 723°C. As we go down the perlite, the fineness and hardness of the steel improve. The nose of the TTT diagram moves in the right direction with an increasing percentage of carbon in steel. The start of the pearlite is marked by the left curve and the end of the pearlite by the right curve. Martensite is produced in the lower region, coarse and fine pearlite is produced in the upper right region and the upper and lower bainites are produced in the middle region. In between the two curves, there are austenite, ferrite and pearlite. Depending on the cooling rate, different transformation products are obtained in the diagram.

3.3 COMMON HEAT TREATMENT PROCESSES AND THEIR APPLICATIONS

Heat treatment is the term that describes an operation, or combination of operations, which involves the controlled heating and/or cooling of a metal or alloy in the solid state in order to modify the existing structure deliberately and/or bring about a change in its properties. However, the International Federation for the Heat Treatment of Materials (IFHT) has defined this term as "a process in which the entire object, or a portion thereof, is intentionally submitted to thermal cycles and, if required, to chemical or additional physical actions in order to achieve desired (change in the) structures and properties."

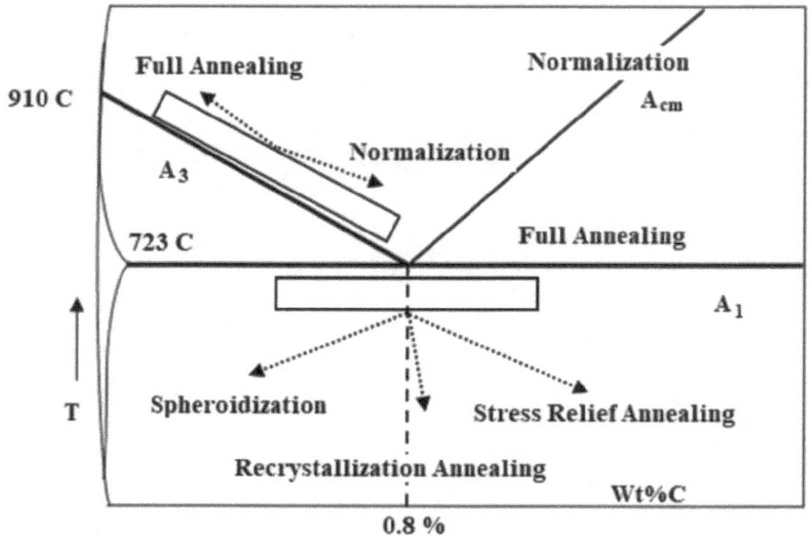

FIGURE 3.4 Common heat treatment processes.

3.3.1 Types of Heat Treatment Processes

Common types of heat treating methods include annealing, hardening, quenching and stress relieving, each of which has its own unique process to produce different results as explained in the following sections. Figure 3.4 shows common heat treatment processes.

1. Annealing processes

These include (a) full annealing; (b) homogenise annealing, solution treating, or austenitising; (c) spheroidising, critical range annealing, or subcritical annealing; (d) stress-relief annealing; (e) process annealing; (f) recrystallisation annealing; (g) isothermal annealing; (h) dehydrogenation annealing, (i) blueing and bright annealing; (j) box annealing; and (k) continuous annealing.

2. Normalising

3. Through hardening and tempering processes

These include (a) water-, oil-, or air-quenching and tempering; (b) time-quenching and tempering; (c) isothermal quenching and tempering (e.g., patenting); (d) austempering; and (e) martempering.

4. Other through-hardening processes

These include (a) precipitation (age) hardening; (b) dispersion hardening; (c) maraging; (d) thermomechanical treatment; and (e) order–disorder reactions.

5. Thermal surface hardening treatment (i.e., without compositional change)

This includes (a) flame hardening; (b) induction hardening; (c) laser hardening; and (d) electron-beam hardening.

6. Thermochemical surface hardening treatment (i.e., with compositional change)

(a) Austenitic thermochemical treatment:
- Carburising, solid, liquid, gas, vacuum, plasma, fluidised bed
- Carbonitriding, liquid, gas, plasma
- Cyaniding

(b) Ferritic thermochemical treatment:
Nitriding, liquid, gas, plasma
Nitrocarburising, liquid, gas

7. Other surface diffusion treatments

(a) Siliconising
(b) Chromising
(c) Boronising
(d) Ion implantation, etc.

8. Miscellaneous heat treatments

These include (a) patenting; (b) stiffening temper; (c) malleabilising; (d) cold, subzero, or deep cryogenic treatments; (e) restoring stainless characteristics; (f) developing magnetic characteristics; (g) improving electrical properties; and (h) relieving embrittlement.

3.3.2 Annealing

Annealing describes a heating process above a certain temperature, which is to be sustained at that temperature for a specified period of time, followed by cooling at a predetermined rate (usually in the furnace) to room temperature in order to achieve or restore desirable properties such as maximum ductility, low strength and improved machinability and cold-workability.

The annealing process is carried out primarily in order to obtain the following properties:

- Softening the steels;
- Boosting machinability;
- Relieving internal stress caused by any prior treatment (rolling, forging, extrusion, uneven cooling);
- Removing the coarseness of the grains;
- Producing a fully stable system.

It is applied to castings, forgings, cold-worked sheets and wires.

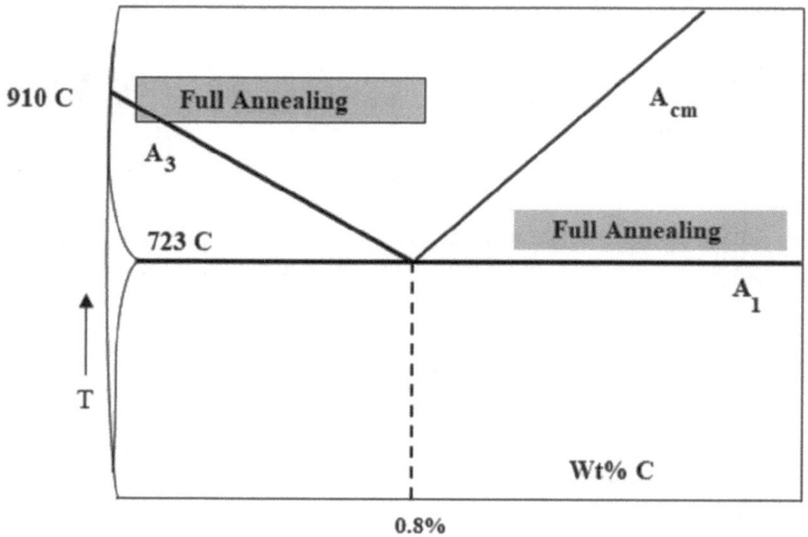

FIGURE 3.5 Full annealing process.

The process consists of:

- Heating of steel to a certain predetermined temperature;
- Soaking at a constant temperature for a sufficient time to allow the required adjustments to occur;
- Cooling at a very slow predetermined rate.

1. Full annealing

Figure 3.5 shows the full annealing process. The purpose of full annealing is to:

- Minimise internal stress caused by cold working, welding, etc.;
- Decrease hardness and improve ductility;
- Refine the size of the grain;
- Improve machinability;
- Make the steel ready for further heat treatment.

Process

- Steel is heated above A3 (for hypo-eutectoid steels) and A1 (for hypereutectoid steels and held down for a certain period of time, then the furnace is cooled to produce coarse pearlite. A1 is the lower critical temperature and A3 is upper critical temperature.
- Coarse pearlite has low hardness but high ductility.
- In the case of hypereutectoid steels, the heating does not take place above Acm (γ/γ+cementite phase field boundary) to prevent a continuous network of proeutectoid cementite along the previous austenite grain boundary.

2. Process annealing

This is also referred to as sub-critical annealing or recrystallisation. Its purpose is to:

- Soften the part to restore ductility;
- Remove internal stresses developed in the casting by welding or by prior heat treatment.

Process

Steel is heated to a temperature of between 600–650°C, held at that temperature, and then cooled in the air or the furnace. This process results in a high degree of softening due to the elimination of stress from pearlite. There is no phase shift and ferrite and pearlite actually rearrange themselves to cause softening of the materials.

3. Isothermal annealing

This process is suitable for small rolled and forged components but not for large ones. It is quicker than full annealing and saves a lot of time.

Purpose

- To achieve a stable structure;
- To save the time required for heat treatment.

Process

- The process is similar to ordinary annealing, but it is first rapidly cooled in the air or by blowing in the furnace at a temperature of 600–700°C;
- The steel is kept at this temperature for a certain amount of time until it is quickly cooled in air.

4. Spheroidise annealing

The method of creating a globular pearlite structure is known as spheroidising or spheroidising annealing.

Purpose

- Enhancing the machinability of steel;
- Minimising hardness;
- Avoiding the risk of cracking during cold working.

Process

In general, this procedure is applied to hypereutectoid steels. Steel is heated just above the low critical temperature (740–770^{0}C) and held for the necessary time

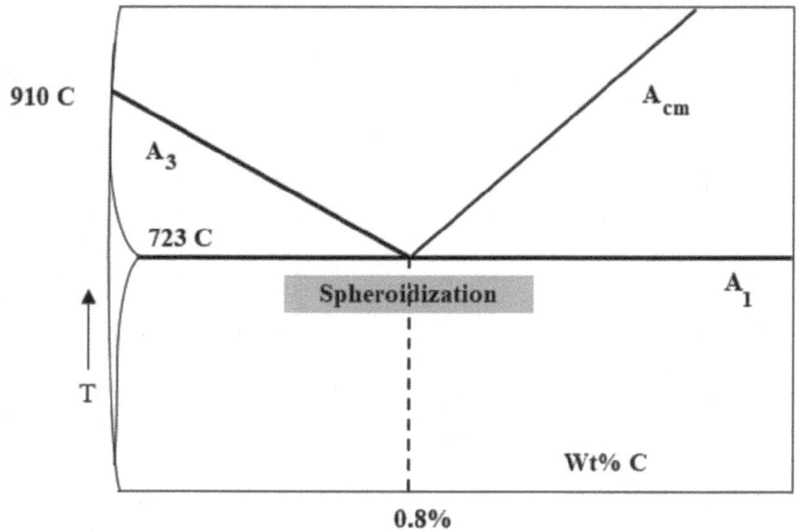

FIGURE 3.6 Spherodising.

and cooled very slowly to 600°C in the furnace. Further cooling is carried out in still air. The rate of cooling ranges from 20 to 25°C per hour. It should be noted that heating well above Acm would produce lamella instead of granular cementite (Figure 3.6).

5. Homogenising

This is often referred to as diffusion annealing.

Purpose

- To remove the non-uniformity of the castings, this is induced by coring. Coring is a change in the composition from the centre to the surface of the grain.
- To strengthen the steel structure.

Process

The steel is heated as quickly as possible to a temperature of up to 1150°C and maintained at that temperature for sufficient time to allow the diffusion to take place. It is then air cooled to 800–850°C in 6–8 hours. Complete annealing is performed after homogenisation to refine the grain structure.

3.3.3 Normalising

Normalising consists of heating the steel to a temperature 40–50°C above the upper critical points (that is, the *Ac*3 or *Ac*cm temperatures), holding it there for a period depending on

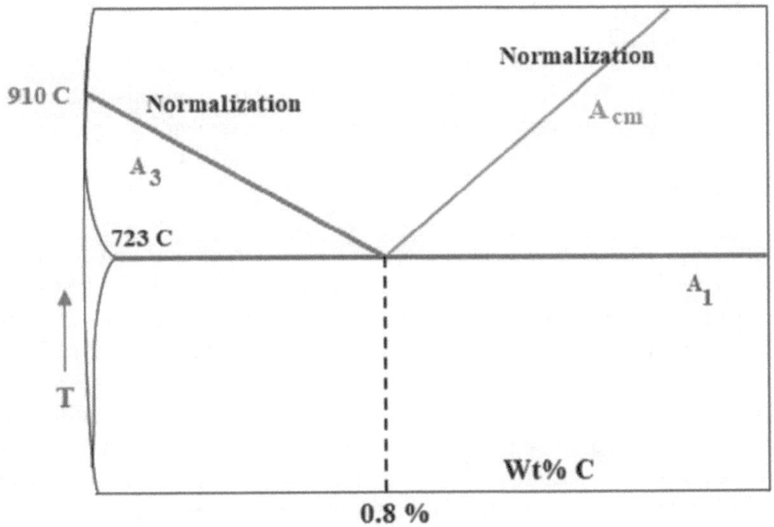

FIGURE 3.7 Normalising.

the dimensions of the part and type of steel being treated until it is completely austenitised, followed by cooling, in still air, free of drafts, to room temperature (Figure 3.7).

Purpose

- Remove the coarse-grained structure;
- Eliminate internal stresses that could have been caused by work;
- Boost the mechanical properties of steel;
- Increase the strength of medium-carbon steel to a certain degree (in comparison with annealed steels);
- Increase the machinability of low-carbon steels;
- Boost the welding structure.

Normalisation is also used as a final heat treatment for products that are required to work at relatively high stress levels.

Process

Heating of the metal to temperatures within the normalisation range, typically between 40°C and 50°C above Ac3 (for hypoetectoid steels) and Acm (for hypereutectoid steels); keeping for a limited time at this temperature (about 15 minutes) it can be air-cooled. Normalised steels have higher yield points, tensile strength and impact strength than if they are annealed, but the ductility and machinability gained by normalising would be much lower.

3.3.4 Hardening and Quenching

Quench hardening is a mechanical process in which steel and cast iron alloys are strengthened and hardened. These metals consist of ferrous metals and alloys. This is done by heating the material to a certain temperature, depending on the material.

- **Hardening**

By hammering, bending and cold rolling of iron materials and non-ferrous metals reinforcement of the metal structure is accomplished, this is called "cold straining." Another method is "age-hardening" of light metals that is carried out after the heat treatment is finished by storing the materials for a period of several days. Steels with a carbon content below 0.35% may be hardened on their surfaces when carbon is added to the steel from outside by means of a special process (case hardening). The most critical type is "hardening by means of heat treatment" by which the structure of steels is systematically modified. This form of heat treatment is used for the manufacture of hard and wear-resistant steels for specific purposes. When the steel is heated, the properties of the steel are changed, depending on the carbon content.
Conditions for hardenability include:

Carbon content:	Above 0.35% – hardenable
	Below 0.35% – not hardenable

The lower the carbon content, the higher must be the hardening temperature.

- **Objectives**

- To improve mechanical properties, like strength, ductility elasticity, toughness, etc.;
- To enable the metal to cut other metals;
- To develop desired hardness.

Hardening process consists of four phases.

First phase: Heating of steel at temperatures above A1 for hypereutectoid steels by 500^{0}C and above A3 for hypoeutectoid steels.
Second phase: For homogeneous austenisation, keeping the steel components for a reasonable soaking time.
Third phase: Cooling of hot steel components at a rate that exceeds the critical cooling rate of the steel at room temperature or below room temperature.
Fourth phase: Tempering of the martensite to achieve the hardness required.

- **Quenching**

The rapid cooling of a metal in a liquid bath during heat treatment is known as quenching, e.g., steel is heated above its critical temperature and plunged into water to cool it, creating an incredibly rigid, needle-shaped structure known as "martensite." The rate at which the heat is consumed by the quenching bath has various effects on the hardness of the metal. Cold clean water is used as a quenching medium, although the addition of salt increases the hardness considerably. Oil offers the strongest balance between hardness, toughness and distortion. Special soluble oils are used as a quenching media. The hardened pieces have good tensile strength, but low ductility, durability and impact strength.

3.3.5 Tempering

The hardened steel is not readily suitable for engineering applications. It possesses the following three drawbacks.

- Martensite obtained after hardening is highly brittle in nature and can be subject to failure of the structural parts by cracking.
- Martensite formed by austenite extinguishing is subjected to the creation of high internal stresses in hardened steel.
- The structures to be recovered after hardening consist of martensite and retained austenite.

All phases of austenite are metastable and can transition to stable phases, resulting in improvements in the dimensions and properties of the steel in operation over time. The method of tempering can reduce these issues. The hardened steel is heated to a high temperature up to a lower critical temperature (A1) and can be soaked at this temperature and then cooled, typically at a slower rate.

Purpose

- Reduction of internal stresses generated during the previous heating process;
- The reduction in hardness produced during the hardening process;
- To give the metal the correct structural condition to stabilise the structure.

Process

The first phase of temperature increase is from room temperature to 200°C. Steel tempering reactions, containing less than 0.2% carbon, vary significantly from steels containing more than 0.2% carbon. The structure at this point is referred to as tempered martensite, which is a double-phase mixture of low tetragonate martensite and ε-carbide. The second phase of the temperature is about 200–300°C. In the second phase of tempering, the preserved austenite is converted into lower bainite (the bainite carbide is ш-carbide). The third stage of temperature is between 200–350°C. At this point of tempering, ε-carbide dissolves in the matrix, and low tetragonal martensite loses the carbon and hence the tetragonality to become ferrite. The fourth stage of tempering temperature is about 350–700°C. Cementite growth and spheroidisation, as well as ferrite recovery and recrystallisation, occur.

In the centre, the optical microstructure consists of equiaxed ferrite grains with coarse cementite spheroidal particles, and then the structure is called globular pearlite or spheroidised cementite. This structure is perhaps the most stable of all ferrite–cementite aggregates and the softest with the highest ductility and greatest machinability possible.

Tempering is done after hardening for the following reasons.

- After hardening, as the metal is taken from the quenching medium, it is very hard but brittle and there are many inequalities present in the metal structure. Tempering is performed to restore ductility and reduce ductility.
- The procedure involves reheating the metal below the critical point, retaining it for a long time and then slowly cooling it down.

Tempering should be conducted immediately after quenching in order to relieve hardening strains. The temperature at which the tempering is carried out changes with the carbon content of the metal and the mechanical properties of the finished article. In the case of lathe tools, chisels etc., it is necessary only for cutting edges to be hardened, which can be hardened and tempered in one operation. The entire tool is heated to a hardening temperature and quenching can be carried out for only for the cutting edge. The entire tool is considered as quenched when the cold end is brightly ribbed, the heat from the unquenchable part is tempered, and the colour is satisfactory.

3.4 SURFACE HARDENING PROCESSES: GAS CARBURISING, NITRIDING, CYANIDING

Sometimes special characteristics are required in metal such as a hard outer surface and soft, tough and more strength-oriented core or inner structure of metal. This can be obtained by a case hardening process. It is the process of carburisation, i.e. saturating the surface layer of steel with carbon or some other substance, by which the outer case of the object is hardened whereas the core remains soft. It is applied to very low-carbon steel. It is performed for obtaining hardness and wear resistance on the surface of metal and higher mechanical properties with higher fatigue resistance, strength and toughness in the core. The following describes the case hardening processes.

- **Surface hardening processes**

Requirements for a significant range of mechanical components, including gears, cams, drive worms, brake drums, and gear shafts, include to have a surface that is resistant to wear and tear. It consists of a soft core, so it must have shock resistance as well as be tough. Plain carbon steel as well as alloy steels cannot fulfil both requirements, i.e. a hard surface as well as a tough core that is resistant to shock. Steel containing 0.1% carbon is tough, whereas steel containing 0.8% C is very hard and brittle. Both these properties can be achieved by the process of case hardening. Case hardening is the heat treatment process used to produce a hard wear-resistant carbon-rich case (surface layers) on a tough and soft core of steel. Low-carbon steel is used for the case hardening processes except in induction and hardening, where medium-carbon steel or high-carbon steel is used. The processes generally employed for case hardening are as follows:

- Carburising;
- Cyaniding;
- Nitriding;
- Carbonitriding;
- Flame hardening;
- Induction hardening.

3.4.1 Carburising

- **Process**

The process of enrichment of carbon in the surface layer of low-carbon steel to produce a hard case is called carburising. Figure 3.8 shows packing components for solid carburizing. Carburising is also referred to as cementation. Generally, many machined elements use low-carbon steel. For the carburising process these components are packed with carburising mixture in a steel box. The carburising mixture contains 50–70% charcoal, 5–15% barium carbonate, 2–15% calcium carbonate and 3–13% sodium carbonate. A layer of the carburising mixture of almost 25 mm thickness is

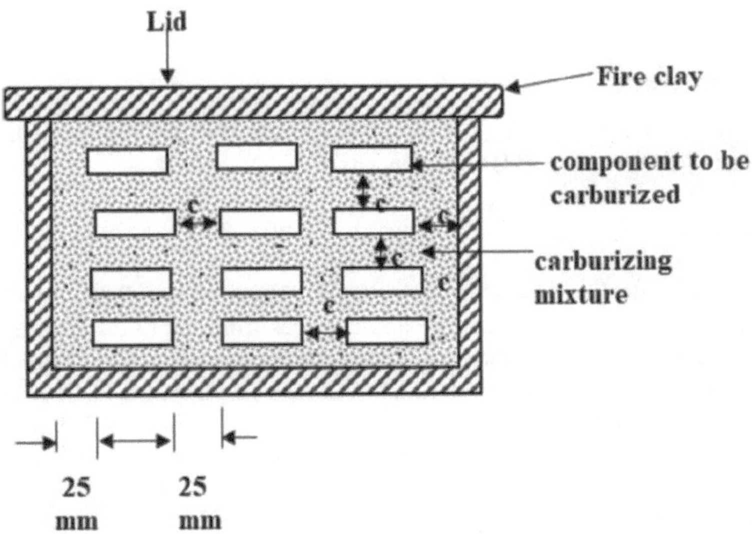

FIGURE 3.8 Packing components for solid carburising.

placed at the bottom. Then, the components are placed so that no component touches another or the sides of the box. The box is covered and the lid tightly sealed to avoid the entry or escape of gases.

The portion or surface of the component which does not need to be case hardened is protected by an electroplating process so that it does not absorb carbon. These boxes are placed in a furnace which it can be heated to a temperature in the range of 900–980°C for about 6–8 hours. The temperature and time of heating depend upon the depth of the case required. After heating for a specified time, the box is held for cooling with the components remaining inside the furnace. As austenite continues to absorb carbon at high temperatures, the percentage of carbon on the surface increases. The depth of the case obtained varies from 1 mm to 1.5 mm with the carbon content on the outer surface at 1.1–1.2%.

3.4.2 Cyaniding

The method of producing a case with hard-wear resistance with a hard core of low-carbon steels with the use of a liquid cyanide bath is called cyaniding.

- **Process**

The mixture of cyanide (20–50% sodium cyanide and 40% sodium carbonate) is heated to between 870–930°C and a wire basket with the parts is immersed in a molten cyanide bath. The soaking time requirement varies, depending on the case depth, but usually is from 10 minutes to 3 hours. Nitrogen in atomic form can dissolves on the surface and increases in hardness as an effect of nitride formation. In the nitriding process, the portion of the surface of the parts to be softened and coated with specific materials is not affected by the molten cyanide bath. Careful handling of cyanide is required as these salts are very poisonous.

3.4.3 Nitriding

The heat treatment process that creates a heavy-wear resistant nitride coating on a hard core of low-carbon steel is known as nitriding.

- **Process**

This process is ideal for 1% aluminium, 1.5% chromium and 0.2% molybdenum steels. In these steels, the amount of carbon ranges from 0.2 to 0.5%. The process consists of heating machined and heat-treated components at a temperature of 500°C for 40–90 hours in a gas container in which ammonia gas is circulated. The basic condition for the operation is close adherence to a temperature of 500°C. The part may be cooled in the oven after the supply of ammonia has been transferred. If ammonia vapours come into contact with steel, they are disassociated. $NH_3 = 3H + N$ and the nascent nitrogen thus formed are diffused to the surface of the hard nitride workpiece (Figure 3.9).

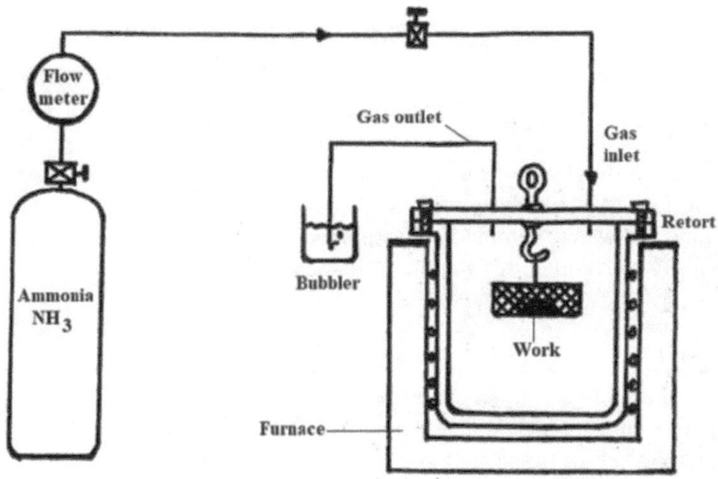

FIGURE 3.9 Nitriding process.

3.4.4 Carbonitriding

The method of adding carbon and nitrogen to the surface of steel to make a hard case is called carbonitriding.

- **Process**

Hydrocarbons, ammonia and carbon monoxide gases are used for the carbonitride process. At a temperature of 800–875°C for 6–10 hours carbonitriding is carried out, with the depth of the case being 0.5 mm. Low-carbon steels (steels used for carburising) are treated with the carbonitriding process. Nitrogen in the surface layer of the steel increases its hardenability and permits hardening in oil quenching. The risks of distortion and cracking are also minimised. The portion of components which is not to be carbonitrided is protected by copper plating.

3.5 INDUCTION AND FLAME HARDENING

Induction hardening involves using induced electrical currents to very rapidly generate heat via hysteresis, usually in a workpiece made from medium- to high-carbon steel. Flame hardening uses oxy-fuel burners to heat the workpiece via conduction.

3.5.1 Induction Hardening

The inductive heating method for surface hardening is known as inductive hardening.

- **Process**

High-frequency current is transmitted through the inductor blocks that surround the surface to be hardened without being directly affected. The inductor block current induces current on the surface of the metal around the block. The induced eddy current and the loss of hysteresis in the surface material have an effect on the heat available. When the surface to be hardened is heated for a reasonable amount of time, the circuit is opened and the water is immediately sprayed onto the surface for quenching. It is widely used for the hardening of crank shafts, cam shafts and axles.

Benefits:

- The time required for this process is less;
- Deformation is reduced;
- Hardening can be monitored by controlling the current;
- The depth of the hardening can be controlled.

Disadvantages:

- High costs of equipment;
- High cost of maintenance;
- The method is suitable only for large-scale production.

3.5.2 Flame Hardening

In this process, the flame of an oxyacetylene torch is used for heating the metal which is then immediately quenched, called flame hardening. Four methods are generally used in flame hardening by industry:

- Stationary (spot): Torch and work are stationary;
- Progressive: Torch moves over a workpiece;
- Spinning: Torch is stationary while the workpiece rotates;
- Progressive-spinning: Torch moves over a rotating workpiece.

The surface to be case-hardened is heated by means of an oxyacetylene torch for ample time and the water sprays which are integrally attached to the heating system are used to quench it. Heating is usually carried out for sufficient time to increase the temperature of the surface of the specimen above the critical temperature. As the target temperature is reached immediately, the spraying of the water is initiated. In mass production work, the gradual hardening of the surface is carried out where the flame is organised together with the in-progress quenching.

Advantages:

- The selective surface can also be hardened on very large parts;
- There is less distortion than in traditional approaches.

Disadvantages:

- The temperature cannot be regulated precisely;
- Hardening is limited to pieces that are affected by wear.

3.6 HEAT TREATMENT PROCESSES IN THE AUTOMOTIVE INDUSTRY

Automotive heat treatment is essential for maintaining the structural integrity of modern cars. Metal components vary in type and quality, and different heat treatment methods are required for lightweight aluminium body parts, along with high-strength engine and drivetrain components. There are many types of automotive heat treatments performed in the industry, and some of the more common methods of application as they relate to the parts are discussed below.

- **Powertrain**

A car's engine/transmission is made up of safety-critical, fast-moving parts that require durability, surface hardness for wear resistance, and strong fatigue strength when working in high-heat and corrosive environments.

Induction hardening, low-pressure carburising, ferritic nitrocarburising, carbonitriding and gas carburising are typical processes used in a powertrain. The process chosen is determined by the type of steel, the necessary properties, the surface finish and the dimensional tolerances. The following are examples of typical treated parts: shafts, pinions, gears, parking pawls, valves, retainers, hubs, support arms and pistons.

- **Drivetrain**

Drivetrain components are subjected to heavy operating loads and may be exposed to harsh environmental conditions. Many sections are hardened to achieve the necessary strength, while others are hardened in wear-prone areas. Additional procedures may be carried out to provide corrosion resistance, such as: induction hardening, furnace hardening, carburising and ferritic nitrocarburizing, which are common heat treatment processes for drivetrain components. The following are examples of typical treated parts: wheel hubs, differential pins, axle shafts, differential cases, pins, gears, hub annulus, friction plates and gear carriers.

- **Brake systems**

When it comes to critical systems on any vehicle, the braking system is probably the most relevant. Failure of any part of the brake system will have significant repercussions, therefore material and processing specifications must be strictly followed. Unless properly secured, brake system components will deteriorate due to the brake hydraulic fluid. The following are examples of typical treated parts: brake shoe, calliper piston and various stampings.

The brake components must be resistant to wear and corrosion. Induction hardening, nitriding and other case hardening methods are also used in the braking system.

- **Suspension and chassis**

The frame, suspension and steering components must be durable, with strong wear resistance and corrosion resistance due to exposure to water and other road surface fluids. Furthermore, as with many automotive parts, cost must be kept as low as possible for competitive reasons. Induction hardening, ferritic nitrocarburising and continuous furnace quenching and tempering are examples of common processes used. The following are examples of typical treated parts: ball studs, stampings, torsion bars, fasteners, seating, arm levers and pins.

3.7 SUMMARY

Strength, toughness, hardness, ductility, malleability and wear resistance are important properties of materials which have a major impact by alloying elements and the heat-treatment processes employed. Heat treatments are an important process through which modifications in the microstructure are carried out to improve the properties as per the application requirements such as crankshafts, camshafts, dies and moulds. A graphical representation of the relationships between environmental constraints (e.g. temperature and sometimes pressure), composition and regions of phase stability, ordinarily under conditions of equilibrium is called a phase diagram. A car's engine/transmission is made up of safety-critical, fast-moving parts that require durability, surface hardness for wear resistance and strong fatigue strength when working in high heat and corrosive environments.

3.8 REVIEW QUESTIONS

1) Define the following terms:
    (a) Phase equilibrium
    (b) Phase
    (c) Solute
    (d) solid solution
    (e) Interstitial solution
    (f) Substitutional solution
    (g) Heat treatment
    (h) Annealing.
2) State the properties of a phase.
3) State the purpose of annealing, normalising and hardening.
4) State the methods used for the hardening.
5) List the objectives of heat treatment.
6) State the different types of annealing processes.
7) Describe the following processes:

(a) Annealing
(b) Carburising
(c) Tempering
(d) Normalising
(e) Hardening
(f) Quenching.

8) Describe the following processes:
   (a) Carburising
   (b) Cyaniding
   (c) Induction hardening
   (d) Nitriding
   (e) Flame hardening.
9) Describe the Fe Fe_3C diagram.

4 Moulding and Casting

LEARNING OBJECTIVES

- To understand the basics of moulding casting process, their terminologies.
- To understand the details of pattern making, their materials, types and their applications as per requirements.
- To learn the casting and moulding processes, as well as their various types, the defects, remedies and applications.
- To get acquainted with aspects of the casting process's design.
- To understand the applications castings in vehicular applications.

4.1 INTRODUCTION

Casting is the method of pouring molten metal into a mould with the desired form cavity and allowing it to solidify. When the desired metal object has solidified, it is removed from the mould by breaking or disassembling the mould. The solidified object is nothing but the casting. This process can give intricate parts strength and rigidity that no other manufacturing process can match. The metal is poured into a heat-resistant mould. Sand is the most generally used material because it can withstand the high temperature of molten metal. Casting processes are classified into two broad categories:

1. Casting with expendable moulds (sand casting)
2. Casting in a permanent mould (die casting)

The pattern is a physical representation of the casting used to create the mould. Moulds are created by packing a readily formed aggregate material like moulding sand, around the required pattern. When the pattern is removed, its imprint serves as the cavity of the mould, which is eventually filled with metal to form the casting. A pattern is simply a replica of the components that will be produced by the casting process, with certain modifications as needed, such as the addition of various pattern allowances and core print provision. If a casting part is made to be hollow or with a specific type of cavity, an additional type of pattern known as a core can be used. The internal cavities of the casting parts can be created using core. There are various

DOI: 10.1201/9781003367321-4

materials used for the pattern making, its design, and construction, all of which affect the quality of the cast products. The total cost of casting is directly related to the cost of the pattern and related equipment. The following are some advantages of the casting process:

- It allows for the most design freedom in terms of shape, size and product quality.
- Casting gives cast components uniform directional properties and improved vibration damping capacity.
- Complex and uneconomical shapes that are difficult to produce using other methods can be easily produced using the casting process.
- Because a product obtained by casting is one piece, metal joining processes are not required.
- Casting process can be used for very heavy and bulky parts that are difficult to fabricate. It also produces machinable parts.
- The casting process is mechanised and is commonly used for mass production of components.

4.2 PATTERN MAKING

A pattern creates a mould cavity for the purpose of casting. A pattern may provide projections known as core prints if the casting includes a core and must be hollow. Runner, risers and gates, used to supply molten metal in the mould cavity, may form a part of the pattern. Proper pattern with finished and smooth surfaces reduces casting defects. A proper construction of pattern directly reduces the overall cost of the casting products.

4.2.1 Types of Pattern

The type of pattern to be used for a specific casting depends on several factors, such as the bulk of casting, the type of moulding process, the quantity or quality of castings needed and the expected complexity of moulding the typical form. A pattern's primary functions are as follows:

(i) To prepare a mould cavity of appropriate shape and size for the purpose of making a casting.
(ii) To make seats for the cores in the mould where the cores can be placed, in order to make a cavity in the casting. These seats in the mould are referred to as core prints.
(iii) Creating the parting line and parting surfaces in the mould.
(iv) To reduce casting defects.
(v) To aid in the positioning of a core prior to the ramming of moulding sand.
(vi) It should reduce overall casting costs.

The following pattern types are widely used in the automobile industry.

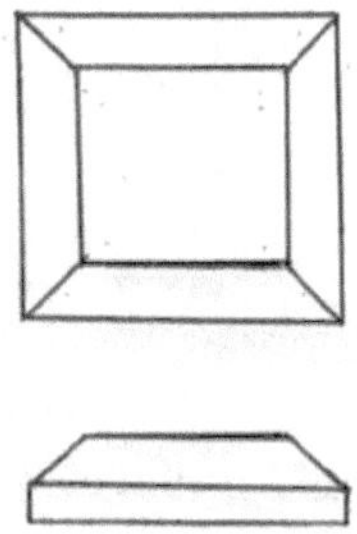

FIGURE 4.1 Single piece pattern.

1. Single-piece pattern
2. Two-piece pattern or split pattern
3. Multi-piece pattern
4. Match plate pattern
5. Sweep pattern
6. Pattern with loose piece
7. Follow board pattern
8. Gated patterns
9. Cope and drag pattern
10. Skeleton pattern
11. Segmental pattern

1. Single-piece pattern

This pattern is made in one piece and does not include joints or partition of loose pieces. It is the simplest pattern. Figure 4.1 shows single piece pattern. Depending on the shape, it can be moulded in one or two boxes. This pattern is the cheapest, but its use is limited to a small amount of manufacturing, since it necessitates a large number of manual operations.

2. Two-piece pattern or split pattern

This can be used when it is impossible to create a pattern by a single pattern. Two-piece patterns may be used in these situations. Figure 4.2 shows split pattern. They are rendered in two sections which are connected to the line of separation called the parting line. While moulding, one part of the pattern is contained by the drag and the other by the cope.

3. Multi-piece pattern

Casting with a more complicated design than the above requires a pattern in more than two sections to allow quick and easy moulding and removal of the pattern. These patterns can consist of three, four or more parts depending on their design.

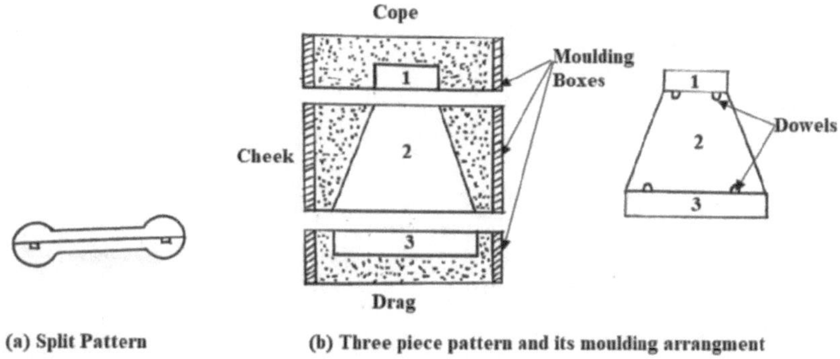

FIGURE 4.2 Spilt patterns.

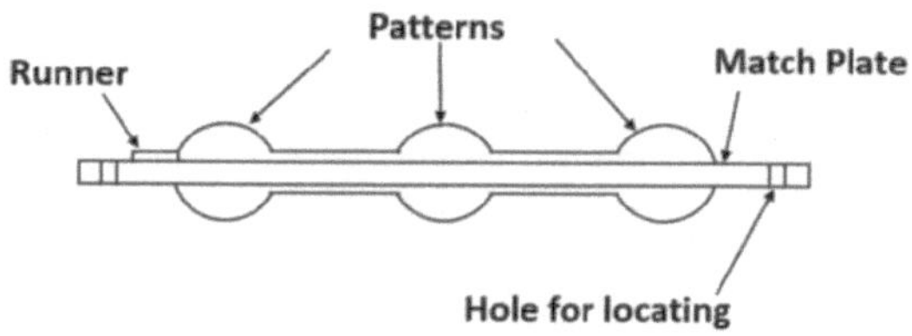

FIGURE 4.3 Match plate pattern.

4. Match plate pattern

These types of patterns are widely used where a high production rate of small and precision casting is required on a large scale. The construction costs of these types of patterns are very high, but this can be easily compensated for by a higher production rate, high dimensional precision and a minimum requirement of machining of the casting. Figure 4.3 shows match plate pattern. The pattern is made of two pieces, one piece mounted on one side and the other on the other side of the plate. The plate can be made of wood, steel or aluminium. Aluminium is favoured due to its lightness and low cost.

5. Sweep pattern

Sweeps can be used advantageously for the preparation of large symmetric casting moulds, in particular circular cross-section moulds. This results in a substantial saving in time, labour and material. Figure 4.4 shows sweep pattern. The equipment consists of a base, placed in the mass of the sand as per requirement, a wooden template known as sweep and vertical spindle. The sweep is rotated around the axis of the spindle to form the required cavity. The sweep and the spindle are withdrawn, leaving the base in the sand. The separately prepared core is placed in the mould, the gates are cut and the mould is ready to pour.

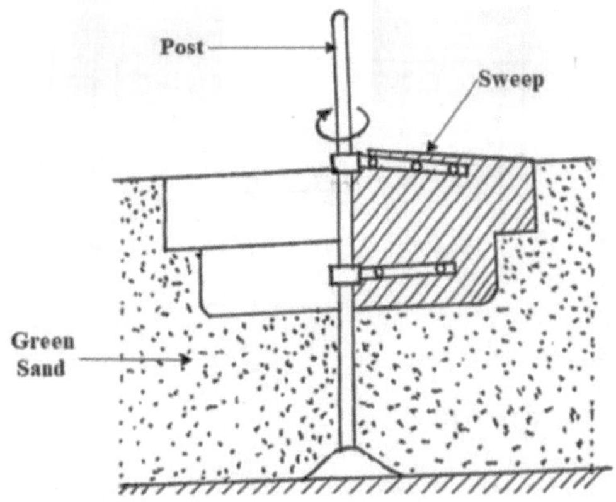

FIGURE 4.4 Sweep pattern.

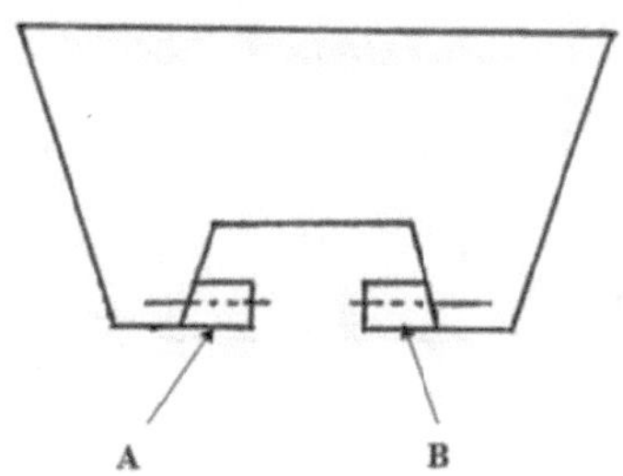

FIGURE 4.5 Loose piece pattern.

6. Pattern with loose pieces

Some patterns are designed to have loose parts, so that they can be quickly removed from the mould. Figure 4.5 shows loose piece pattern. These pieces form an integral part of the pattern throughout the moulding process. After the mould is completed, the pattern is removed leaving the parts in the sand, which are then removed separately through the cavity created by the pattern.

7. Follow board pattern

Wooden boards called follow boards are used to support a pattern during moulding. Such single-piece patterns which have an odd shape or very thin wall require a follow board. Figure 4.6 shows follow board pattern. The follow board has a projection conforming to the inside shape of the thin-walled pattern to support it throughout the moulding process. If such assistance is not offered, the pattern can be broken due to less wall thickness during ramming.

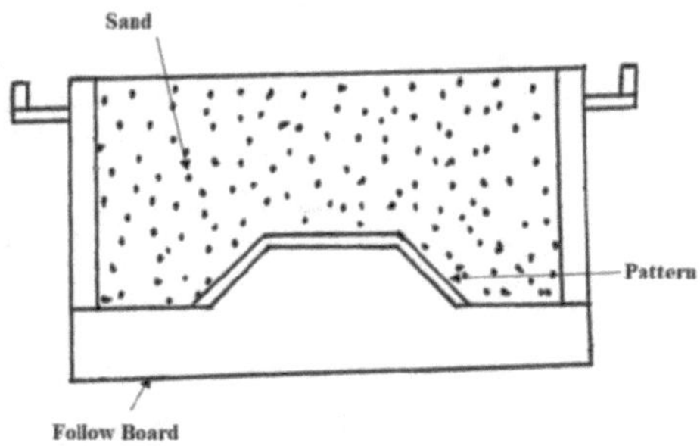

FIGURE 4.6 Follow board pattern.

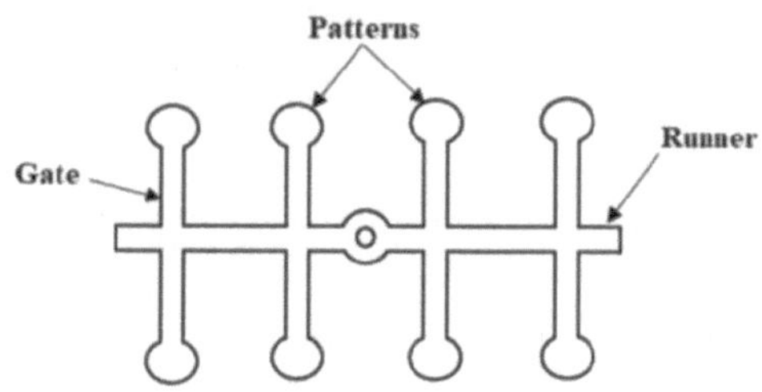

FIGURE 4.7 Gated patterns.

8. Gate pattern

The groups of patterns attached to the gate are called gated patterns (Figure 4.7). The material used for the gated pattern may be wood or metal and it can be used for the mass production of small casting elements.

9. Cope and drag pattern

It is very difficult for a single individual to manage in cases of heavy casting production. Different cope and drag patterns may be used to solve this challenge. They are rendered in half, divided by a convenient joint line.

10. Skeleton pattern

Patterns for very large castings would require a quantity of tremendous of wood for a complete pattern. In such situations, a skeleton pattern can be used to assess the general content and size of the desired castings. Figure 4.8 shows skeleton pattern.

12. Segmental pattern

The principles of working of both the segmental and sweep pattern are similar. The key difference between them is that a sweep revolves continuously to produce the

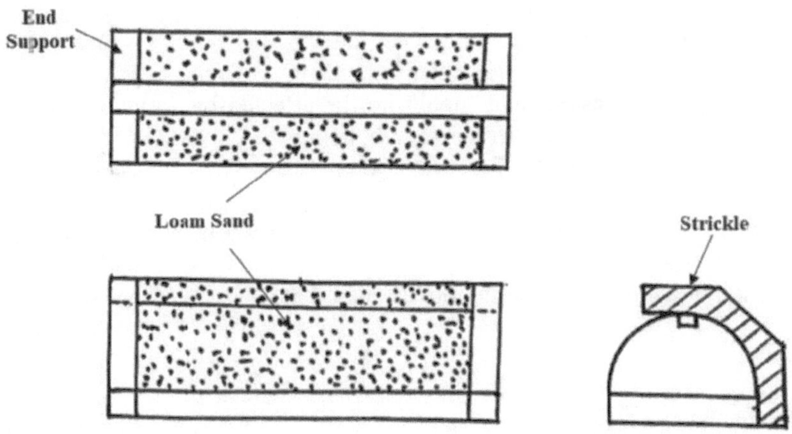

FIGURE 4.8 Skeleton pattern.

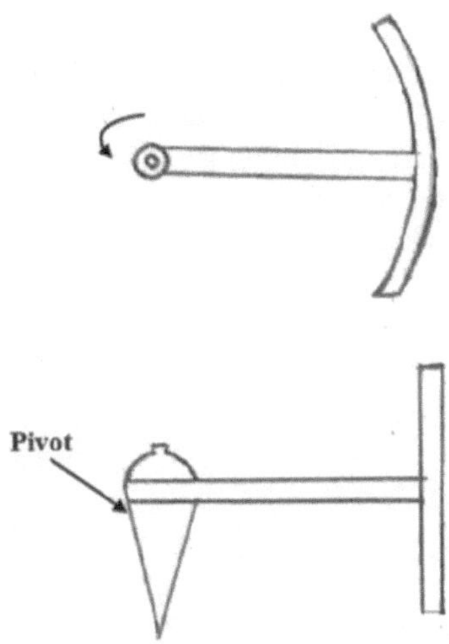

FIGURE 4.9 Segmental pattern.

required shape of the component and in the case of segmental pattern is a portion of the solid pattern itself and the mould is prepared in parts by it. It is mounted on a central pivot and it completes one portion of the mould and then moves to the next portion. Figure 4.9 shows segmental pattern.

4.2.2 Pattern Materials

To be suitable for use, the pattern material should be:

1. In a position for easy work, joining and shape
2. Light weight

3. Resistant to abrasion and wear
4. Hard and durable strong
5. Corrosion resistant and also chemical reaction resistant
6. Stable in dimensions and should not be affected by variations in environmental factors like temperature and humidity
7. Available in abundant amounts at affordable cost

The common materials used for making patterns are wood, metal, plastic, plaster, wax or mercury.

Some important pattern materials are discussed next.

1. Wood

Wood is the most widely used material for pattern making. It is inexpensive, readily available in abundance, simple to fix and easy to produce in various forms using resin and glues. It is very light and can create a very smooth surface. Wood can maintain its surface by applying shellac coating for longer life of the pattern. However, considering the above properties, it is prone to shrinkage and warping, and its existence is limited due to the fact that it is highly affected by moisture from the moulding sand. After some use, it warps and wears out easily as it has less resistance to sand abrasion. It cannot withstand rough handling and is soft compared to metal. In light of the above qualities, wooden patterns are favoured only when the number of castings to be produced is smaller. The main varieties of wood used in pattern-making are shisham, kail, deodar, teak and mahogany.

Advantages of wooden patterns

- It can be easily worked.
- It is widely available.
- It is light in weight.
- It is cost effective.
- It can be easily repairable.
- It can be joined easily.
- Wooden laminated patterns are strong.
- Good surface finish can be easily achievable.

Disadvantages

- It is very prone to moisture.
- It tends to warp.
- It wears out quickly due to sand abrasion.
- It is weaker as compared to metallic patterns.

2. Metal

Metallic patterns are favoured if the number of castings needed is large enough to justify their use. These patterns are not as much influenced by moisture as by the wood pattern. The wear and tear of this pattern is much less and thus it has a longer life. In addition, metal is simpler to form the pattern with reasonable precision, surface finish and complex shapes.

- It can tolerate corrosion and handling for longer periods of time.
- It has an outstanding strength-to-weight ratio.
- Higher cost, higher weight and rusting propensity are the key drawbacks of metallic patterns.
- It is preferable to manufacture castings in large quantities of the same design.
- The metals widely used for pattern making are cast iron, brass and bronze and aluminium alloys.

3. Plaster

Plaster of Paris can be cast very easily into any shape. It has a very high compressive strength and can be used to make patterns of smaller sizes with close dimension control. It has the property that it expands on solidification. In the case of proper selection of plaster it neutralises the effect of shrinkage automatically.

4. Plastics and resins

Plastic is used to produce a pattern due to the properties of lighter weight more strength with less wear, stronger finish and lower shrinkage during the melting process. Thermosetting resins and phenolic resins are the plastic used for pattern making.

5. Wax

Wax is used mainly in investment castings. It has a decent surface finish and high dimensional precision, much lower cost and is only used to create small patterns. Paraffin wax, shellac wax and microcrystalline wax are widely used.

4.3 MOULDING

The moulding process refers to the method and materials used in the creation of the mould. The term casting process has a broader meaning, often encompassing the moulding process, the method of introducing metal into the mould cavity, or all processes involved in the casting's production.

Certain characteristics are shared by moulding processes, including:

(1) The application of a pattern.
(2) A granular refractory and binder-containing aggregate mixture.

(3) A method for shaping the aggregate mixture around the pattern.
(4) Aggregate hardening or bonding while in contact with the pattern.
(5) The pattern is removed from the hardened aggregate mould.
(6) The mould and core pieces are assembled to form a complete mould, and metal is then poured into the mould.

4.3.1 Classification of Moulding

Moulding processes are classified in several ways. They are broadly classified based on the method used or the mould material used.

4.3.1.1 Classification based on the material used in the mould

(a) Sand moulding

Moulding processes which use sand aggregate to make the mould can be called sand castings and the most commonly used. Whatever metal is poured into sand moulds, the finished product is known as a sand casting.

1. Green sand mould: Of all the sand-casting processes, moulding with green sand is the most common.
2. Dry sand mould: Dry sand moulds are made with moulding sand that is still green. After the mould has dried, the sand mixture is slightly modified to favour good strength and other properties.
3. Skin-dried mould: The effect of a dry-sand mould can be obtained in part by drying the mould surface to a depth of 1/4 to 1 in. Torches or electrical heating elements directed at the mould surface can be used to dry the skin. Skin-dried moulds must be poured as soon as possible after drying to prevent moisture from the undried sand from penetrating the dried skin.
4. Cement bonded sand mould: A silica sand mixture containing 8–12% cement and 4–6% water is used. When creating the mould, the cement-bonded sand mixture must first harden before the pattern is removed. After that, the mould is allowed to cure for about 3–5 days. When the metal is poured, heat causes the water of cement crystallisation to be driven off, and thus steam must be allowed to pass through the sand via its porosity and appropriately distributed vent holes. This method is typically used to produce large castings with intricate shapes, precise dimensions and smooth surfaces. The only drawback is the lengthy time required for the moulding process.
5. Carbon-dioxide mould: This method is widely used for rapidly hardening green sand moulds and cores. The mould-making procedure is similar to that of conventional moulding, with the exception of the mould material, which is pure dry silica sand free of clay, 3–5% sodium silicate as a binder, and a moisture content of less than 3%. A small amount of starch can be added to improve the green compression strength, and a very small amount of coal dust, sea coal, dextrin, wood floor, pitch, graphite and sugar can be added

to improve moulding sand collapsibility. The prepared moulding sand is rammed around the pattern in the mould box, and the mould is prepared using any standard technique.

6. The shell mould: Shell mould casting is a relatively new invention (Germany during WWII) in moulding techniques for mass production and a smooth finish. It is a method of creating a thin mould around a heated metallic pattern plate. The moulding material is a combination of dry, fine silica sand (clay content should be kept to a minimum) and 3–8% thermosetting resin, such as phenol formaldehyde or silicon grease.

(b) Plaster moulding

In plaster moulding, the mould material is gypsum or plaster of Paris. To this plaster of Paris, additives such as talc, fibres, asbestos, silica flour and others are added to control the mould's contraction characteristics as well as the settling time. The plaster of Paris is used as slurry. This plaster slurry is poured over a flask containing a metallic pattern.

(c) Metallic moulding

Because of their long life, metallic moulds are also known as permanent moulds. The metallic mould can be reused a number of times before being discarded or rebuilt. Heat-resistant cast iron, steel, bronze, anodised aluminium, graphite or other suitable refractoriness materials are used to make permanent moulds. The mould is divided into two halves to facilitate the removal of the casting from the mould. The metallic mould is commonly referred to as a die, and the metal is introduced into it by gravity.

Its classification is according to the method used

1. Bench moulding
   Bench moulding is moulding done on a bench at a convenient height to the moulder. It is used for small work.
2. Floor moulding
   Floor moulding refers to the moulding done on the foundry floor and is used for all medium and large castings.
3. Pit moulding
   Very large moulds in a pit or cavity cut in the floor are used to accommodate very large castings. The pit acts as a drag. Because pit moulds can resist pressures created by hot gases, this greatly reduces pattern expenses.
4. Machine moulding
   Machine moulding refers to the process of making moulds using a machine. Small, medium and large moulds can be made using a variety of machines. Machine moulding is usually faster and more uniform than bench moulding.

4.3.2 Green Sand Moulding

Moulding with green sand is the most common sand-casting process. Green moulding sand is a plastic mixture of sand grains, clay, water and other materials that can be used in moulding and casting processes. Because of the moisture present, the sand is referred to as "green", and it is distinguished from dry sand. The basic steps in green sand moulding are as follows:

1. Pattern preparation. The majority of green sand moulding is done with match plate or cope and drag patterns. When only a few castings of a type are to be made, loose patterns are used. In simple hand moulding, a loose pattern is placed on a mould board and surrounded by a suitable-sized flask.
2. Making the mould. Moulding necessitates the ramming of sand around the pattern. As the sand is packed, it gains strength and becomes rigid within the flask. Ramming can be done by hand. Cope and drag are moulded in the same way, but the cope must accommodate the sprue. The gating-system parts of the mould cavity are simply channels for the molten metal to enter.
3. The core setting. Cores are set into the mould cavity to form the internal surfaces of the casting after the cope and drag halves of the mould is made and the pattern is withdrawn.
4. Closing and weighting with the cores set, the cope and drag are closed. To keep the cope from floating when the metal is poured, it is usually weighted down or clamped to the drag. Because of the nature of green sand moulding and moulding sands, the process has both advantages and disadvantages.

The advantages are as follows:

1. A high level of adaptability in the manufacturing process. Moulding and related operations can be carried out using mechanical equipment. Furthermore, green sand can be reused many times by reconditioning it with water, clay and other materials. Moulding can be a fast and repetitive process.
2. Green sand moulding is typically used to direct the metal directly from the pattern to the mould ready for pouring.
3. Economically, green sand moulding is typically the least expensive method of moulding.

The following are some drawbacks to using green sand moulding:

1. Some casting designs necessitate the use of other casting processes. Thin, long projections of green sand in a mould cavity are washed away by the molten metal or may not be mouldable at all. Cooling fins on air-cooled engine cylinder blocks and heads are one example. The mould's strength is then increased.
2. When certain metals and castings are poured into moulds containing moisture, they develop defects.

3. The dimensional accuracy and surface finish of green-sand castings may be inadequate.
4. Large castings necessitate greater mould strength and erosion resistance than is available in green sands.

4.3.3 Core

Core castings are frequently required to have holes, recesses and other features of varying sizes and shapes. Cores can be used to obtain these impressions. As a result, where coring is required, provisions should be made to support the core within the mould cavity. This is accomplished through the use of core prints. The core print is an additional projection on the pattern that serves as a seat in the mould for the sand core. A core is a sand shape or form that creates the contour of a casting for which no moulding provision has been made in the pattern. The core could be made of sand, plaster, metal or ceramics. The core is an obstruction that, when placed in the mould, prevents the molten poured metal from filling the space occupied by the core, resulting in hollow casting. Cores are used as inserts in moulds to create design features that are difficult to achieve through simple moulding. Functions of core include:

- Providing a means of forming the main internal cavity for hollow casting; and providing an external undercut feature.
- To obtain deep recesses in the casting, cores can be inserted.
- Cores can be used to increase the strength of the mould and as part of the gating assembly.
- It can be used to improve the surface of the mould as well as form part of the green sand mould.

The following are the core characteristics of a dry sand core, which must have the following properties.

It must be strong enough to support itself without breaking, it must have high permeability and refractoriness, it must have a smooth surface to ensure a smooth casting, and it must have high collapsibility to aid in the free contraction of the solidifying metal. It should contain ingredients that do not produce mould gases.

Core applications: The core and its shape increase the adaptability of moulding processes and operations. Cores are used in addition to recess forming and casting holes as follows:

- Cores are used to make moulds.
- Cores can function as strainers, gates and pouring cups.
- Cores are used to increase output from a match plate pattern.
- In the centrifugal casting process, cores can be used as core moulds.
- They can also be used as a slab core to increase casting output from a single mould.

Types of cores

Various types of cores of various designs and sizes are used in various ways in foundry work. A general way to categorise them is based on their shapes, sizes and positions in the ready-made moulds. Their primary classifications are as follows:

1. Horizontal core
2. Vertical core
3. Hanging core
4. Balanced core
5. Ramp up core
6. Kiss core.

4.4 CASTING

Casting processes are frequently chosen over other methods of manufacture for the following reasons:

- Casting can produce complex shapes with internal cavities or hollow sections.
- Very large parts can be produced in one piece.
- Casting can use materials that would be difficult or uneconomical to process otherwise.
- Casting can be economically competitive with other manufacturing processes.

The casting process consists essentially of:

(a) Pouring molten metal into a mould that is patterned after the part to be manufactured
(b) Allowing it to solidify and
(c) Removing the part from the mould.

As with all manufacturing processes, an understanding of the underlying science is required for the production of high-quality, cost-effective castings as well as the establishment of proper mould design and casting practice techniques.
The following are important considerations in casting operations:

- Flow of molten metal into the mould cavity
- Solidification and cooling of the metal in the mould
- Effect of mould material.

The following are some of the most common casting processes:

a. Sand mould casting
b. Plaster mould casting
c. Metallic mould casting
    I. Permanent mould casting

II. Slush casting
III. Pressure dies casting

d. Centrifugal casting
e. Investment casting
f. Continuous casting
g. CO_2 mould casting
h. Ceramic mould casting.

4.4.1 Permanent Mould Casting

This method is also known as gravity die-casting. Figure 4.10 shows details of permanent mould casting process. The primary distinction between permanent mould casting and sand casting is that, in this case, the mould is permanent and is not destroyed or recreated after each cast. High-resistant fine-grained alloys, iron and steel are commonly used in the manufacture of permanent moulds. Pouring in permanent moulds is simply done by gravity and is thus referred to as gravity pouring. Casting stages include: a permanent mould made up of several blocks joined together. The mould is first preheated to a temperature of 400°C using some means. Following the mould, a refractory coating is applied to the mould cavity surfaces, runner and riser, and so on.

The casting is poured once the mould temperature has been reached. Cores are removed as the metal begins to solidify; otherwise, it may shrink onto the metal's surface.

The mould is then cleaned by blowing, coated with refractory coating, cores are assembled, and the mould is closed again for pouring.

- **Advantages**

  - It is a quick process, and the moulds last a long time.
  - A better surface finish is possible.
  - Less skilled operators are required, as is less floor space.

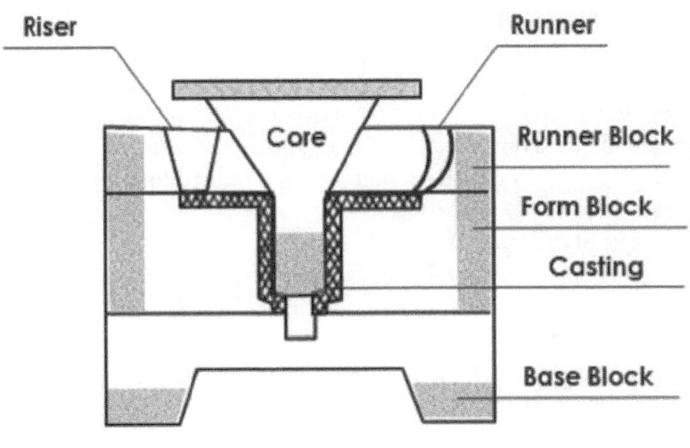

FIGURE 4.10 Permanent mould casting.

Limitations

- Moulds are much more expensive; this method is not suitable for small-scale production; and the shape and weight of the casting are limited.
- Gates, runners and risers are not movable.

Applications

Hydraulic brake cylinders are among the components produced by this method.

4.4.2 Centrifugal Casting

Centrifugal casting is the process of pouring molten metal into a rotating mould cavity caused by centrifugal acceleration. Impurities such as slag and sand, which are lighter, travel towards the rotating mould's centre axis, keeping the main casting free of defects.

Centrifugal casting processes can be classified as:

(i) True centrifugal casting
(ii) Semi-centrifugal casting
(iii) Centrifuging casting.

(i) True centrifugal casting

The true centrifugal casting method is used to create objects that are symmetrical about their axis but hollow on the inside, such as pipes. The mould is rotated about an axis, which can be horizontal, vertical or inclined. By using centrifugal force, the molten metal is pushed to the mould's walls, where it solidifies into a hollow cylinder. The wall thickness of hollow castings is determined by the amount of poured metal. Figure 4.11 shows details of true centrifugal casting process. The machines that rotate the mould can have either a horizontal or vertical axis of rotation. Horizontal axis machines are used to cast longer pipes, such as water supply and sewer pipes, whereas vertical axis machines are used for casting short tubes.
True centrifugal casting has the following advantages:

(i) Increased output
(ii) High metal utilisation efficiency due to the elimination of sprues and risers
(iii) The castings produced have a high density, refined fine-grained structure and superior mechanical properties
(iv) A low rejection rate
(v) The castings produced by this method have small machining allowances
(vi) This process can cast ferrous and nonferrous metals
(vii) It is a quick and cost-effective method.

(ii) Semi-centrifugal casting

The difference between semi-centrifugal casting and true centrifugal casting is that the mould cavity is completely filled with molten metal in semi-centrifugal casting process (Figure 4.12).

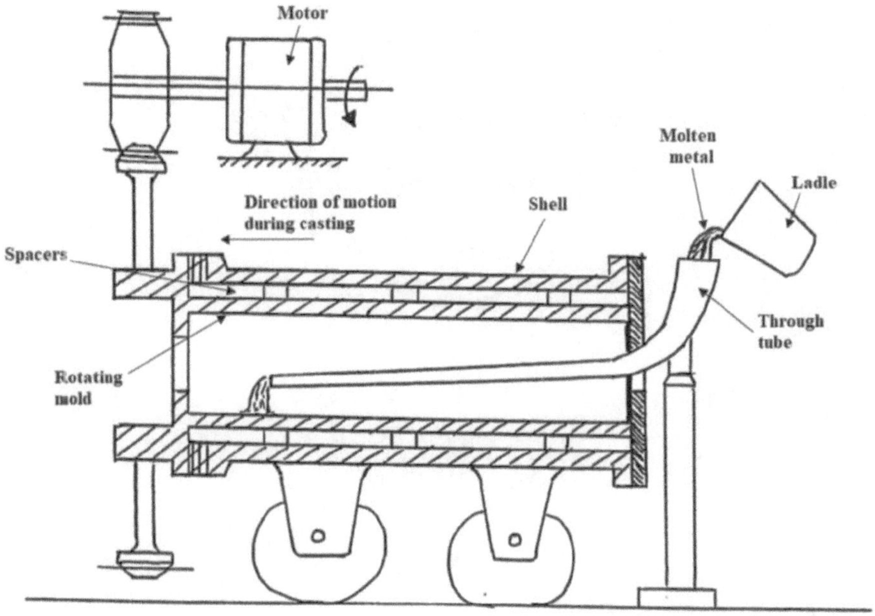

FIGURE 4.11 True centrifugal casting.

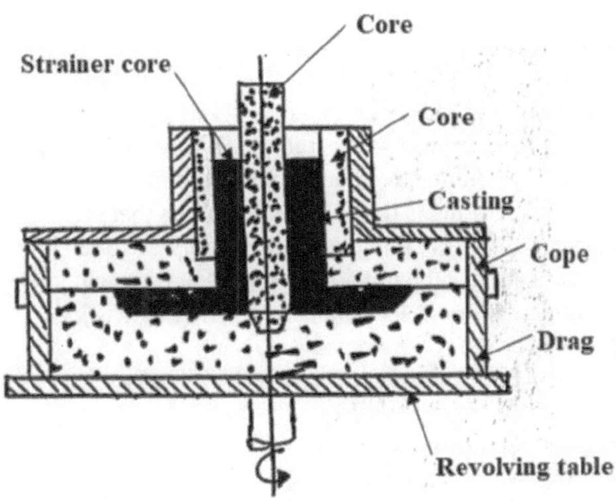

FIGURE 4.12 Semi-centrifugal casting.

Semi-centrifugal casting produces parts that are symmetrical about their axis but may or may not be hollow. The spinning speeds are not as fast as they would be in a true centrifugal casting. Impurities are not effectively separated from the metal due to the slower speed. A semi-centrifugal casting process is shown in Figure 4.12

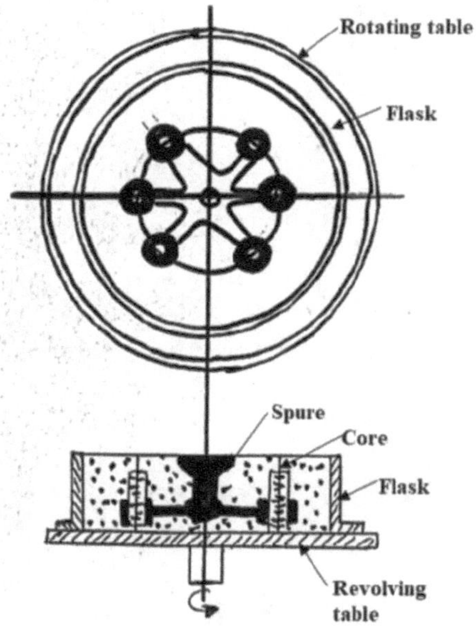

FIGURE 4.13 Centrifuging casting.

The semi-centrifugal casting method is employed for making large castings which are axis symmetrical. Examples include cast track wheels for tanks, tractors and the like, pulleys, spoked discs, gears, propellers, etc.

(iii) Centrifuging casting

The centrifuging casting method consists of a series of mould cavities arranged in the shape of a circle. Radial gates connect these cavities to a central down sprue. The mould is then filled with molten metal and rotated around the sprue's central axis.

In other words, each casting is rotated around an axis that is different from (shifted from) its own centre axis. As a result, the mould cavities are filled under high pressure. Centrifugal forces ensure that the mould fills evenly. It is typically used to create intricately shaped castings. Figure 4.13 depicts a centrifuging casting.
Centrifuging casting has the following advantages:

- It ensures the filling of the mould cavity properly.
- It ensures uniform thickness with a smooth surface.
- It removes impurities from the casting.
- It is used to produce small intricate shapes.
- It is used for long castings such as pipes.

Centrifuging casting's limitations include:

- The unit is quite pricey.
- Expensive upkeep.
- To achieve uniform thickness, uniform speed or rotation is required.

4.4.3 Pressure Die Casting

High-pressure die casting (HPDC) is an extremely efficient manufacturing method for producing various product forms. The process forces molten metal at high speed and high pressure into a closed steel die cavity. The die has a stationary and moving half, both of which are mounted to the die casting machine's platens.

In pressure die-casting, molten metal is poured by pressure into a metal mould known as the die. Because the metal solidifies under pressure, the casting conforms to the die cavity in shape and surface finish. The pressure is generally obtained with the help of compressed air or hydraulically. The pressure varies from 70 to 5000 kg/cm^2.

The main types of die-casting machines are:

A. Hot chamber die-casting
B. Cold chamber die-casting.

The principal difference between the two methods is determined by the location of the melting pot. In the hot chamber method, a melting pot is included with the machine and the injection cylinder is immersed in the molten metal at all times. The injection cylinder is operated by either hydraulic or air pressure, which forces the metal into the dies to form a casting. meanwhile, the cold chamber machine consists of a separate melting furnace and metal is introduced into the injection cylinder by hand or mechanical means.

A. Hot chamber die-casting

In this method, metal is forced into the mould and pressure is maintained during solidification either by a plunger or by compressed air. Figure 4.14 shows the main parts of the hot chamber die-casting machine. The plunger acts inside a cylinder formed at one end of the goose neck type casting submerged in the molten metal. Near the top of the cylinder, for entry of molten metal, a port is provided. When the bottom of the plunger is above the port, the cylinder is connected to the melting pot through this port. This downward stroke of the plunger closes this port, cuts off the supply of metal and applies pressure on the metal present in the goose-neck to force it into the die cavity through the injecting nozzle.

After some time, the plunger is raised up, causing the remaining molten metal in the nozzle and channel to fall back into the casting. Before the end of the upward stroke, the plunger uncovers the port, through which a greater amount of molten metal enters into the cylinder. Then the dies are opened and the casting is ejected. These machines are generally used for producing castings of low melting point metals like zinc, tin and lead.

B. Cold chamber die-casting

The machine consists of a separate furnace for melting the metal. Figure 4.15 shows the main parts of the cold chamber die-casting machine. The metal is melted separately in the furnace and transferred to the cold chamber using a small hand ladle. After

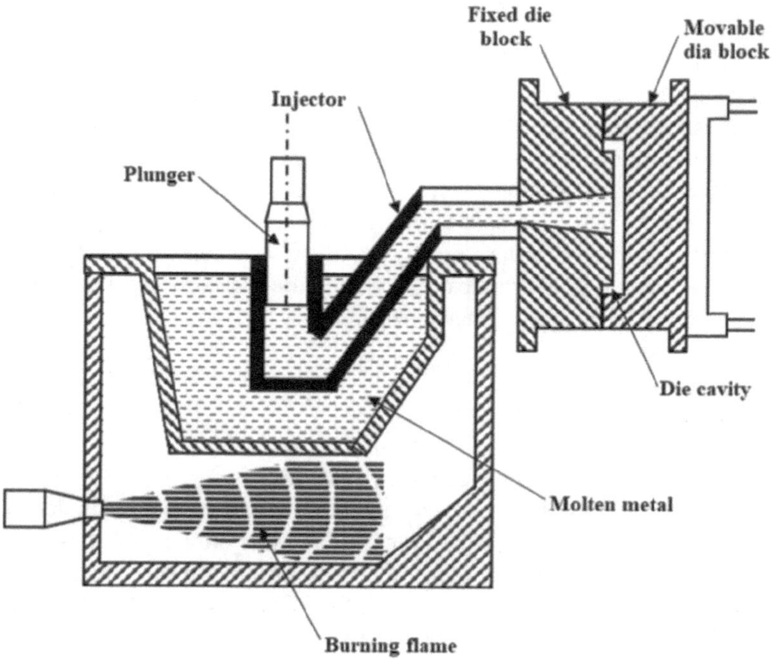

FIGURE 4.14 Hot chamber die casting.

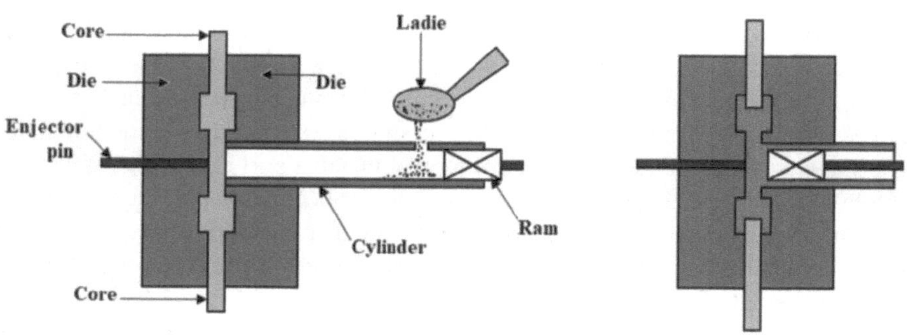

FIGURE 4.15 Cold chamber die casting machine.

closing the die, the molten metal is poured into the horizontal chamber through the metal inlet. To force the metal into the die, the plunger is pushed forwards hydraulically. After solidification, the die is opened and the casting is ejected. Cold chamber machines are mainly used for making castings in aluminium, brass and magnesium. The life of these machines is longer, because the melting unit is separated from the working parts. However, the life of the die is less because the machine involves very high pressure, i.e. about 200–2000 kg/cm^2.

Advantages of pressure die casting include:

- High production rates are possible. Economical for large production quantities.
- Close tolerances up to ± 0.076 mm on small parts is possible. Good surface finish can be obtained. Thin sections up to 0.5 mm can be cast.

Limitations of pressure die casting include:

- Only small parts can be made.
- Only non-ferrous alloys and metals can be commercially cast.
- Due to the high cost of equipment and dies, the process is economical only for mass production.
- Due to entrapped air, the die castings are porous, which reduces the mechanical properties of the component.

Applications of pressure die casting include:

- Household equipment such as decorative parts, mechanical parts of mixers, fans, vacuum cleaners, washing machines, can openers; refrigerators, etc.
- Industrial equipment such as motor housing, crane parts, motor, rotor fan, impeller wheel, etc.
- Automotive parts such as windshield frames, window channels, bodies of fuel pump and carburettor, handles, rear-view mirror parts, brake shoe (Al), etc.
- Toys such as pistols, electric trains, model aircraft, automobiles, etc.
- Other parts such as taps, valves, burners, fire alarm systems, telephone sets, speakers, staplers, typewriters, etc.

4.5 POURING (GATING DESIGN)

The metal is poured or pumped into the mould cavity after it has been melted. We now go through the challenges that we encountered and how we overcame them by employing an effective gating design. A successful gating design ensures proper metal distribution in the mould cavity without undue temperature loss, turbulence or entrapping gases and slag. If the liquid metal is poured slowly, it will take a long time to fill the mould, and solidification will begin before the mould is fully filled. This can be prevented by using too much superheat, but this can create a problem with gas solubility. On the other hand, if the liquid metal strikes the mould cavity at an extremely high velocity, the mould surface can be eroded. As a result, a compromise must be made in order to achieve the optimum velocity. A gating system's configuration is determined by the metal and mould compositions. To avoid dross (e.g., oxides) in easily oxidised metals with low melting points, such as aluminium, an elaborate gating design is needed. However, in the case of cast iron, a short route for the liquid metal is chosen to prevent a high pouring temperature. A ceramic mould's gating nature varies greatly from that of a permeable sand mould. Gating designs are roughly categorised into three types: (i) vertical gating, (ii) bottom gating and (iii) horizontal gating. The

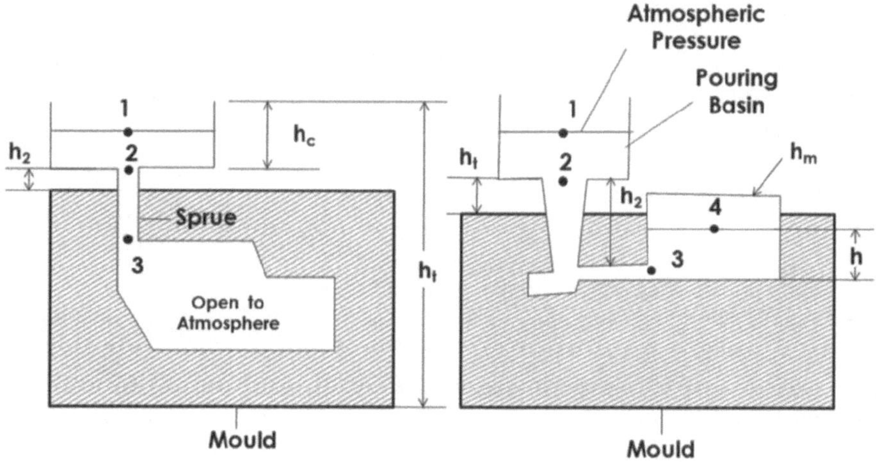

FIGURE 4.16 Simple vertical and bottom gating.

molten metal is poured vertically to fill the mould with ambient pressure at the base of vertical gating. Bottom gating, on the other hand, fills the mould with liquid metal from bottom to top, preventing the splashing and oxidation associated with vertical gating. Figure 4.16 depicts a straightforward vertical gating system as well as a bottom gating design. Additional horizontal sections are introduced in the horizontal gating method for optimal distribution of the liquid metal with reduced turbulence. Simple calculations based on principles of fluid flow can lead to an estimate of the time taken to fill up a mould. We illustrate this for the two designs in separate figures. The integrated energy balance equation on the basis of per unit mass flow, more commonly known as Bernoulli's equation, will be used. For example, in Figure 4.16a it is assumed that the pressure at points 1 and 3 is equal (i.e., $p_1=p_3$) and that level 1 is maintained constant. Thus, the velocity at station 1 (v_1) is zero. Moreover, the frictional losses are neglected. Then, the energy balance equation between points 1 and 3 gives

$$gh_t = \frac{v_3^2}{2}$$

or

$$v_3 = \sqrt{2gh_t}$$

where g is the acceleration due to gravity and v_3 is the velocity of the liquid metal at the gate, subsequently referred to as v_g. Therefore, the time taken to fill up the mould (t_f) is obtained as

$$t_f = \frac{V}{A_g v_3}$$

where A_g and V are the cross-sectional area of the gate and the volume of the mould, respectively.

In Figure 4.16b, applying Bernoulli's equation between points 1 and 3, we get

$$gh_t = \frac{p_3}{\rho_m} + \frac{v_3^2}{2}$$

where ρ_m is the density of the liquid metal, p_3 is the gauge pressure at station 3, and h_t is again assumed to be constant. Further, applying Bernoulli's equation between points 3 and 4, with the assumptions that v_4 is very small and all the kinetic energy at station 3 is lost after the liquid metal enters the mould, we can write

$$\frac{p_3}{\rho_m} = gh$$

From these equations, the velocity of the liquid metal at the gate we obtain is

$$v_g = v_3 = \sqrt{2g\left(h_t - h\right)}$$

The above equation gives the velocity of a jet discharging against a static head h, making the effective head $\left(h_t - h\right)$. Now, for the instant shown, let the metal level in the mould move up through a height d_h in a time interval d_t; A_m and A_g are the cross-sectional areas of the mould and the gate, respectively. Then,

$$A_m dh = A_g v_g dt.$$

Using above equations we get

$$\frac{1}{\sqrt{2g}} \frac{dh}{\sqrt{\left(h_t - h\right)}} = \frac{A_g}{A_m} dh$$

At $t = 0$, $h = 0$ and at $t = t_f$ (filling time), $h = h_m$. Integrating above equation between these limits, we have

$$\frac{1}{\sqrt{2g}} \int_0^{h_m} \frac{dh}{\sqrt{\left(h_t - h\right)}} = \frac{A_g}{A_m} \int_0^{t_f} dt$$

or

$$t_f = \frac{A_m}{A_g} \frac{1}{\sqrt{2g}} \sqrt{h_t} - \sqrt{h_t - h_m})$$

If a riser (reservoir to account for shrinkage from the pouring temperature) is used, the pouring time t_f should also account for the time required to fill the riser. Normally, open risers are filled to the level of the pouring sprue; thus, the time required to fill the riser is calculated in the equation with A_m replaced by A_r (riser cross-section) and h_m replaced by h_t.

4.5.1 Aspiration Effect

When making a mould with a permeable material (e.g., sand), care should be taken to ensure that the pressure in the liquid metal stream does not fall below ambient pressure anywhere in the stream. Otherwise, the gases produced by baking the organic compounds in the mould would mix with the molten metal stream, resulting in porous castings. This is known as the aspiration effect. Using Figure 4.16b and Bernoulli's equation between points 2 and 3, we obtain

$$gh_2 + \frac{p_2}{\rho_m} + \frac{v_2^2}{2} = \frac{p_3}{\rho_m} + \frac{v_3^2}{2}$$

where p and v are the pressure and velocity, respectively, of the liquid metal at stations 2 and 3. If the pressure at point 3 is atmospheric, i.e. $p_3 = 0$, the $p_2 = -p_m g h_2$ as $v_2 = v_3$. Hence, the design in Figure 4.16a is not acceptable. To avoid negative pressure at point 2 (to ensure positive pressure throughout in the liquid column), the sprue should be tapered, the ideal shape of which can be determined as follows.

If, in the limiting case, p_2 is equal to zero, then, from the equation,

$$\frac{v_3^2}{2} = gh_2 + \frac{v_2^2}{2}$$

From the principle of continuity of flow, $A_2 v_2 = A_3 v_3$, where A is the cross-sectional area. Thus,

$$v_2 = \frac{A_3}{A_2} v_3 = Rv_3,$$

where $\frac{A_3}{A_2} = R$. From the above two equations

$$\frac{v_3^2}{2g} = h_2 + \frac{R^2 v_2^2}{2g}$$

or

$$R^2 = 1 - \frac{2gh_2}{v_3^2}$$

Again, $v_3^2 = 2gh_t$ (applying Bernoulli's equation between points 1 and 3, with $p_1 = p_3 = 0$ and $v_1 = 0$). Substituting this in the equation we have

$$R^2 = 1 - \frac{h_2}{h_t} = \frac{h_c}{h_t}$$

or

$$R = \frac{A_3}{A_2} = \sqrt{\frac{h_c}{h_t}}$$

This can easily be seen to be the shape of a freely falling stream when $v_2 = \sqrt{2gh_c}$ and $v_3 = \sqrt{2gh_t}$.

As a result, the sprue profile should ideally be as shown by the solid lines. Figure 4.17 shows ideal actual shape of the sprue. When the pressure in the stream is just above atmospheric pressure, a straight tapered sprue (shown by the dashed lines) is safer (pressure is above atmospheric everywhere except points 2 and 3) and easier to build. Figure 4.18 shows principle of avoiding vacuum generation.

Another scenario in which the aspiration effect is present is when the flow direction abruptly changes. As illustrated in Figure 4.18a, due to the momentum effect,

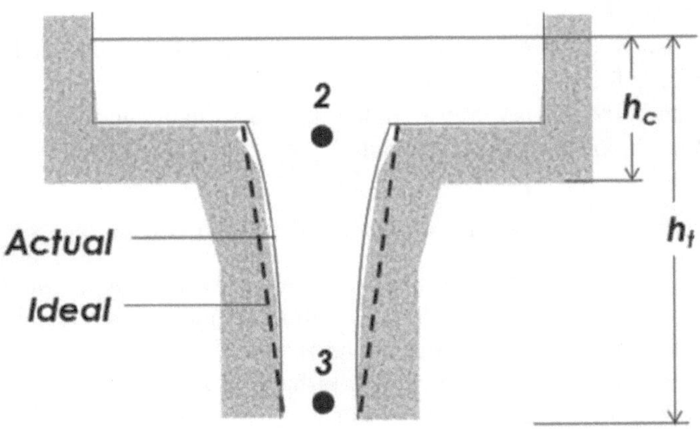

FIGURE 4.17 Ideal actual shapes of sprue.

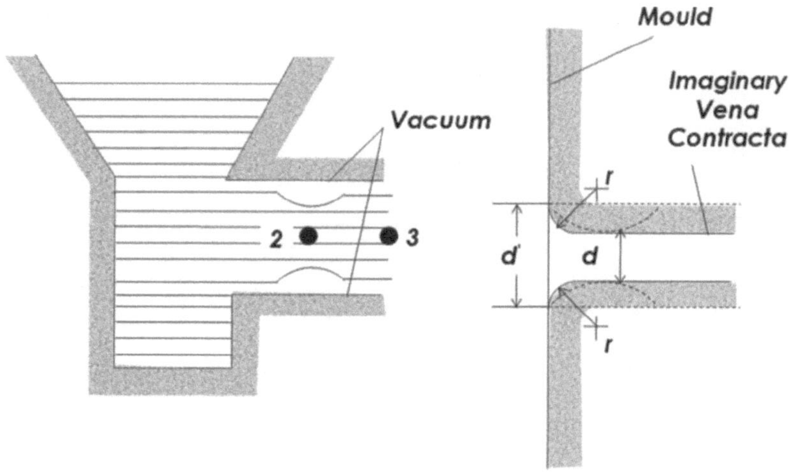

FIGURE 4.18 Principle of avoiding vacuum generation.

the liquid metal stream contracts around a sharp corner. This has nothing to do with gravity acceleration in vertical gating. The constrictive region depicted in Figure 4.18a at station 2 is an example of a constrictive region that is referred to as vena contracta. To avoid the formation of a vacuum around station 2, the mould is made to fit the vena contracta, as shown in Figure 4.18b. This avoids a sharp change in flow direction. If the runner diameter is d and the entrance diameter is d', then normally d'/d is kept at a value close to 1.31. This equals $r \approx 15d$.

The common items employed in a gating design to prevent impurities in the casting are as follows:

(i) Pouring basin: This decreases the eroding force of the direct metal stream from the furnace. A pouring basin can also be used to maintain a constant pouring head.
(ii) Strainer: In the sprue, a ceramic strainer eliminates dross.
(iii) Splash core: A ceramic splash core mounted at the sprue's end reduces the eroding force of the liquid metal stream.
(iv) Skim bob: This is a trap installed in a horizontal gate to keep heavier and lighter impurities out of the mould.

4.5.2 Effects of Friction and Velocity Distribution

We assumed that the velocity of a liquid metal in the sprue and gate is uniform across the cross-section. In fact, the velocity of a fluid in contact with any solid surface is zero and reaches its maximum at the conduit's axis. The velocity distribution within the conduit is determined by the conduit's shape and the nature of the flow (i.e., turbulent or laminar). Furthermore, in our previous discussion, we believed there were no frictional losses. Frictional losses are often present in real fluids, particularly when there is a sudden contraction or enlargement of the flow cross-sections. In the following discussion, we will modify the equations we have already developed in light of these two factors, namely velocity distribution and friction. The nonuniform velocity distribution can be accounted for by changing the kinetic energy term in the integrated energy balance equation by replacing the (velocity)2 term by $\frac{\overline{v}^2}{\beta}$, where $\overline{v}$ is the average velocity and β is equal to 0.5 for laminar flow and approximately equal to 1 for turbulent flow.

The energy loss due to friction in a circular conduit (on the basis of per unit mass) is given by

$$E_{f_1} = 4f\frac{L}{D}\frac{\overline{v}^2}{2}$$

where

$\overline{v}$ = average velocity,

D, L = diameter and length, respectively, of the conduit and

f = friction factor.

The value of f is affected by the roughness of the conduit as well as the quality of the flow. As a result, when applying the integrated energy equation between two points, say, 1 and 2, in that order in the direction of flow, E_{f1} should be added to station 2's energy. The equations give the value of f for a smooth conduit.

$$f = \frac{16}{Re} \text{ for laminar flow } (\text{Re} < 2000), \tag{4.16}$$

$$\frac{1}{\sqrt{f}} = 4\, log_{10}(R_e\sqrt{f} - 0.4 \text{ for turbulent flow} (\text{Re} > 2000), \tag{4.17}$$

where Re is the Reynolds number. For the range 2100 < Re < 10^5, the equation can be simplified to the form

$$f = 0.0791\ (\text{Re})^{-1/4} \tag{4.18}$$

Frictional losses may also occur as a result of a gradual shift in flow direction, such as in a 900 bend or other similar fittings. In this case, the frictional losses are compensated for by using an equation with an analogous (L/D) factor for the bend. The frictional loss (per unit mass) caused by a sudden expansion or contraction of a flow area is denoted as

$$E_{f_2} = \frac{1}{2}e_f\overline{v}^2$$

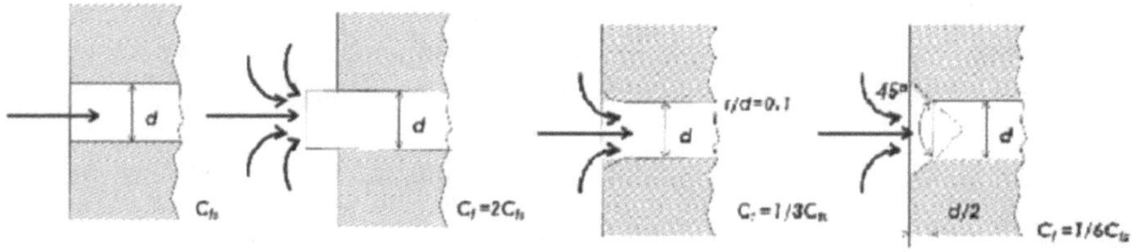

FIGURE 4.19 Values of e_f for some types of charges in flow geometry.

where v is the average velocity of the fluid in the smaller cross-section and e_f is the friction loss factor, which depends on the flow area-to-Reynolds number of the flow ratio. The length and diameter of the smaller flow cross-section must also be considered for laminar flow. The value of e_f is determined by whether the flow area expands or contracts in the direction of flow. Figures 4.19a and 4.19b display the values of e_f for a sharp shift in the flow cross-section for sudden expansion and contraction, respectively. The Reynolds numbers of the flow have been determined in these figures based on the average flow velocity in the smaller flow cross-sections, and l and d are the length and diameter of the smaller cross-sections, respectively. Figure 4.19 shows the values of e_f for certain other forms of changes in flow geometry; e_f refers to the values for a sharp shift in geometry of equal initial and final dimensions (i.e., the values of e_f can be obtained from Figures 4.19a and 4.19b. Returning to Figure 2.6a, let us revise the review provided in Section 4.5 to account for the effects of friction and velocity distribution. Using the integrated energy balance equation between points 1 and 3, we get after accounting for the loss due to a sudden contraction at 2:

$$\frac{p_1}{\rho_m} + 0 + gh_t = \frac{p_3}{\rho_m} + \frac{v_3^2}{2\beta} + E_{f_1} + E_{f_2}$$

where (v_3) is the average fluid velocity in the sprue. With $p_1 = p_3$ and using above equations we get

$$2gh_t = v_3^2\left(\frac{1}{\beta} + e_f + 4f\frac{l}{d}\right)$$

where d is the diameter of the sprue and l is the length of the sprue (=h_2 in Figure 4.16a). Thus, Equation (4.2) can be modified as

$$v_3 = C_{D\sqrt{2gh_t}}$$

where the discharge coefficient

$$C_D = \left(\frac{1}{\beta} + e_f + 4f\frac{l}{d}\right)^{-\frac{1}{2}}$$

In this analysis, we have neglected the fluid velocity (and hence the loss) between points 1 and 2. If the sprue also has a bend or fitting, then E_{f1} is modified as

$$E_{f_1} = 4fv_3^{\,2}\left[\frac{l}{d} + \left(\frac{L}{D}\right)_{eq}\right]$$

[The values of $\left(\frac{L}{D}\right)_{eq}$ for various types of fittings are listed in standard tables.]

In such a case, the discharge coefficient C_D is, finally, given by

$$C_D = \left(\frac{1}{\beta} + e_f + 4f([\frac{l}{d} + \left(\frac{L}{D}\right)_{eq}\right)^{-\frac{1}{2}}$$

4.6 SOLVED EXAMPLES

1. A casting of size 100 cm × 100 cm × 30cm was filled by top and bottom gates with manometric height in pouring basin to be 30 cm. Compare the time to fill the casting by different gates. The area of gate is 5 cm^2.

Solution

Casting area = 100 × 100 = 10^4 cm^2

Casting height, H=30 cm

Manometric height, h_t = 30 cm

Gate area, A_g = 5 cm^2

Top gate: $t_{f1} = \dfrac{AH}{Ag\sqrt{2ght}} = \dfrac{10^4 \times 30}{5 \times \sqrt{2 \times 981 \times 30}} = 247.3$ s

Bottom gate: $t_{f2} = \dfrac{2 \times 10^4}{5 \times \sqrt{2 \times 981}}[\sqrt{30} - \sqrt{30 - 30}] = 494.6$ s

2. Two gating design for a mould of 50 cm × 25 cm × 20 cm are shown in Figure N4.1. The cross-section area of the gate is 5 cm^2. Determine the filling time for both designs.

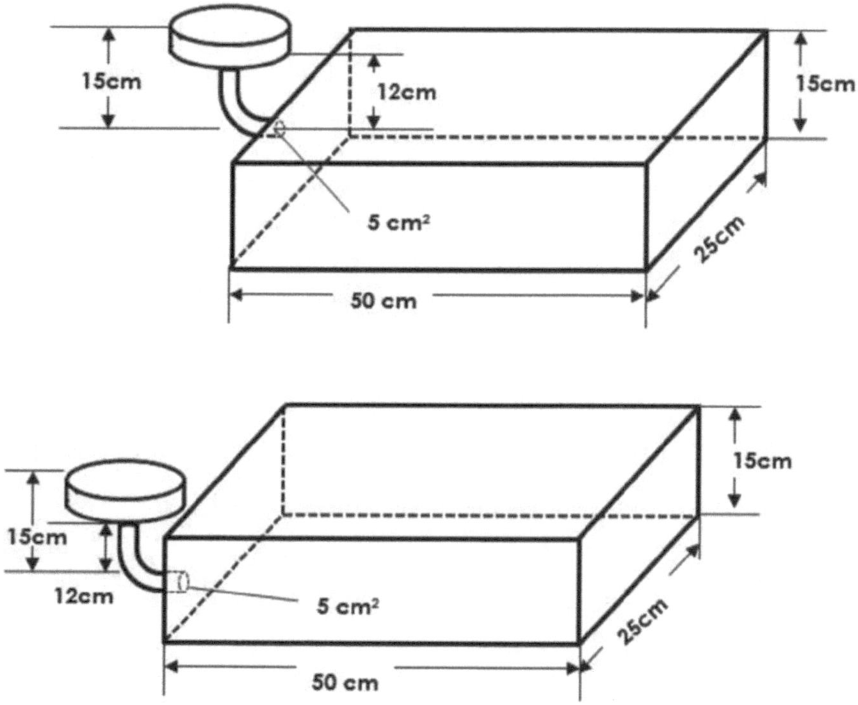

Figure N4.1 Top and bottom gating.

Solution

Top gating: $h_t = 20$ cm

Velocity at the gate, $v_3 = \sqrt{2ght} = \sqrt{2\times 981\times 20} = 198.09$ cm/sec

Volume of the mould, $V = 50 \times 25 \times 20\ cm^3$

Cross-sectional area of gate, $A_g = 5\ cm^2$

Time of filling, $t_f = \dfrac{V}{Ag\,v^3} = \dfrac{50\times 25\times 20}{5\times 198.09} = 25.24$ sec

For bottom gating: $t_f = \dfrac{Am}{Ag}\sqrt{\dfrac{2}{g}}\ (\sqrt{ht} - \sqrt{ht-h}\,)$

In this case, $h_t = h = 20$ cm

Area of mould, $A_m = 50 \times 25\ cm^2$

Cross-section of gate, $A_g = 5\ cm^2$

$$t_f = \frac{50\times 25}{5}\sqrt{\frac{2}{981}(20)} = 50.48\,\text{sec}$$

3. A cube and a sphere made of cast iron (each of volume 1331 cm^3) were cast under identical conditions. The time taken for solidifying the cube is 5 s. What is the solidification time (in s) for the sphere?

Solution

Time of solidification (T) = $k\left[\frac{\text{Volume}}{\text{Surface area}}\right]^2$

$$\left[\frac{Tsphere}{Tcube}\right] = \left[\frac{Acube}{Asphere}\right]^2 = \left[\frac{6a2}{4\pi r\wedge 2}\right]^2 \text{ --------- } (4.25)\ [\because V_{sphere} = V_{cube}]$$

$V_{cube} = a^3 = 1331\ cm^3$

$\Rightarrow a = 11cm$

Similarly,

$V_{sphere} = \frac{4}{3}\pi r^3$

$\Rightarrow r = 6.8238cm$

Putting in Equation (4.25), we get

$$\frac{Tsphere}{Tcube} = \left[\frac{6\times 11\wedge 2}{4\text{À}\times(6.8238)\wedge 2}\right]^2$$

$T_{sphere} = 7.696$ s

4. For sand casting a steel rectangular plate with dimensions 80 mm × 120 mm × 20 mm, a cylindrical riser has to be designed. The height of the riser is equal to its diameter. The total solidification time for the casting is 2 minutes. In Chvorinov's law for the estimation of the total solidification time, the exponent is to be taken as 2. For a solidification time of 4 minutes in the riser, what is the diameter (in mm) of the riser?

Solution

Casting size = $80 \times 120 \times 20\ mm^3$

$(t_s)_{casting} = 2$ min

$(t_s)_{riser} = 4$ min

$(t_s)_{casting} = k\left(\frac{V}{A}\right)^2$

$$2 = k\left(\frac{80\times 120\times 20}{2(80\times 120+120\times 20+20\times 80)}\right)^2$$

$k = 0.040138\ min/mm^2$

Riser: $(t_s)_{riser} = k\left(\frac{V}{A}\right)^2$

$4 = 0.040138\left(\frac{V}{A}\right)^2$

$$\left(\frac{V}{A}\right) = 9.98$$

$$\frac{V}{A} = \frac{\frac{\pi}{4} d \wedge 2h}{2\frac{\pi}{4} d^{2h} + \pi dh}$$

For riser if h = d *then* $\frac{V}{A} = \frac{d}{6}$

$$\frac{d}{6} = 9.98$$

$$h = d = 59.88 \text{ mm}$$

5. A cylindrical riser of 6 cm diameter and 6 cm height has to be designed for a sand casting mould for producing a steel rectangular plate casting of 7 cm × 10 cm × 2 cm dimensions having a total solidification time of 2 minute. What is the total solidification time (in minute) of the riser?

Solution

$$V_{casting} = 7 \times 10 \times 2 = 140 \text{ cm}^3$$

$$A_{casting} = 2(7 \times 10 + 7 \times 2 + 2 \times 10)$$

$$= 2 \times 10 = 208 \text{ cm}^2$$

$$\left(\frac{V}{A}\right)_{casting} = \frac{140}{208} = 0.673 \text{ cm}$$

$$\left(\frac{V}{A}\right)_{Side\ Riser} = \frac{\frac{\pi}{4} d \wedge 3}{\frac{\pi}{2} d^2 + \pi d \wedge 2} = \frac{d}{6} = 1 \text{ cm}$$

$$\frac{tR}{tC} = \frac{\left(V/A\right) SideRiser \wedge 2}{\left(V/A\right) Casting \wedge 2} = \left(\frac{1}{0.673}\right) \wedge 2$$

$\Rightarrow t_R$ =4.42 minutes (for bottom riser)

4.7 DEFECTS IN CASTING: CAUSES AND REMEDIES

Casting defects are the undesirable irregularities that appear in the casting during the metal casting process. Defects in the cast metal can be caused by a variety of factors or sources. In this section, we go over all of the major types of casting defects. Some of the defects produced may be ignored or tolerated, while others must be eliminated for the parts to function properly.

Characteristics

1. Porosity of gas: blowholes, open holes and pinholes
2. Defects caused by shrinkage: shrinkage cavity
3. Defects in mould material: cut and washes, swell, drops, metal penetration, rat tail
4. Pouring metal defects include: cold shut, misrun and slag inclusion
5. Metallurgical defect: hot tears and hot spots.

1. Shift or mismatch

This is a defect caused by misalignment of the upper and lower parts of the casting, as well as misalignment of the core at the parting line. The causes are:

(1) Improper upper and lower part alignment during mould preparation.
(2) Flask misalignment (a flask is a type of tool used to contain a mould in metal casting. It can be square, round, rectangular, or any other shape that is convenient.)
  (i) Proper alignment of the pattern or die part moulding boxes
  (ii) Double-check the flask alignment.

2. Swell

The molten metal pressure causes the mould cavity to enlarge, resulting in localised or overall enlargement of the casting. Causes are:

(i) Defective or improper mould ramming.

Remedy:

(i) The sand should be rammed evenly and properly.

3. Blowholes

Blowholes are formed when gases entrapped on the surface of a casting due to solidifying metal form a rounded or oval cavity. These defects are always present in the mould's cope.

The following are the causes:

(i) Too much moisture in the sand.
(ii) The sand has low permeability.

(iii) The sand grains are far too fine.
(iv) Rammed sand that is too hard.
(v) Inadequate ventilation is provided.

Treatments:

(i) The sand's moisture content must be controlled and maintained at the desired level.
(ii) Sand with a high permeability should be used.
(iii) Sand with the proper grain size should be used.
(iv) Enough ramming should be done.
(v) An adequate venting system should be provided.

4. Drop

A drop defect occurs when the upper surface of the sand cracks and sand pieces fall into the molten metal. The following are the causes:

(i) Soft ramming and low sand strength.
(ii) Inadequate molten metal fluxing. Fluxing refers to the addition of a substance to molten metal in order to remove impurities. Impurities from the molten metal can be easily removed after fluxing.
(iii) Inadequate sand projection reinforcement in the cope.

Remedies:

(i) Strong sand should be used in conjunction with proper ramming (neither too hard nor soft).
(ii) Molten metal should be properly fluxed so that impurities in the molten metal can be easily removed before pouring it into the mould.
(iii) Adequate reinforcement of the cope's sand projections.

5. Metal penetrating

These casting defects manifest themselves as an uneven and rough surface of the casting. When the sand grain size is large, the molten metal fuses into the sand and solidifies, resulting in a metal penetration defect. The following are the causes:

(i) It is caused by the sand's low strength, large grain size, high permeability, and soft ramming.
(ii) As a result, the molten metal penetrates the moulding sand, resulting in a rough or uneven casting surface.

Remedy:
This defect can be eliminated by using sand with high strength, small grain size, low permeability and soft ramming.

6. Pinholes

These are very small holes, approximately 2 mm in diameter, that appear on the surface of the casting. This flaw occurs as a result of hydrogen gas dissolution in molten metal. When the molten metal is poured into the mould cavity and begins to solidify, the solubility of the hydrogen gas decreases and it begins to escape, leaving behind a small number of pinholes. The following are the causes:

(i) The use of sand with a high moisture content.
(ii) Molten metal absorbs hydrogen or carbon monoxide gas.
(iii) Pouring steel from wet ladles or insufficient gasification

Remedies:

(i) Lowering the moisture content of the moulding sand.
(ii) Proper fluxing and melting procedures should be followed.
(iii) Increasing sand permeability.
(iv) Performing rapid solidification.

7. Shrinkage cavity

Shrinkage cavity refers to the formation of a cavity in a casting as a result of volumetric contraction.

The following are the causes:

(i) Uneven or erratic solidification of molten metal.
(ii) The pouring temperature is excessively high.

Remedies:

(i) This defect can be eliminated by employing the directional solidification principle in mould design.
(ii) Appropriate use of chills (an object used to promote solidification in a specific portion of a metal casting) and padding.

8. Cold shut

It is a type of surface defect, and a line can be seen on the surface. When molten metal enters the mould through two gates, and these two streams of molten metal meet at a junction with low temperatures, they do not fuse and solidify, resulting in a cold shut (appear as line on the casting). It appears as a crack with a rounded edge. Causes include the following:

(i) Inadequate gating system
(ii) A low melting point
(iii) Inadequacy of fluidity.

Remedies:

(i) An improved gating system
(ii) The correct pouring temperature.

9. Misrun

Misrun occurs when molten metal solidifies before completely filling the mould cavity, leaving a space in the mould. Causes are:

(ii) Low molten metal temperature, which reduces fluidity.
(iii) Inadequate gating system and too thin section.

Remedies:

(i) Increasing the molten metal's pouring temperature increases fluidity.
(ii) An appropriate gating system.
(iii) Avoiding a section that is too thin.

10. Inclusion of slag

This defect occurs when molten metal containing slag particles is poured into the mould cavity and solidifies. The following is the cause:

(i) The presence of molten metal containing slag.

Remedy:

(i) Before pouring the molten metal into the mould cavity, remove any slag particles from it.

11. Hot tears or hot cracks

When the metal is hot, it is weak, and as the molten metal cools, the residual stress (tensile) in the material causes the casting to fail. The failure of casting in this case appears as cracks and is referred to as hot tears or hot cracking. Causes are:

(i) Inadequate mould design.
(ii) Removal of residual stress from the casting material.

Remedy:

Proper mould design can easily eliminate these types of casting defect.

12. Hotspot vs. hardspot

When an area of the casting cools faster than the surrounding materials, this is referred to as a hotspot defect. The term "hotspot" refers to areas of the casting that are harder than the surrounding area. It is also known as a hard spot. The cause is:

(i) Rapid cooling of an area of the casting faster than the surrounding materials.

Remedies:

(i) This flaw can be avoided by using proper cooling techniques.
(ii) By altering the metal's chemical composition.

13. Sand holes

This refers to the holes made on the casting's external surface or within the casting. It happens when loose sand washes into the mould cavity and fuses into the interior of the casting or when the molten metal is poured quickly. The following are the causes:

(i) Sand ramming loosely.
(ii) Rapidly pouring the molten metal into the mould causes the sand in the mould to wash away, resulting in the formation of a hole.
(iii) Inadequate mould cavity cleaning.

Remedies:

(i) Proper sand ramming.
(ii) Carefully pour molten metal into the mould.
(iii) Sand holes are eliminated by thorough cleaning of the molten cavity.

14. Dirt

The embedding of dust and sand particles in the casting surface results in a dirt defect. The following are the causes:

(i) Mould curse as a result of improper handling and sand washing (a sloping surface of sand that spreads out by stream of molten metal).
(ii) The presence of slag particles in molten metal.

Remedies:

(i) Proper mould handling to avoid crushing.
(ii) Enough fluxing should be performed to remove slag impurities from molten metal.

15. Sponginess or honeycombing

This is an external defect in which a number of small cavities are present in close proximity in the metal casting. Causes include:

(i) It is caused by dirt and scurf that are mechanically held in suspension in the molten metal.
(ii) As a result of insufficient skimming in the ladle.

Remedies:

(i) Prevent dirt and scurf from entering the molten metal.
(ii) Avoid sand wash.
(iii) Remove slag materials from the molten metal using proper ladle skimming.

16. Warpage

This is an unintentional and undesirable deformation of the casting that occurs during or after solidification. The dimension of the final product changes as a result of this defect. The causes are:

(i) Because different sections solidify at different rates, this causes stresses in adjacent walls, resulting in warpage.
(ii) Large and flat sections, as well as intersecting sections like ribs, are more prone to casting defects.

Remedies:

(i) It can be avoided by creating large areas with wavy, corrugated construction, or by adding enough rib-like shape to provide equal cooling rates in all areas.
(ii) Proper casting designs can more effectively reduce these defects.

17. Fins

Fins are thin metal projections that are not considered to be a part of the casting. It usually happens at the mould or core section parting. The causes are:

(i) Incorrect mould and core assembly.
(ii) Inadequate mould weight or improper flask clamping may result in the formation of fins.

Remedies:

(i) Correct mould and core assembly.
(ii) There should be enough weight on the top part of the mould for the two parts to fit tightly together.

4.8 AUTOMOTIVE MOULDING

The automotive parts industry is the foundation and an important part of development of the automotive industry. There are about 30,000 parts in a whole vehicle, which can be divided into automotive engine system parts, automotive body system parts, chassis system parts, electrical and electronic equipment and general-purpose parts according to the functions. According to the material classification, it can be divided into metal parts and non-metal parts, of which metal parts account for about 60–70%, and non-metal parts account for about 30–40%. Automotive mould is a term used in the automotive industry and is considered as the "mother of the automotive industry". Moulds are used in the manufacture of more than 90% of automotive parts. Generally, it takes more than 1,000 to 1,500 sets of stamping dies for a car, which accounts for about 40% of the total output value of the entire car. There are about 800 sets of plastic moulds, about 300 pairs of sheet metal moulds and about 100 sets of die-casting moulds. Products in the automotive and vehicle industries must reach the highest quality to provide safe usage. Most of the parts of the braking system and many other parts are made by casting. Injection moulding is used to make parts of vehicles' exteriors, especially plastic parts. Aluminium alloys and steel, as well as other materials, are used for these purposes (plastics, polystyrene, rubber, etc.). Casting moulds must be given special attention in order to improve efficiency and shorten the time it takes to produce a new product. In certain cases, prototype cast metal parts must be checked before they can be approved for mass production.

4.9 METAL CASTING APPLICATIONS IN THE AUTOMOTIVE INDUSTRY

The most common application of casting in the automobile industry is vehicle manufacturing. The automobile industry is a major application of metal casting. Metal casting is a massive market for cast products made from ferrous, nonferrous and alloyed metals. There are numerous processes for efficiently producing automobile parts and accessories that are light in weight, inexpensive to manufacture and have a long life. The most common casting processes for producing automobile parts are zinc die casting and aluminium die casting. The weights of these parts are typically one-third that of steel parts and the parts have good thermal properties. Pistons, engine blocks, valve covers, wheels, transmission housing, carburettors, fan clutches and other major automobile components are cast. Magnesium alloy and titanium casts have grown in popularity due to their superior corrosion resistance and the constant need to make automobiles lighter and stronger. Because of their excellent thermal conductivity, which prevents thermomechanical damage, and lower density,

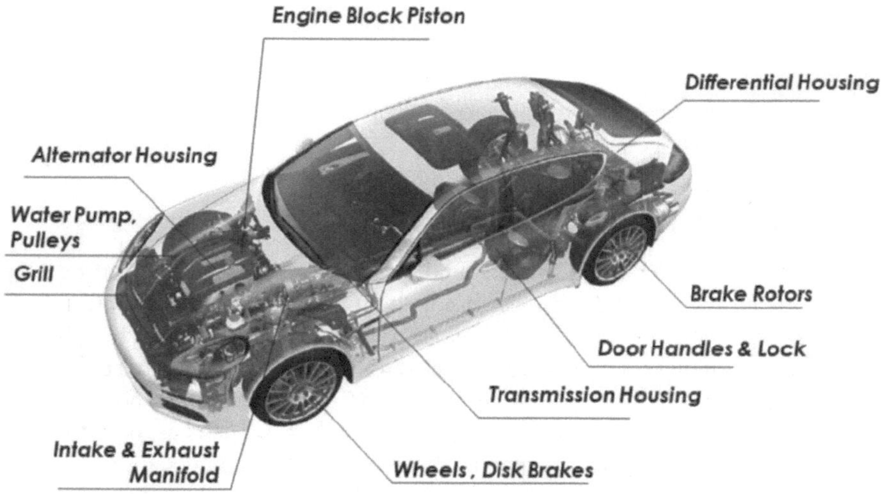

FIGURE 4.20 Cast parts in a vehicle.

Al-Si alloys are commonly used for casting internal combustion engine components. The automobile industry provides a plethora of casting process applications that are constantly being worked on and improved in order to meet the industry's established standards. Zinc die casting can produce an aesthetic appearance while also allowing for the creation of unique shapes with a high durability tolerance. Zinc alloy has excellent anticorrosive properties, making it an excellent choice for the automotive industry. Figure 4.20 shows cast parts in a vehicle.

Following are some of the parts that can be manufactured using this process: parts for sunroofs, transmission components, chassis parts, brake systems and hardware, engine parts, power steering systems, vehicle interior components, fuel intake parts, air-conditioning systems, other mechanical parts, and many more.

4.10 SUMMARY

Casting is the process of pouring molten metal into a mould with a cavity of the desired shape and allowing it to solidify. When the desired metal object has solidified, it is removed from the mould by breaking or disassembling the mould. The pattern is a physical representation of the casting used to create the mould. A core is a sand shape or form that creates the contour of a casting for which no moulding provision has been made in the pattern. The core could be made of sand, plaster, metal or ceramics. Magnesium alloy and titanium casts have grown in popularity due to their superior corrosion resistance and the constant need to make automobiles lighter and stronger. Because of their excellent thermal conductivity, which prevents thermo-mechanical damage, and lower density, Al-Si alloys are commonly used for casting internal combustion engine components. More than 90% of parts in automotive production rely on mould forming. Generally, it takes more than 1,000 to 1,500 sets of stamping dies for a car, which accounts for about 40% of the total output value of the entire car.

4.11 UNSOLVED EXAMPLES

1. A cube-shaped casting solidifies in 5 min. What is the solidification time in min for a cube of the same material, which is six times heavier than the original casting? (**Answer: 16.5 min**)
2. A casting size 400 mm × 200 mm ×140 mm solidifies in 20 min. What is the solidification time for a casting 400 mm × 200 mm × 70 mm under similar conditions? (**Answer: 9.03 min**)
3. A mould cavity of 1200 cm^3 volume has to be filled through a sprue of 10 cm length feeding a horizontal runner. The cross-sectional area at the base of the sprue is 3 cm^2. Consider acceleration due to gravity as 9.81 m/s^2. Neglecting frictional losses due to molten metal flow, what is the time taken to fill the mould cavity in seconds? (**Answer: 2.85 s**)
4. A cylindrical job with diameter of 250 mm and height of 150 mm is to be cast using the modulus method of riser design. Assume that the bottom surface of the cylindrical riser does not contribute as a cooling surface. If the diameter of the riser is equal to its height, then what is the height of the riser (in mm)? (**Answer: 204.54 mm**)
5. A sprue in a sand mould has a top diameter of 20 mm and height of 250 mm. The velocity of the molten metal at the entry of the sprue is 0.5 m/s. Assume acceleration due to gravity as 9.8 m/s^2 and neglect all losses. If the mould is well ventilated, what is the velocity (up to three decimal points accuracy) of the molten metal at the bottom of the sprue. (**Answer: 2.2702 m/s**)

4.12 REVIEW QUESTIONS

(1) State the function of the core.
(2) State the functions of the pattern.
(3) List the types of patterns used in the casting process.
(4) Describe the various defects in casting and their causes and remedies.
(5) Describe pressure die casting with a neat sketch.
(6) Describe the role of casting in the automotive industry.
(7) Describe various types of moulding processes.

5 Forging Process

LEARNING OBJECTIVES

- To recall the properties and related terminologies of forged material.
- To understand some common terms such as billet, press, and die , tools and equipments used and forging methods
- To understand the forging defects, their causes and remedies.
- To learn automotive forging applications and forging sequences.
- To understand the metallurgical spectrum of the forging and similar alloys.

5.1 INTRODUCTION

Forging is the forming process of a red hot metal through the application of uniform pressure (i.e., press forging) or the impact of sudden blows (i.e., hammer forging) and makes use of the plasticity characteristics of the material. In this deformation process the work is squeezed between two dies, using either impact or gradual pressure to convert the part into the required shape. Metals like steel can be converted into the required shape in a cold state but heating of the metal reduces the yield point and makes the deformation easier. Forging of the component may be done by hand or machine. Machine forging consists of the use of dies and normally is used in mass production. Hand forging or blacksmithing can be utilised in cases where the quantity of the production is small and in the cases where work is very specialised. Normally this process can be carried out with the hand and using some special hand tools. Therefore, blacksmithing is a manufacturing process in which the metal can be heated and formed to achieve its requirements with the use of blacksmith tools by using hand tools as well as power hammers.

In smithing, smaller parts are formed by heating them in an open fire or hearth. Shaping is performed under hand supervision by means of hand tools. This work is carried out in a smithy shop. Forging refers to the manufacturing of medium- and large scale heavy parts on a large scale by means of closed heating furnaces, heavy hammers, forging presses and machines.

DOI: 10.1201/9781003367321-5

5.2 FORGING AND FORGEABILITY

The ability of a metal to undergo deformation without cracking by means of forging is what experts call the forgeability of a metal. With this characteristic, metals can easily be formed with low force and resistance. For instance, closed die forging involves a process wherein the dies move toward each other by covering the metal. The heated raw material is placed in the bottom die to produce a design or shape according to specifications.

5.2.1 Forging

Forging is the regulated plastic deformation of metals at elevated temperatures into a predetermined size or shape by compressive forces exerted through some kind of die by a hammer, press or upsetting machine. It effectively involves changing the form and section of a steel specimen by pressing or hammering it at temperatures of around 1000°C, when the steel is fully plastic and will flow under pressure.

Forging, by definition, requires the shaping of metal by the application of impact or pressure; however, the primary distinction between forging methods is the rate at which the energy is applied to the workpiece. The forging hammer deforms at a relatively fast rate, while the hydraulic press deforms at a relatively slow rate. A bar, billet or blank is typically used as the raw material for forging.

Forging deforms the initial crystals and precipitates many of the constituents at high temperatures, which then become soluble in the solid iron on freezing, enhancing the material's local homogeneity. Forging is commonly used for components that need high strength, resistance to shock or vibration, and uniform properties.

The insoluble segregates and non-metallic inclusions, on the other hand, cannot be dispersed by forging, but their undesirable effects are reduced in relation to the properties parallel to the main direction of flow of the material throughout forging.

The structure created by the flow of these separated phases in the direction of working is known as grain fibre, or a flow line fibre structure. Properties such as elastic limit and ultimate tensile strength are gradually improved with pronounced grain flow up to a limit in a direction parallel to the principal grain flow.

The properties in the transverse direction can initially increase due to preliminary working of the metal, but as the new directional grains become more defined, the properties in the transverse direction rapidly decline and may be lower than those exhibited by a cast structure.

- **Temperature range for forging**

Forging should be done at a high enough temperature that the steel can be easily deformed by applied loads. It should be higher than the recrystallisation temperature to prevent work hardening (and therefore depends on the carbon content of the steel), but not so high that it causes excessive grain growth and burning. For steel with a carbon content of 0.2%, the suitable range is 1150–1300°C, and for steel with a carbon content of 0.7%, the range is 1000—1500°C.

- **Characteristics of the forged parts**

(a) It refines the metal structure by filling cavities and breaking up huge grain formations.
(b) Forged components have good strength because they have directional properties.
(c) Mechanical properties such as percentage elongation, percentage reduction of area and shock and vibration resistance are enhanced.
(d) Cracks and blow holes are kept to a minimum.

5.2.2 Forgeability

The ease with which forging is done is called forgeability. Forgeability increases with temperature up to a point at which a second phase, e.g., from ferrite to austenite in steel, appears or if grain growth becomes excessive. The basic lattice structure of metals and their alloys seems to be a good index of their relative forgeability. Certain mechanical properties are also influenced by forgeability. Metals which have low ductility have reduced forgeability at higher strain rate, whereas highly ductile metals are not so greatly affected by increasing strain rates. The pure metals have good malleability and thus good forging properties. The metals having high ductility at cold working temperature possess good forgeability.

The term forgeability can be described as the capacity of a metal to undergo deformation without failure or rupture. For the proper forging process, it is essential to understand the deformation behaviour or trend of the metal to be forged with regard to the resistance to deformation and any adverse effects, such as cracking etc. after heating.

It depends upon the following factors:

(i) Material composition
(ii) Material purity
(iii) Grain size of the metal
(iv) Working temperature
(v) Strain rate and strain distribution throughout the metal
(vi) Number of phases present.

The forgeability is directly proportional to the working temperature, i.e. it increases with increasing temperature up to a point at which the phase changes. Pure metals have good forgeability. There is no commonly accepted standard test for forgeability of metals, however, the following tests can be carried out for a quantitative assignment of the forgeability of a metal.

(i) Upsetting test

As a measure of the forgeability index, the absolute limit of upset ability without cracking or failure is used. This test entails subjecting a series of cylindrical billets of the same dimensions to varying degrees of deformation.

(ii) Notched-bar upsetting test

In this test longitudinal notches or serrations are made prior to upsetting, and the remainder of the test is the same as for the notched-bar upsetting test. This test gives a more reliable index of forgeability.

(iii) Hot-impact tensile test

The impact tensile strength is used as an indicator of the forgeability index in this test. This test is performed using an impact-testing system equipped with a tension-test attachment.

(iv) Hot twist test

Twisting a round, hot bar and counting the number of twists before failure is the measure of the hot twist test. The greater the number of twists, the greater the forgeability of the material.

5.3 ADVANTAGES AND DISADVANTAGES

- **Advantages of forging:**

  1. Strength of the component:
     The forging process minimises failures of the component. In this process the workpiece yields a high strength to weight ratio. As a result it is able to withstand fluctuating stresses which are developed by sudden shock or impact loading.
  2. Conservation of metals:
     There is practically is no or very little material waste.
  3. Weight saving:
     Strong thin-walled parts may be produced without damaging important physical requirements.
  4. Machining time:
     Forging can be made with close dimensional tolerances, which reduces further material removal time for finishing operations and further operations of the products.
  5. Speed of production: a high production rate is easily achievable.
  6. Incorporation in welded structures: parts can be quickly welded due to a fibrous structure.
  7. Maintains uniform and same standards across all sections.
  8. It gives close dimensional tolerances.
  9. Good surface finish is achievable.

- **Disadvantages of forging:**

1. Higher cost of the tool.
2. High maintenance of the tool.
3. No cord holes.
4. Limitations of size and shape.
5. The heat treatment process raises the cost of the product.
6. It is impossible to forge brittle materials like cast iron.
7. Complex forms cannot be produced by forging.

5.4 FORGING TOOLS AND FORGING METHODS: HAND TOOLS AND FORGING OPERATIONS

Some traditional hand forging tools are used to perform forging operations manually. These are also called blacksmith's tools.

The main hand forging tools are as follows: flatter, tongs, fuller, swage, rivet header, punch, hot chisel, anvil, hammers, swage block, set-hammer, drift, blacksmith's gauge, brass, brass scale, and heading tool.

- **Smith's forge or hearth**

Figure 5.1 shows smith's forge or hearth. This has a solid cast iron or steel frame consisting of four leg supports, an iron foundation known as the hearth, a hood at the top and a tuyere opening in the hearth either from the back or from the bottom. The hearth carries coal and is provided with fire brick liners to withstand the high heat generated by the burning of coal. In the absence of this liner, the heat generated, as stated above, would have a direct effect on the metal structure of the hearth, so that the body, in particular the bottom and the surrounding walls, could even melt.

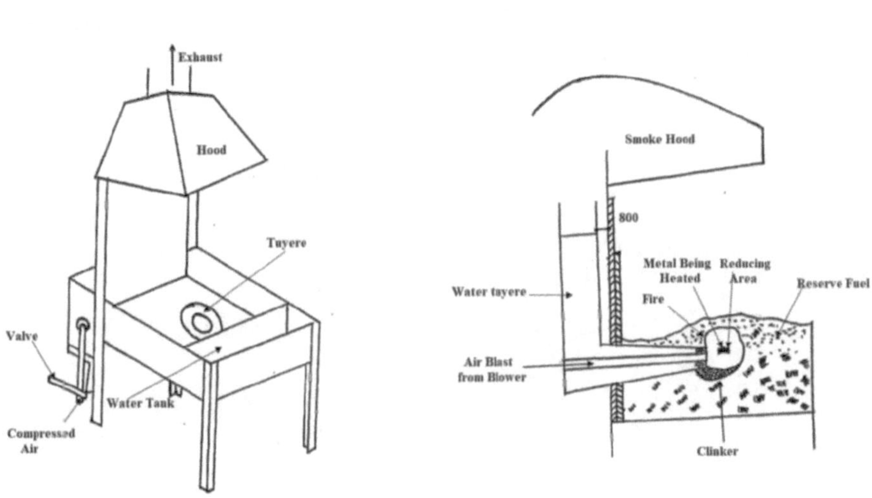

FIGURE 5.1 Smith's forge or hearth.

As a result, the entire system will crumble and the hearth will no longer be useful. Air, under pressure, is supplied by the blower, conveniently located near the forge, through the opening of the tuyere in the hearth. This blower can be either hand-operated or power-driven. The latter is preferable, but the former has no alternative in the absence of the availability of a power supply option. If hand blowers are to be used, they are normally mounted on the back of the forge itself. In the event that power-driven units are to be used, the blower is fitted in one corner of the shop and all the forges are connected to it by means of a well-drawn pipe running underground around the hearths. At the necessary points, auxiliary pipes are used to link the tuyere to the main pipe line. The valve is incorporated in the auxiliary pipe, just before the position where it is attached to the pipe, to regulate the supply of air to the furnace.

The chimney at the top allows smoke and gases created by burning coal to escape as easily as possible. In front of the forge, a water tank is provided that carries water for the purpose of quenching. These hearths can also be used for masonry construction with all attachments such as chimney, tuyere, blower, water tank, etc.

- **Anvil**

In order to carry out successful forging operations, a proper supporting device is needed which should be capable of withstanding the heavy blows rendered during use. An anvil is used for these forging operations. Figure 5.2 shows various parts of anvil. The body of the anvil is commonly made of wrought iron, cast steel or mild steel which is provided with a hardened top that is 20–25 mm thick. The hardened top is in the form of a plate on top of the anvil. Bending of the metal or forming of curved shapes is carried out through the horn or beak. In between the top and the horn a flat step is provided which is used to support the job for the cutting operation and this is known as the chipping block. The flat projecting part at the backside of the anvil is known as the tail. It consists of a square hole to facilitate the square shank of the bottom part of various hand tools such as fullers and swages. It is referred as the hardie hole. Near the hardie hole, a circular hole is provided called the punching hole.

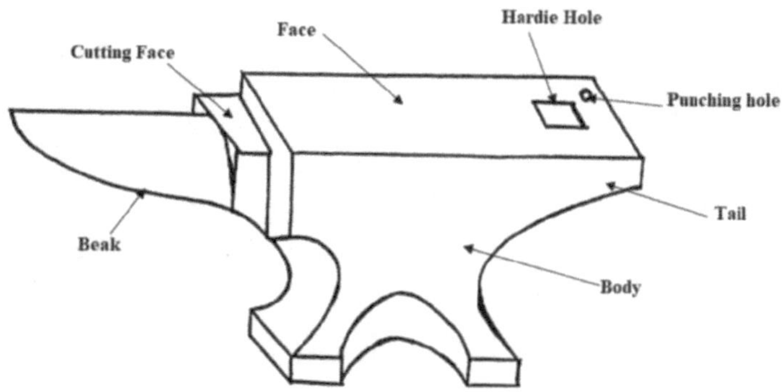

FIGURE 5.2 Anvil.

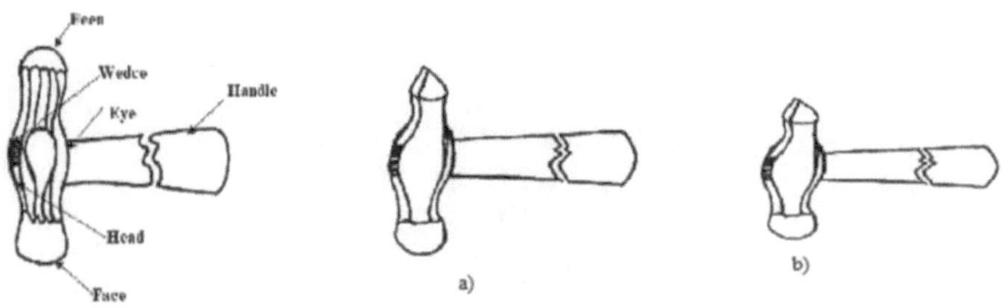

FIGURE 5.3 Types of hammers: (a) cross pin hammer; (b) straight pin hammer.

Since the anvil can be manufactured in the various sizes, the weight of the anvil can be in the range of 50–150 kg. The top face of the anvil is almost 0.75 m from the floor.

- **Hammer**

Hammers can be classified on the basis of size and weight and are used for the forging operation. A smith's hand hammer is a hammer with small size and the sledge hammer is comparatively larger in size and heavier in weight. The smith's hand hammer is normally a small-sized ball peen hammer. Figure 5.3 shows various types of hammers.

All hammers are divided into four parts: the peen, the eye, the cheeks and the face. The top portion of the peen is slightly tapered from the cheeks and rounded to the top. The face is hardened and well-polished and has a subtle rounding around the circular edges so that the metal surface is not harmed by sharp edges when the former is struck by a hammer. The eye is usually oval or elliptical in shape and matches the handle or shaft. These handles are made of shisham wood or bamboo for small hammers, but in the case of sledge hammers the handles are made of solid bamboo.

The steel wedge is often pressed into the handle after it has been fixed to the hammer to prevent the hammer from sliding off the handle during use. A smith's hammer is typically a ball peen hammer or a straight peen sledge hammer of a relatively small scale. Its weight typically ranges between 1.0 and 1.8 kg. A ball peen hammer is used for all general work and its peen is used when light blows are required at a faster speed, such as when fullering a rivet head in a countersunk hole. Sledge hammers are three to four times heavier than hand hammers. They are available in various sizes and weights from 3 to 8 kg. They are used when heavy blows are required in forging and other heavy-duty operations.

- **Swage block**

This is typically a block of cast steel or cast iron bearing a variety of slots of various shapes and sizes along its four sides and through holes from the top to the bottom. It is used as a support for punching holes and for creating various shapes. Figure 5.4 shows swage block. The job to give the desired shape is held on a similar shaped slot, which functions as a bottom swage, and then the top swage is added to it.

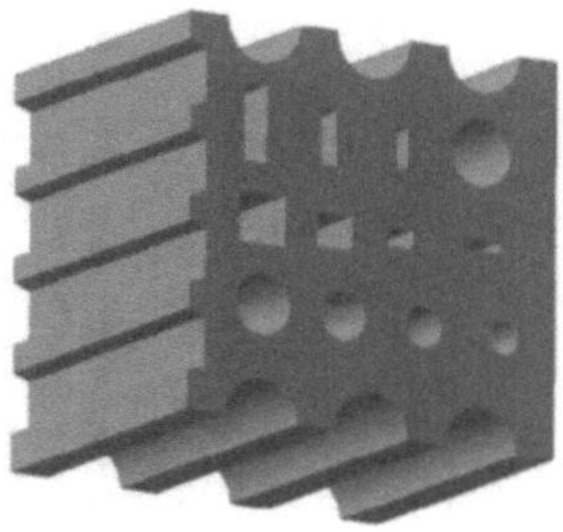

FIGURE 5.4 Swage block.

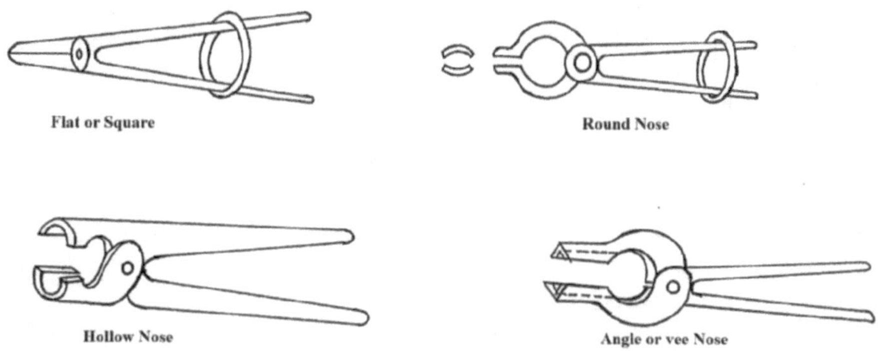

FIGURE 5.5 Tongs.

- **Tongs**

These are used to keep jobs in place and switch over during the forging process. They are made of mild steel. Tongs are typically made in two parts, which are riveted together to create a hinge. The smaller length on one side of the hinge carries the holding jaws, which are rendered in various shapes and sizes to accommodate the corresponding shapes and sizes of the work, and the longer portions on the other side of the hinge form the arms that are held in the hand by the smith. Figure 5.5 shows various types of tongs. The sizes of the tongs differs depending on the size and form of the work to be carried out, but the widely used length of the tongs in hand forging varies from 400 mm to 600 mm, with the opening of the jaws varying from 6 mm to 55 mm.

Tongs are generally named for the inside outline of the jaws.

- Flat tongs are used to grip thin parts and small flat bits.
- Circular hollow pins, with a curved surface inside, are used to carry round work.
- Hollow tongs with square jaws are used to carry square or hexagonal work. Pick-up tongs with shaped jaws allow for quick pick-up of even small sections. They are not used to hold a job.

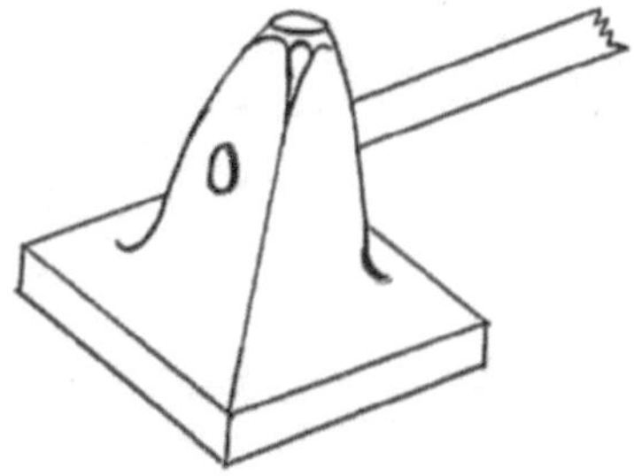

FIGURE 5.6 Flatters.

- **Chisels**

Chisels are used to cut metals in a hot or cold state. Those used to cut metals in a hot state are referred to as hot chisels, and those used to cut metals in a cold state are known as cold chisels. The biggest difference between these chisels is the angle of the cutting edge used. These chisels are provided with the included angle at the cutting edge. A cold chisel has an included angle of 60^0 at the cutting edge and is fully hardened and tempered. The material used for the chisel is made from high-carbon steel. The material used for a hot chisel is medium-carbon steel and no hardening is needed. It is used to cut metal in a plastic state. The angle of cutting edge used is 30^0.

- **Punches**

Punches are tapered instruments made in different shapes and sizes. They are used to make holes in red-hot work. A bigger tapered punch is called a drift. The work is put on the anvil and the punch is pounded through it to about half its size. In is then turned over and the punch is made to pass through. The execution of this operation in two stages prevents the work from being split and completed to the point of bursting.

- **Flatters**

These are often referred to as smoothers. They are made of high-carbon steel and consist of a square body, a handle and a flat square base. They are used for levelling and finishing a flat surface after drawing or some other forging process. Figure 5.6 shows a flatters.

- **Set hammers**

These are made of tool steel and hardened. They are not used to strike. Their construction is similar to that of a flatter, but they are smaller in size and lack an expanded bottom face. They are used to finish corners created by two adjacent right-angle surfaces. Figure 5.7 shows set hammer. The anvil is used to serve the work, and the tool is hammered from the tip.

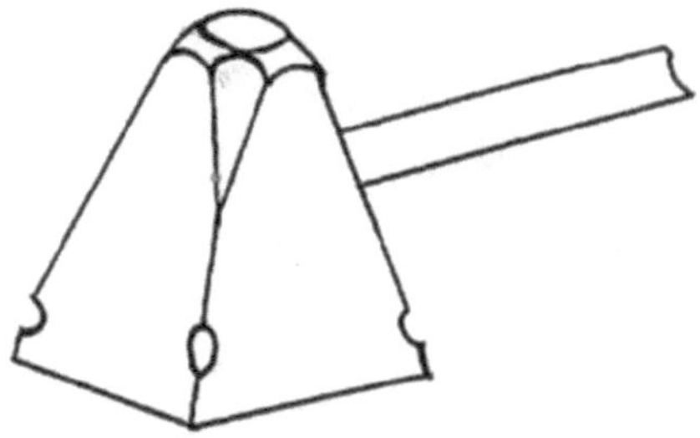

FIGURE 5.7 Set hammers.

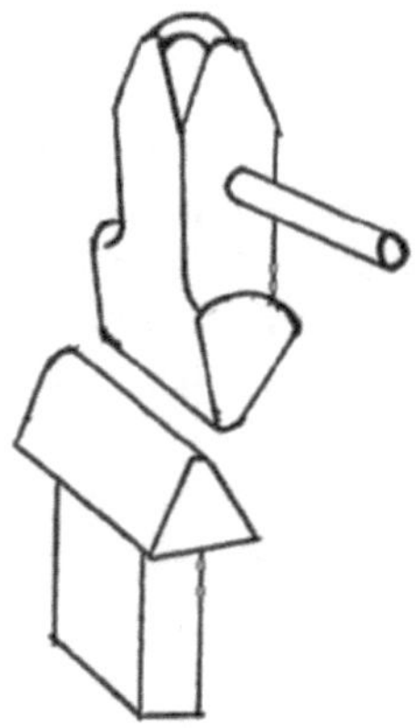

FIGURES 5.8 Fullers.

- **Fullers**

These materials are made of high-carbon steel of varying sizes to accommodate different types of work. Typically used in pairs, they consist of a top and bottom filler. Normally, their working edges are rounded. Figure 5.8 shows fullers, they are used to render the necks by reducing the cross-section of the job and also by drawing them out.

- **Swages**

Like fullers, these are often made of high-carbon steel in two sections, called the top and bottom swages (Figure 5.9). Their working faces have circular grooves to accommodate the scale of the work. They are available in different sizes. The top swage is a handle and the bottom swage is a square shank to match the hardie hole of the anvil during service. They are used to increase the length of the circular rod or to finish the circular surface of the work after forging.

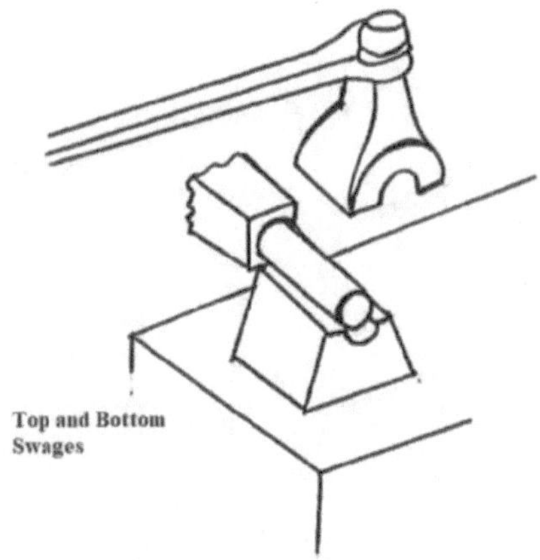

FIGURES 5.9 Swages.

5.5 POWER FORGING: SPRING HAMMERS, DROP HAMMERS

The impact of hand hammer blows will not always be sufficient to affect the proper plastic flow in medium-sized or heavy forging. It also causes fatigue to the hammer user. To have a heavy impact or blow for more plastic deformation, power hammers are generally employed. These hammers are operated by compressed air, steam, oil pressure, springs and gravity. They are generally classified as spring hammers and drop hammers. The capacity of these hammers is given by the total weight. A 100-kg hammer will be one of which the falling pans weigh 100 kg. The heavier these parts and the greater the height from which they fall, the higher will be the intensity of blow the hammer will provide. Power hammers are of different types, e.g. spring power hammers, pneumatic power hammers, etc. These hammers are named due to their construction, according to their way of operation and according to the type of fuel they use for getting the required power for operation. Besides these, a large number of forging presses are also used in forging work. The typical hammers are discussed in the following paragraphs.

- **Spring hammers**

For small forgings, a spring hammer is widely used. This is a light-weight power hammer. Figure 5.10 depicts a typical spring hammer configuration. It is made up of a heavy rigid frame with a vertical projection at the end. This projection serves as a bearing housing inside which the laminated spring oscillates. The back end of this spring has a connecting rod, and the front end has a vertical top that carries weight and travels vertically up and down between fixed guides. At its lower end, the connecting rod is attached to an eccentric sheave, which is then connected to the crank wheel. To operate the hammer, the treadle is pressed downwards, which allows the sheave to rotate through the crank wheel, allowing the laminated spring to oscillate in the bearing. The spring oscillation is responsible for the up and down

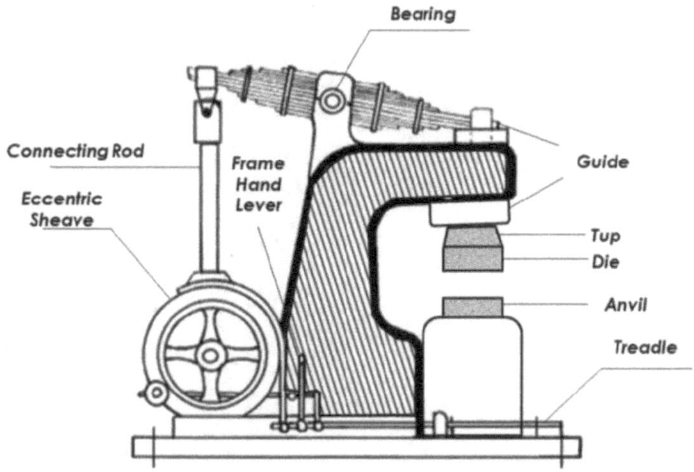

FIGURE 5.10 Spring hammer.

movement of the tup, and thus the necessary blows that are needed on the job to be forged. This mechanical form of hammer is often used on a hand lever to change the stroke of the connecting rod and thus the speed of blows. An eccentric spring hammer is one that uses a rotating eccentric disk to produce vibrations in the spring. It is powered by a foot ring, known as a treadle, located at the bottom and connected to the shaft at the top by a vertical bar with a clutch at one end. The shaft at the top of the hammer is equipped with a pulley and a sturdy disk at the end. A belt from the line shaft or an electric motor drives the pulley. The solid disk at the shaft's end is attached eccentrically to a crank with a laminated spring at its lower end. The weighted nip is suspended on a toggle joint that links the two ends of the laminated spring.

When the foot treadle is pressed, the clutch engages with the shaft and the disk holding the crank begins to rotate, causing variations in the machine's toggle joint. This causes the tup to travel vertically up and down. The speed of the blows is entirely determined by the speed of the driving pulley. Spring hammers are available in a variety of capacities with tup weights ranging from 30 to 250 kg. Those with tup weights of 50–100 kg and blow speeds of up to 300 per minute are commonly used in forging shops. These hammers have a common downside in that their springs are often broken due to extreme vibrations during the forging of jobs in the forging shop.

- **Drop hammers**

Drop hammers are hydraulically driven and are commonly used to form parts by drop hammering a heated bar or billet into a die cavity, as shown in Figure 5.11. A drop forging raises a huge weight and causes it to fall under gravity on dies, allowing the forged part to be compressed. As shown in Figure, the die integrates its form onto the hot work object. Drop hammers are widely used in the forging of copper alloys and steel.

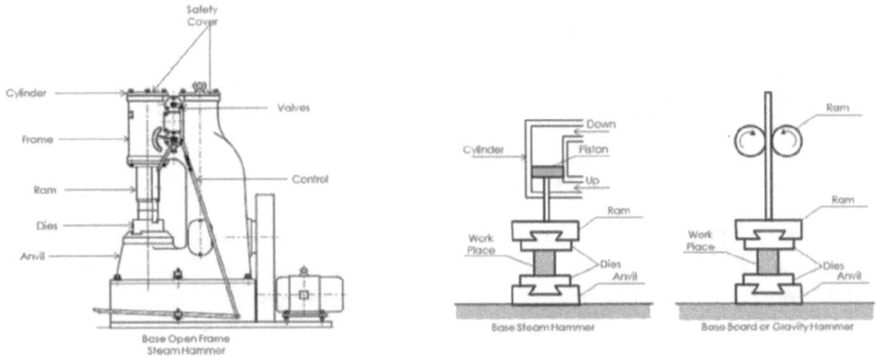

FIGURE 5.11 Drop hammer.

5.6 DEFECTS IN FORGED PARTS AND THEIR CAUSES

Defects occur during forging despite it being one of the best manufacturing processes that gives better mechanical properties. These defects need to be examined and can be prevented. Forging defects can be controlled by careful consideration of work material volume, and good designing of the forging die and the process. To avoid forging defects, care should be taken during the operation and the smith must have experience in forging. Forging defects are explained below.

- **Forging defects**

Although the forging process usually produces higher-quality goods than other manufacturing processes, certain defects are possible if sufficient care is not taken in the design of the forging process. The following is a brief overview of such flaws:

1. A cold shut: This usually happens at corners and at right angles to the surface. It is mostly due to poor die design in which the corner and fillet radii are too short, causing the metal to flow improperly into the corner and result in a cold shut.
2. An unfilled segment: This is similar to a casting misrun in that it happens when metal does not fully fill the die cavity. It is normally caused by inadequate metal or insufficient heating of the metal.
3. Flakes: Essentially, these are internal ruptures. These are caused by insufficient cooling of large forgings and can be remedied by proper cooling practices.
4. Scale pits: These are irregular depressions on the surface of the forging. They are primarily caused by improper cleaning of the forging stock.
5. Improper grain flow: This is caused by an improperly constructed die, which allows the metal flow to deviate from the final intended directions.
6. Internal cracks: These may occur as a result of too rapid a shift in the shape of the raw stock.

7. Die shift: This defect is caused by misalignment of the two die halves, resulting in the two halves of the forging being of incorrect form.
8. Burned and overheated metal: This defect is caused by poor heating conditions and soaking the metal for too long.

5.7 FORGED AUTOMOTIVE COMPONENTS

The demand for forged automotive components is likely to be generated primarily from the automotive ancillary sector which supplies the finished products to automotive original equipment manufacturers (OEMs). Increasing demand for a better driving experience in terms of smooth gear shifting and progress in acceleration, and improvements in the installation of automotive parts, are triggering the global forged automotive component market. It is expected to increase demand for light commercial vehicles and passenger cars in domestic as well as export market, enhancing the global forged automotive component market during the forecast period.

It is anticipated that fluctuating oil and gas prices may hinder the global forged automotive component market during the forecast period. Furthermore, slowdown in automobile production and the availability of an alternative metal-forming process is anticipated to hinder the market. On the other hand, increasing demand for electric vehicles with the chassis and ancillary parts made of polymer materials is expected to cause a decline in the demand for the global forged automotive component market during the forecast period.

- **Connecting rod**

For decades, forged connecting rods have been regarded as the quality standards of the internal combustion engine. Over the years, production methods of connecting rods have constantly changed. This included forging, hammer forging, press forging, metal powder forging and casting.

In the early years of the automotive industry development, hammer forging was considered one of the best forging methods. However, in the middle of the twentieth century, due to the higher requirements for quality and tighter tolerances, there was a need to change the forging method. Therefore, manufacturers began to use a mechanical press to replace forging hammer. The most effective way to reduce the manufacturing cost was to improve material utilisation. Increasing requirements for weight and size tolerance promoted the use of the powder forging method. Another advantage besides tolerance was that it enabled the connecting rod cover to be forged together with the rod. This allowed the cover to be opened from the connecting rod, and thus manufacturing costs were reduced.

In the late twentieth century, the forging method was developed into "open" forged steel connecting rods. This became one of the strengths of powder forging. The strength of the forged steel connecting rod is higher than that of the powder metallurgy rod. Due to this, the engine compression ratio is increased. As a result, the high strength of the forged connecting rod was more satisfactory. This forging method could meet both the strict requirements of weight and size tolerance. Figure 5.12 shows forging sequence for connecting rod.

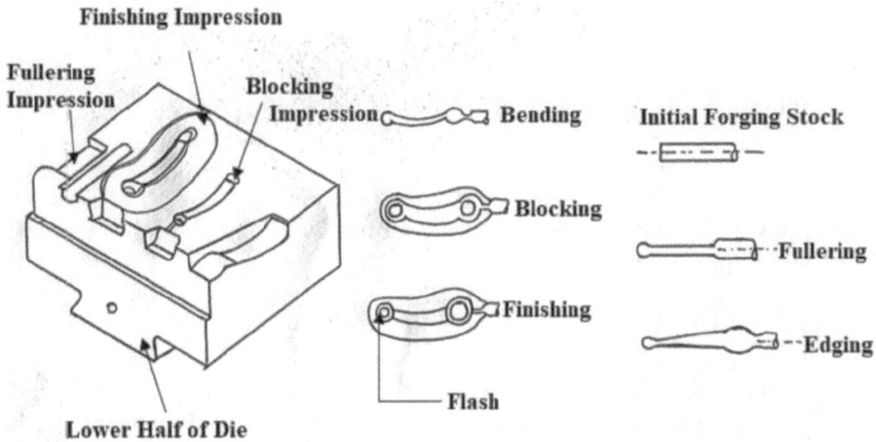

FIGURE 5.12 Steps in the forging sequence of a connecting rod.

(i) Fullering impression

Fullering is the very first step towards reducing the stock to the desired size. The completion of the process is achieved in the fuller impression of the die. The heated stock is first put in a fuller impression and then hammered once or twice to obtain local metal distribution on the surface of its cross-section.

(ii) Edging impression

The stock is then moved to the edging print where the metal is redistributed along its length in order to better fill the finishing die cavities. In this process, the exact amount of metal is "gathered" to some predetermined cross-sections and reduced to some others. Edging is an important move and is typically accomplished through a series of blows, along with the turnover of the metal, as necessary.

(iii) Bending

The next operation is bending, which may or may not be required, depending on the design and shape of the product. The bending operation may be accomplished without a bending impression, but then the grain flow path may not match the bending form, resulting in a weak grain strength flow. Thus, in order to enhance the flow of grain, the bending impression in the die is often integrated.

(iv) Impression blocking

The blocking operation is often referred to as a semi-finishing operation. It is the penultimate step before the process is over. The blocking operation helps to minimise tool wear in the finishing impression. More than one blocking impression can be used for complex shapes.

(v) Finishing impressions

This is the impression where the actual shape is made. In this process, the excess metal will form a flash in the parting line that surrounds the forging.

(vi) Operation of trimming

Finally, the extra flash present around the forging is trimmed in order to achieve optimal forging.

- **Crankshaft**

Forged crankshafts, often referred to as forged cranks, provide strength and reliability that far surpass any cast or turned bar stock, making forged crankshafts the standard for any use that demands strength, consistency or quality. The method of producing crankshafts by forging is explained below.

(i) Stock is redistributed and size is increased at certain places and reduced at others by rolling forging.
(ii) After preliminary roll forging operation the stock is forged again.
(iii) This stock is then forged in the first impression or blocking die.
(iv) The final shape is given to the forging in the next blocking die.
(v) Then the finished part is trimmed in the blanking die to eliminate excess metal or flash.

5.7.1 Forging Sequence for Camshaft

According to the conventional cold forging method, a blank for a camshaft is axially compressed in a forging die so that cams are formed in sequence, first, at portions near the ends, and then, at inner and central portions. Figure 5.13 shows forging sequence for the production of camshaft.

(i) Stock is redistributed and size is increased at some locations and decreased at others by roll forging.
(ii) After preliminary roll forging, the stock is again roll forged.
(iii) This stock is then forged in the first impression or blocking die.
(iv) The final shape is given to the forging in the next blocking die.
(v) Then the finished part is trimmed in the blanking die.

5.8 METALLURGICAL SPECTRUM OF FORGING

One major benefit of hot working is the ability to detect and upgrade mechanical properties. Most of the porosity, directionality and segregation found in cast forms is eliminated by hot-rolling or hot-forging. As a result, the "wrought" product is more ductile and harder than the unworked casting. During the forging of a bar, the metal grains become greatly elongated in the direction of flow. As a result, the hardness

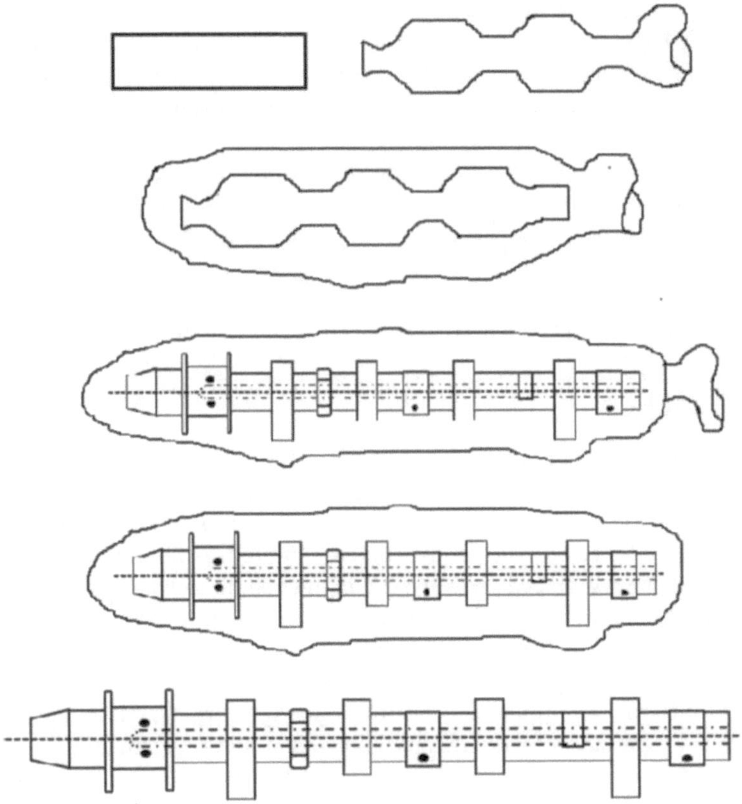

FIGURE 5.13 Forging sequence for the production of a camshaft.

of the metal increases significantly in this direction, while decreasing slightly in directions transverse to the flow. A good forging design involves making sure that the flow lines in the finished component are aligned in the direction of maximum stress when the element is put to use.

In alloy selection and process design, a metal's resistance to thinning and fracture during cold-working operations is critical. The best alloys for stretching operations are those that harden with pressure (strain harden), such as the copper–zinc alloy used in cartridges and the aluminium–magnesium alloys used in beverage cans, which both exhibit greater strain hardening than pure copper or aluminium.

Another useful property that can be regulated by processing and composition is the plastic anisotropy ratio. When a sheet segment is bent (i.e., elongated) in one direction, the thickness and width must shrink in order for the volume to remain stable. The thickness and width of an isotropic sheet exhibit equal strain, but if the grain orientation of the sheet is right, the thickness will shrink about half as much as the width. Due to the fact that thinning induces early fracture, this plastic anisotropy increases the deep-drawing properties of sheet material with the best grain orientation.

Defects in the metal, which often consist of non-metallic inclusions such as oxides or sulphides embedded in the metal during grinding, may cause workpiece fracture during shaping. Proper manufacturing procedures can aid in the prevention of such inclusions. A lap is another form of defect that occurs when a part of a metal piece is inadvertently folded over on itself but the two sides of the fold are not properly welded together. If a force tending to open this fold is applied during the forming process, the metal may fail at the lap.

The ability of different metals to withstand strain varies greatly. The amount of shape change that can be achieved in a single forming step is often restricted by the tensile ductility of the metal. Metals with a face-centred cubic crystal structure, such as copper and aluminium, are by far more ductile in such operations than metals with a body-centred cubic structure. Processes that primarily apply compressive stresses rather than tensile stresses are used to avoid early fracture in the latter type of metal.

5.9 SUMMARY

Forging is the forming process of a red-hot metal by applying uniform pressure (i.e., press forging) or the impact of sudden blows (i.e., hammer forging). The term forgeability can be as described the capacity of a metal to undergo deformation without failure or rupture. The forging process minimises failures of a component. In this process the workpiece yields with a high strength to weight ratio. As a result of this it is able to withstand fluctuating stresses which are developed by sudden shock or impact loading. One major benefit of hot working is the ability to detect and upgrade mechanical properties. Most of the porosity, directionality and segregation found in cast forms is eliminated by hot-rolling or hot-forging. Although the forging process usually produces higher-quality goods than other manufacturing processes, certain defects are possible if sufficient care is not taken in the design of the forging process.

5.10 REVIEW QUESTIONS

(1) Describe the forgeability of a metal.
(2) List the advantages and disadvantages of the forging process.
(3) List the various forging tools used in forging operations.
(4) Describe the various press machines used in forging operations.
(5) State various forging defects and their causes.
(6) Describe metallurgical aspects of the forging process.

6 Welding Process

LEARNING OBJECTIVES

- To understand the fundamentals of the welding process their terminologies, tools and equipments.
- To comprehend the classification of welding and allied process in vehicular applications in detail.
- To study the types of weld joints and their design aspects.
- To get details of automotive applications of welding process and their standard procedures.

6.1 INTRODUCTION

Welding is the method of fusing two metals that can be identical or dissimilar. It joins various metals/alloys without or with pressure and without or with the use of a filler metal. Certain welding processes are performed solely by heat, with no pressure applied; others by a mixture of heat and pressure; and still others by only pressure, without external heat being supplied. Welding is typically done on parts made of the same metal, although certain welding operations may be used to join metals that are not the same. Welding processes are classified into two types:

- Fusion welding and
- Solid-state welding.

1. **Fusion welding**
   In fusion welding methods, heat is employed to melt the base metals. A filler metal is applied to the molten pool in certain fusion welding operations to accelerate the process and gives bulk and strength to the welded joint. An autogenous weld is a fusion-welding process in which filler metal is not used.
2. **Solid-state welding**
   Solid-state welding refers to a joining process in which coalescence happens as a result of applying pressure alone or pressure and heat in combination. If

DOI: 10.1201/9781003367321-6

heat is employed, the temperature of the process is kept below the melting point of the metals being welded. There is no usage of filler metal.

The following is a general classification of welding and related processes.

(A) Welding processes

1. **Oxy-fuel gas welding processes**
    - Air-acetylene welding
    - Oxy-hydrogen welding
    - Oxy-acetylene welding
2. **Arc welding processes**
    - Submerged arc welding
    - Carbon arc welding
    - Shielded metal arc welding
    - Plasma arc welding
    - Atomic hydrogen welding
    - Electro-slag welding
    - Gas metal arc welding
    - Gas tungsten arc welding
    - Electro-gas welding
    - Stud arc welding
3. **Resistance welding processes**
    - Seam welding
    - Spot welding
    - Resistance butt welding
    - Projection welding
    - Percussion welding
    - Flash butt welding
    - High-frequency induction welding
    - High-frequency resistance welding
4. **Solid-state welding processes**
    - Cold pressure welding
    - Forge welding
    - Explosive welding
    - Friction welding
    - Diffusion welding
    - Thermo-compression welding
5. **Thermit welding processes**
    - Pressure thermit welding
    - Thermit welding
6. **Radiant energy welding processes**
    - Electron beam welding
    - Laser welding

(B) Allied processes

1. **Metal joining or metal depositing processes**
    - Brazing
    - Soldering
    - Adhesive bonding
    - Braze welding
    - Surfacing
    - Metal spraying
2. **Thermal cutting processes**
    - Arc cutting
    - Gas cutting

- **Advantages of welding**

    - Welding is a more cost-effective and time-efficient process than other methods (riveting, bolting, casting, etc.)
    - Welding, when done properly, produces permanent joints with strength equal to or greater than base metal.
    - A variety of metals and alloys, both identical or dissimilar, can be joined together by the process of welding.
    - The expense of general welding equipment is not prohibitive.
    - Portable welding equipment is readily accessible.
    - Welding allows for a great deal of design flexibility.
    - Welding can connect welding jobs through spots, as well as continuous pressure tight seams, end-to-end as well as in a variety of other configurations.
    - Welding can also be automated.

- **Disadvantages of welding**

    - It causes residual stresses and workpiece distortion.
    - Welded joints require stress relief and heat treatment.
    - Welding emits hazardous radiation (light), gases and spatter.
    - Jigs and fixtures may also be needed to carry and place the welded pieces.
    - Edge preparation of welding jobs is required prior to the welding.
    - A skilled welder is needed for the production of high-quality welding.
    - Since the composition of the welded joint differs from that of the parent metal, heat during welding causes metallurgical changes.

6.2 CONCEPT OF WELDING AND WELDABILITY

Welded structures are put together using five different types of joints: butt, lap, corner, T, and edge joints. Butt joints are created by welding the members' end surfaces or edges together. Lap joints are made by welding two overlapping surfaces together. T joints are made by welding two surfaces that are at right angles to each other. Corner joints are created by welding the two edges of surfaces that are at right angles to each

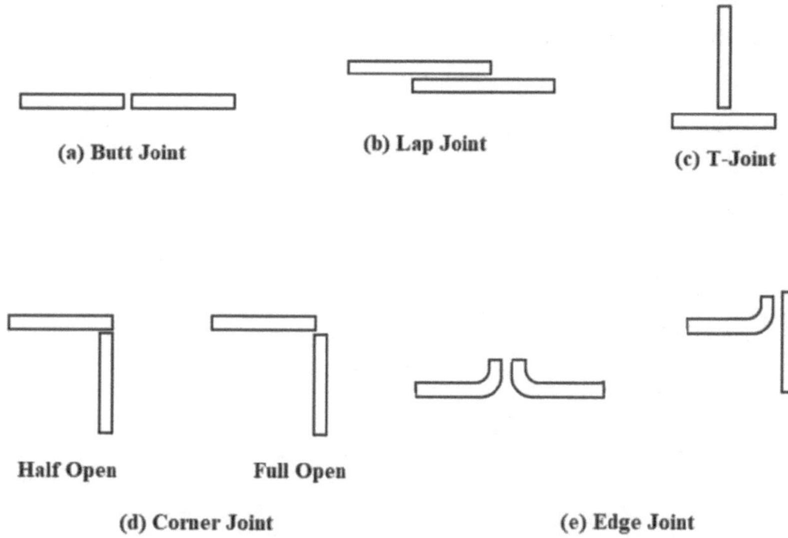

FIGURE 6.1 Basic welded joints.

other. Edge joints are created by welding two edges that have a portion of their surface parallel to each other. Figure 6.1 shows basic welded joints

Types of welds

- A "bead" weld is one in which the filler metal is deposited at a joint where the two neighbouring surfaces are in the same plane.
- A "bead" is a single run of welded metal.
- A "fillet" weld is one where the filler metal is deposited at the intersection of two intersecting surfaces, such as a T or lap joint.
- A "groove" weld is one in which the filler material is deposited in a groove created by the edge preparation of one or both members.
- A "plug" or "slot" weld is one in which a hole is formed through one of the welded parts, and the filler material is then deposited into this hole and fused with the mating portion. Figure 6.2 shows types of welds.

Weldability

A metal's weldability is commonly characterised as its ability to be welded into a particular structure that has certain properties and characteristics and will meet service requirements satisfactorily.

Weldability is affected by many factors, including:

(1) The welding mechanism
 Certain metals or metal combinations that weld easily in one method are difficult to weld in another. Stainless steel, for example, is easily welded by most AW processes but is considered a tough metal for oxyfuel welding.

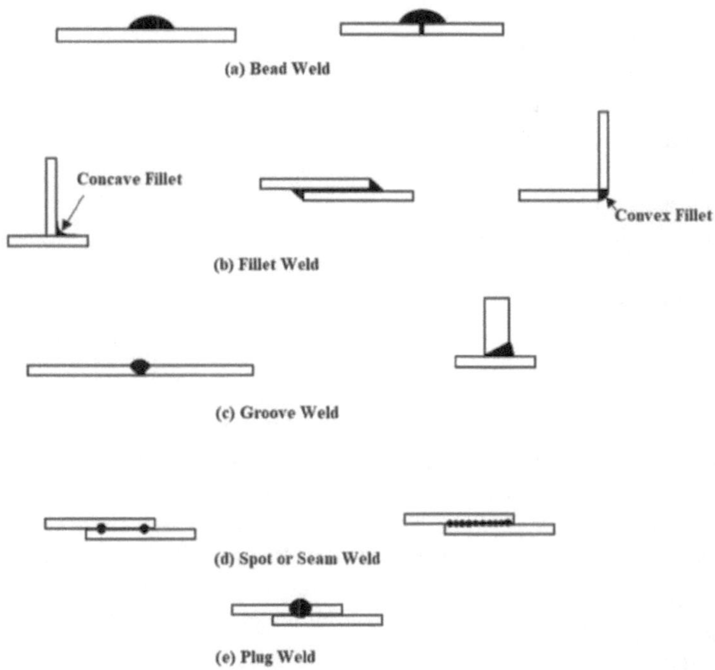

FIGURE 6.2 Types of welds.

(2) Properties of base metals
The properties of the base metal have an impact on welding efficiency. Melting point, thermal conductivity and the coefficient of thermal expansion are all important properties. Metals with a high thermal conductivity appear to transfer heat away from the weld zone, making them difficult to weld (e.g., copper). The welded assembly suffers from distortion due to the metal's high thermal contraction and expansion.

(3) Filler metal
When using a filler metal, it must be matched with the base metal(s). In general, elements mixed in a liquid state that solidify to form a solid solution will not cause a problem. If the solubility limits are exceeded, weld joint brittleness can occur.

(4) Surface characteristics
The surface conditions of the base metals may have a negative impact on the process. Moisture, for example, can cause porosity in the fusion zone. Oxides and other solid films on metal surfaces may prevent proper contact and fusion.

6.3 ARC WELDING: ARC WELDING EQUIPMENT, ELECTRODES

An arc welding method is one in which an electric arc is used to weld base metals between an electrode and a workpiece or between two electrodes. The various arc welding processes are as follows:

- Carbon arc welding
- Flux cored arc welding
- Shielded metal arc welding
- Gas metal arc welding
- Gas tungsten arc welding
- Plasma arc welding
- Electroslag welding
- Atomic hydrogen welding
- Stud arc welding
- Electrogas welding

Process

An electric arc is a discharge of electric current through a circuit gap sustained by the existence of a thermally ionised column of gas (referred to as plasma) through which current flows. In an AW operation, the electrode makes contact with the work and then quickly separates from it by a short distance to start the arc. The electric energy produced by the arc produces temperatures of 5500°C or higher, which is hot enough to melt any metal. A molten metal pool forms towards the tip of the electrode, consisting of base metal(s) and filler metal (if one is used). Filler metal is applied during the procedure in most arc-welding processes to improve the volume and strength of the weld joint. The molten weld pool solidifies in its wake as the electrode moves along the joint. The electrode is moved relative to the job by either by a manual welding or by machine welding, automatic welding or robotic welding. One of the most difficult aspects of manual arc welding is that the accuracy of the weld joint is dependent on the individual welder's expertise and work ethic. Figure 6.3 shows details of arc welding.

Benefits

- Shielded metal arc welding (SMAW) can be performed in any location while maintaining the highest weld reliability.

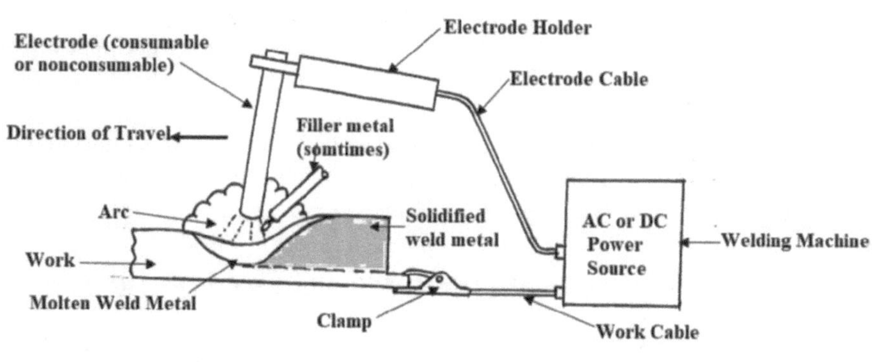

FIGURE 6.3 Arc welding.

- MMAW is the most basic of all arc welding processes.
- Since a wide range of electrodes is available, this welding process has a plethora of applications.
- A wide variety of metals and alloys can be easily welded.
- The method is ideal for hard facing, metal resistance and other applications.
- Joints (for example, between nozzles and shells in a pressure vessel) that are difficult to weld by automated welding machines can be easily achieved by flux shielded metal arc welding.
- The MMAW welding equipment is compact and reasonably priced.

Limitations

- As compared to MIG and TIG welding, the chances of slag entrapment and other associated defects are higher with flux coated electrodes.
- Because of the slag gases and particles, the arc and metal transfer are not as visible, making welding control more complicated in this process than in MIG welding.
- Mechanisation is difficult due to the small duration of each electrode and the brittle flux coating on it.
- When welding long joints (for example, in pressure vessels), the weld should be advanced with the next electrode as one electrode finishes.
- A defect (such as slag inclusion or inadequate penetration) may occur at the location where welding is restarted with the new electrode if not properly cared for.
- Since stick electrodes are used, the process is slower than MIG welding.

Applications

1. Today, almost all widely used metals and alloys can be welded by this method.
2. Shielded metal arc welding is used for both fabrication repair and maintenance tasks.
3. The procedure has applications in
   Bridge and building construction
   Automotive and aerospace industries
   Construction of air receivers, tanks, boilers, and pressure vessels
   Ship construction
   Piping
   Joining penstock.

6.3.1 Arc Welding Equipment

1. Source of power for arc welding

Electric arc welding employs both direct current (DC) and alternating current (AC), each with its own set of applications. DC welding power is usually supplied by generators powered by electric motors or, in the absence of electricity, by internal combustion engines. Transformers are commonly used for AC welding supply in almost all arc welding applications where mains electricity is available. They must reduce the standard supply voltage (200–400 volts) to the open circuit welding voltage (50–90 volts). The following factors affect power source selection:

- The type of electrodes to be used, as well as the metals to be welded
- Availability of a power source (AC or DC)
- Required production
- Cycle of duty
- Effectiveness
- Initial and operating costs
- Floor space available
- Equipment adaptability.

2. Cables for welding

Welding cables are needed for current transmission from the power source to the load. Welding cables are required for conduction of current from the power source through the electrode holder, the arc, the workpiece, and back to the welding power source. These are insulated copper or aluminium cables."

3. Holder for electrodes

The electrode holder is used to manually hold the electrode when applying current to it. They are usually matched to the size of the lead, which is then matched to the arc welder's amperage production. Electrode holders are available in a variety of sizes ranging from 150 to 500 amps.

4. Electrodes for welding

An electrode is a wire or rod made of metal or alloy, with or without coatings. Between the electrode and the workpiece, an arc is formed.

5. Handheld screen

A handheld screen used for eye defence and weld bead supervision.

6. Hammer for chipping

The chipping hammer is used to strike the slag and scrape it.

7. Brush made of wire

A wire brush is used to clean the weldable surface.

8. Protective clothing

To avoid direct heat exposure to the body, the operator wears protective clothing such as an apron.

6.3.2 Electrodes

In AW processes, electrodes are graded as consumable or non-consumable and consumable electrodes supply the filler metal. These electrodes are available in two main types: rods (also known as sticks) and wire. Welding electrodes are rated as follows:

- Electrodes that are consumable
- Electrodes in their purest form
- Electrodes with coatings
- Electrodes that are not consumable
- Electrodes made of carbon or graphite
- Electrodes made of tungsten.

The consumable electrode is made of various metals and alloys. As an arc is formed between the electrode and the workpiece, the end of this electrode begins to melt. As a result, the consumable electrode serves as a filler metal. Bare electrodes are made of a metal or alloy wire that has no flux coating. Coated electrodes have a flux coating that melts when an electric arc is struck. When melted, this coating performs a variety of functions, including joint protection from atmospheric pollution, arc stabilisers, and so on.

Electrodes that are not consumable

Non-consumable electrodes are composed of high melting point materials such as carbon, pure tungsten or alloy tungsten, among others. During welding, these electrodes do not melt. However, because of oxidation and vaporisation of the electrode material during welding, the electrode length decreases over time. Non-consumable electrodes are usually made of copper-coated carbon or graphite, pure tungsten, or thoriated or zirconiated tungsten.

6.4 GAS WELDING: TYPES OF FLAMES, GAS WELDING EQUIPMENT

Oxyfuel–gas welding (OFW) is a broad term for any welding process that generates a flame by combining a fuel gas with oxygen. The heat used to melt the metals at the joint is produced by a flame. Acetylene is used in the most popular gas welding method, known as oxyacetylene–gas welding (OAW), which is commonly used for structural metal fabrication and repair work.

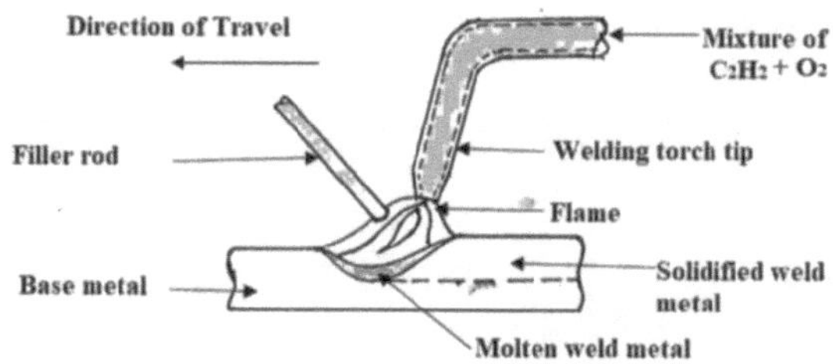

FIGURE 6.4 Oxyfuel–gas welding.

Gas welding process is much easier than arc welding. During this phase, all of the equipment is carefully connected. Pressure regulators attach the gas and oxygen cylinders to the welding torch. The pressure of the gas and oxygen supplied to the torch is adjusted so that they are properly combined. Figure 6.4 shows an oxyfuel gas welding process.

A striker ignites the flame. The torch's tip is pointed downward. The flame is now operated by valves located on the welding torch. Depending on the welding state, the flame is set to natural flame, carburising flame or oxidising flame. The welding torch now moves along the line where the joint is to be formed. This will melt the interface element and permanently join them.

Advantages

- It is easy to use and does not necessitate a highly skilled operator.
- The cost of equipment is low as compared to other welding processes such as MIG, TIG, and so on.
- It can be used on the job site.
- Welding equipment is more compact than other types of welding.
- It can also be used for cutting metals.

Disadvantages

- It has a low surface finish. After welding, this method necessitates a finishing operation.
- Gas welding has a wide heat-affected region, which can cause the mechanical properties of the parent material to alter.
- There is a greater risk of injury due to the naked flame's high temperature.
- It is only suitable for soft and thin boards.
- The metal joining rate is slow.
- There is no shielding field, which leads to further welding defects.

Utilisation

- It is used to connect thin metal plates.
- It can be used to join ferrous and nonferrous metals.

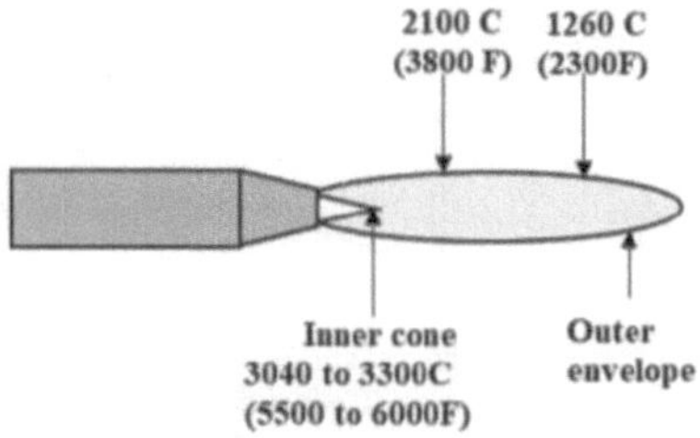

FIGURE 6.5 Neutral flame.

- Gas welding is primarily used in the manufacturing of sheet metal.
- It is commonly used in the automotive and aerospace industries.

Flame varieties

There are three types of flame: neutral flame, carburising flame and oxidising flame. In oxy-acetylene gas welding, the flame is the most important means to control the welding joint and the welding process.

(1) Neutral flame

As the name suggests, this flame contains an equal amount of oxygen and gas fuel by volume. This flame completely burns the fuel and has no chemical effect on the metal to be welded. It emits very little smoke. The flame is divided into two zones. The inner zone is white and has a temperature of about 3100^0C, while the outer zone is blue and has a temperature of about 1275^0C. Figure 6.5 shows a neutral flame.

Applications include stainless steel, cast iron, copper, mild steel and aluminium.

(2) Oxidising flame

As the volume of acetylene in a natural flame decreases or rises, the inner cone tends to vanish, and the flame that results is known as an oxidising flame. It is hotter than natural flame and has two distinct regions. The inner region is very bright white in colour and has a temperature of around 3300^0C. The outer flame is blue in colour and is used to weld oxygen-free copper alloys such as brass and bronze. Figure 6.6 shows an oxidising flame.

Applications include copper base alloy, zinc base metal, brass and bronze.

(3) Carburising flame

This flame contains an excess of fuel gas and chemically reacts with metal to form metal carbide. As a result, this flame should not be used with metals that consume carbon. The flame is smoky and silent, and is divided into three parts. The

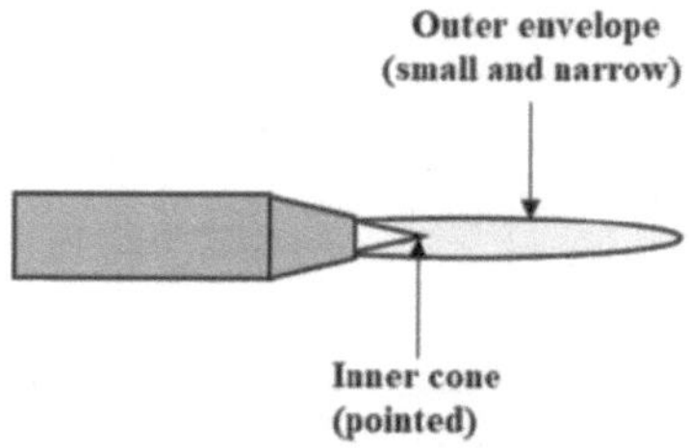

FIGURE 6.6 Oxidising flame.

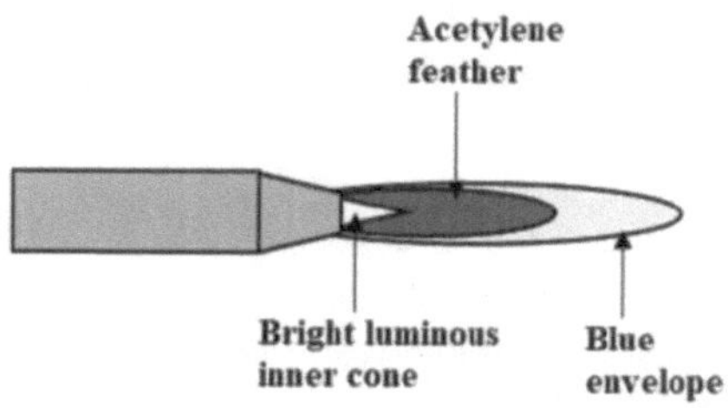

FIGURE 6.7 Carburising (reducing) flames.

inner zone is white in colour, the intermediate zone is red, and the outer zone is blue. The inner zone temperature is approximately 2900^{0}C. Figure 6.7 shows a carburising flame.

Applications include high-carbon steel, nonferrous alloy medium carbon steel and nickel, among other applications.

The following equipment is required for oxy-acetylene welding.

1. Welding torch or blowpipe

This is the tool used to combine two gases in the desired quantities and burn the mixture at the tip of a torch. It has a handle for carrying and two gas inlet connections at the top. Each inlet has a valve that controls the amount of oxygen or other gas pumped in.

A mixture of two gases from two paths is formed, and a flame is produced by igniting the mixture at the torch's tip.

2. Pressure regulator

This device regulates the amount of pressure in the system.

The pressure regulator's purpose is to reduce the pressure from the cylinder and keep it constant regardless of pressure variations at the source.

It is also used to change the gas pressure to the torch. Changing the pressure is as simple as turning the handle on the regulator.

Pressure regulators are classified into two types: single stage and double stage.

3. Hose and connectors

Two hoses are needed to transport oxygen and acetylene separately. They attach the torch to the regulator fixed on cylinders. In general, green colour is used for oxygen and red for acetylene. They should be solid, long-lasting, non-porous, light and flexible. Never apply grease or oil oxygen fittings to prevent explosions. Special hose fittings and connections for attaching to the torch and pressure regulators are supplied.

4. Cylinders

The gases are usually contained under high pressure in steel cylinders. For commercially obtained oxygen, cylinders are made of drawn-steel with no seams and carefully heat treated to develop great strength and durability. The cylinder also has a high-pressure valve and a valve-protector seal. It is charged at a pressure of about 150 kg/cm^2 and a temperature of about 200^0C. A safety fuse plug is also included to release oxygen if the temperature rises. Acetylene cylinders are densely filled with acetone-saturated absorbent filler.

5. Safety goggles

These are important for eye protection. These have colour lenses that protect against harmful heat as well as ultraviolet and infrared rays.

6. Lighter spark

This offers a quick and easy way to light the welding torch. It is made up of a pointed stone and a rough surface that, when rubbed together, create a light.

7. Apron

This protects the operator's clothes from dirt and danger while keeping him alert.

8. Gloves

These are essential for hand protection.

9. Ventilation apparatus

This is especially important when welding in confined spaces. Welding fumes are toxic to the lungs.

10. Welding rods

The welding rod's structure and properties should, in theory, be very similar to those of the base metal. As a result, appropriate welding rods for different non-ferrous and ferrous metals should be selected.

6.5 RESISTANCE WELDING: TYPES AND APPLICATIONS

Resistance welding (RW) refers to a class of fusion-welding processes that use a combination of heat and pressure to achieve coalescence, with the heat produced by electrical resistance to current flow at the to-be-welded junction.

6.5.1 Resistance Welding Types

(1) Resistance spot welding (RSW)

Resistance spot welding (RSW) is a RW method in which opposing electrodes fuse the faying surfaces of a lap joint at a single site. In situations where an air-tight assembly is not needed, the process is used to connect sheet-metal parts with a thickness of 3 mm (0.125 in) or less using a series of spot welds. Figure 6.8a shows details of resistance spot welding.

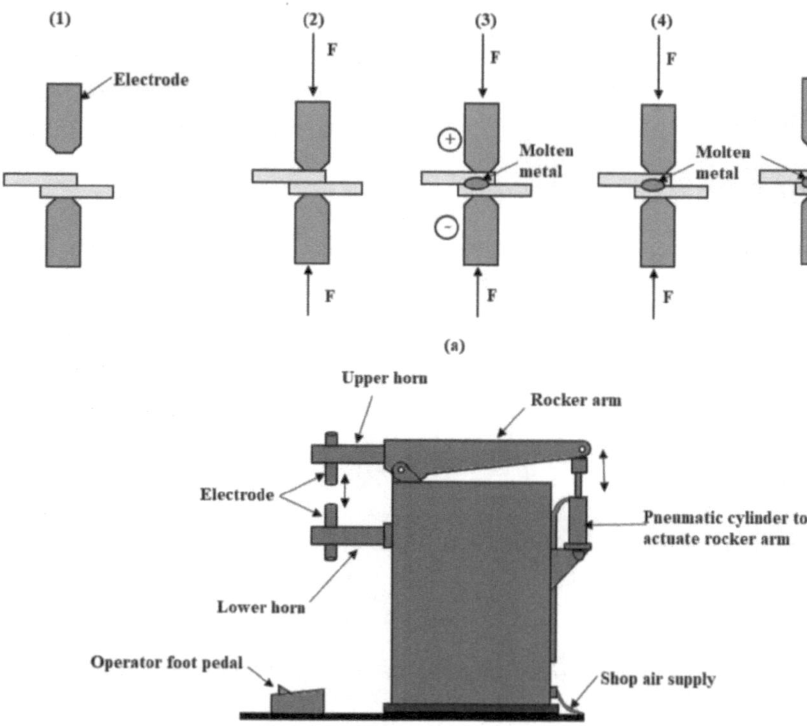

FIGURE 6.8 Resistance spot welding.

The electrode tip determines the size and shape of the weld spot; the most common electrode shape is round, but square, hexagonal and others can be utilised. The weld nugget which is obtained usually has 5–10 mm (0.2–0.4 in) diameter, a heat-influenced zone that extends slightly beyond the nugget into the base metals. If the weld is properly constructed, the strength should be equal to the surrounding metal. Spot-welding operations can be performed using a variety of machines and methods. Portable spot-welding guns and rocker-arm and press-type spot-welding machines are among the tools available. Figure 6.8b shows a rocker-arm spot welder with a stationary lower electrode and a movable upper electrode that can be raised and lowered to load and unload the work.

The upper electrode is mounted on a rocker arm (hence the name), the movement of which is powered by a foot pedal that the worker operates. During the weld cycle, modern devices can be programmed to regulate force and current.

(2) Resistance butt welding (UW)

Resistance butt welding is used to join components of similar cross section by making a weld across the entire section in a single operation. Heat is produced in the weld region by resistance to the passage of the welding current through the parts, which are held under a preset end force. As the material heats, the force forges the soft material to consolidate and complete the joint.

Resistance butt welding is thus a solid-state process. The force across the interface causes deformation which brings the surfaces into sufficiently close contact to make a weld, and there is some expulsion of material which carries oxide film and contaminants out of the joint. Applications include wire and rod joints up to about 16 mm diameter, including chain, and narrow strip joints such as automobile road wheel rims. Types are as follows: (1) flash butt welding and (2) upset butt welding.

Electric resistance butt welding processes are classified into two types: “upset” and “flash.”

During upset welding, the ends of the two sections to be joined together (with the same cross section) are clamped in location in the electrodes. The movable head is pushed towards the fixed head until the work pieces’ abutting surfaces make light contact. The appropriate current is then made to pass through the interface for a predetermined amount of time, while light pressure between the two sections is preserved. After heating the interface to welding temperature (plastic state), the current is turned off and the welding pressure is increased to form a “upset.” To achieve uniform heating of the joint, the two sections to be joined should ideally have the same resistance. Figure 6.9 shows the upset butt welding process.

Applications

It is widely used in the construction of tubular parts, pipes and heavy steel rings, as well as in the joining of small ferrous and non-ferrous strips.

During flash welding, the weldable parts approach each other and come into contact with the current switched on. The butt welding process is as follows:

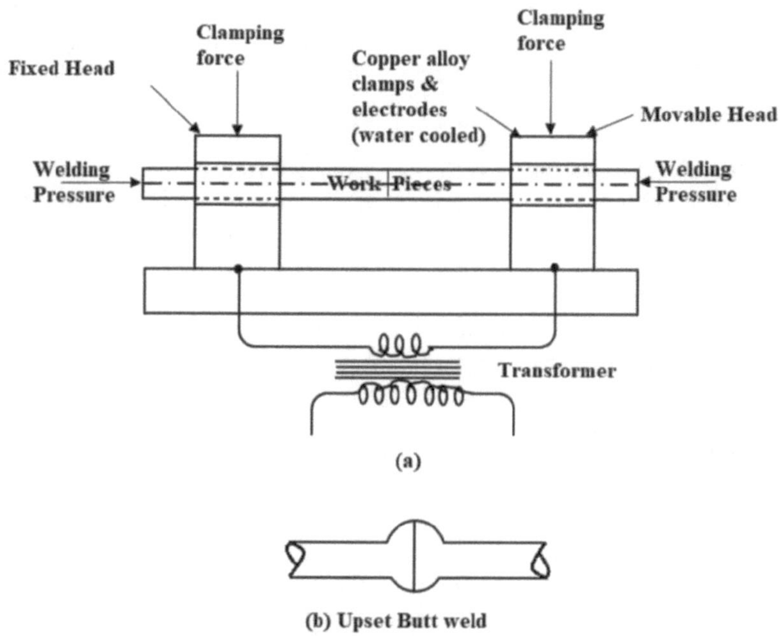

FIGURE 6.9 Upset butt weld.

- After the parts have been correctly arranged and the appropriate current, head speed and time have been selected, the cycle start button is pressed.
- The movable head approaches the fixed head as a result of this.
- Since the abutting surfaces are so close to each other, extremely rapid heating occurs when the surface asperities first make contact.
- Molten metal is violently expelled and burns in air with great force and "sparking" or "arcing," giving the process the term "flash butt."
- During these few seconds, a very thin layer at the interface melts.
- After that, the current is turned off, and the two pieces are quickly pushed together to form a weld. Figure 6.10 shows the flash butt welding process.

This method is used for joining thin-walled tubes, chain links, tools and press-made pieces.

(3) Seam welding

In resistance seam welding (RSEW), rotating wheels replace the stick-shaped electrodes used in spot welding, as shown in Figure 6.11, and a series of overlapping spot welds are made along the lap joint. The process can create airtight joints, and its industrial applications include the manufacture of fuel tanks, vehicle mufflers and a variety of other fabricated sheet metal containers. RSEW is technically the same as spot welding, except that the wheel electrodes add a layer of complexity. Since the

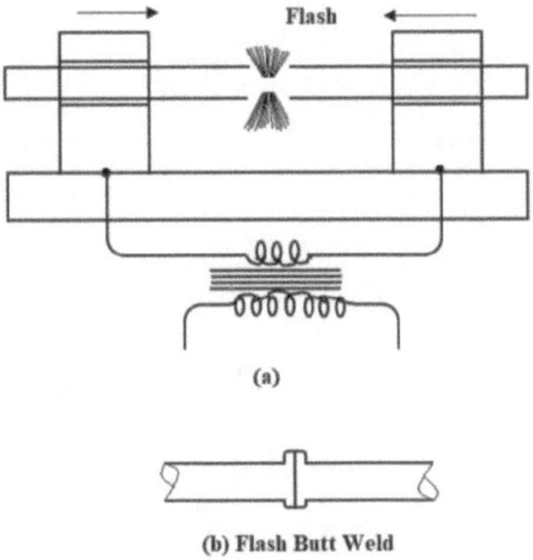

FIGURE 6.10 Flash butt weld.

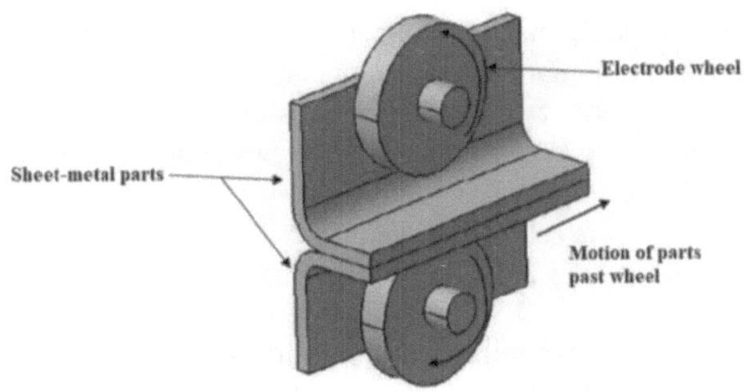

FIGURE 6.11 Seam welding.

procedure is normally done in a continuous rather than discrete fashion, the seams should be straight or uniformly curved. Fixtures are needed to keep the work in place and reduce distortion. Seam-welding devices are similar to press-type spot welders, but instead of stick-shaped electrodes, electrode wheels are used. Cooling the work and wheels is often needed in RSEW, and this is achieved by directing water at the top and bottom of the work component surfaces near the electrode wheels.

(4) Projection welding

High-resistance projection welding (RPW) generates electrical resistance at the junction by embossing one or more projections on one of the surfaces to be welded. For design or strength reasons, the projections may be circular or oval. Localised high temperatures are produced at the projections in contact with the flat mating element.

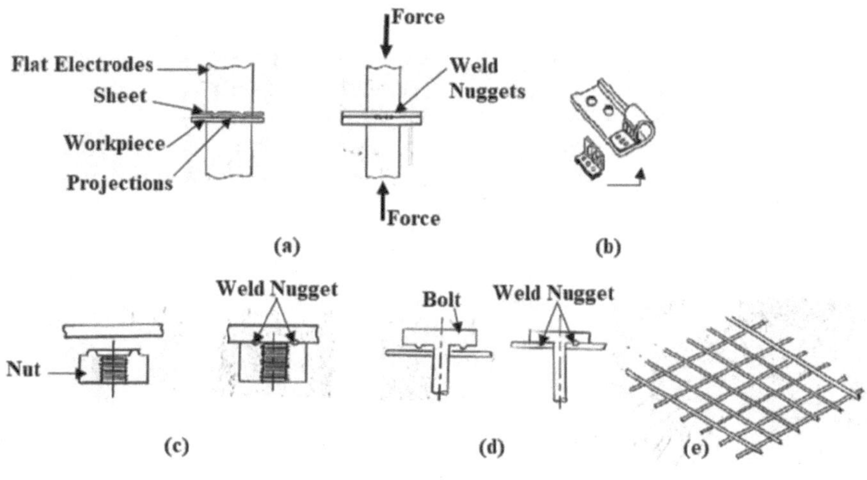

FIGURE 6.12 (a) Resistance projection welding; (b) a welded bracket; (c) and (d) projection welding of nuts or threaded bosses and studs; (e) resistance projection- welded grills.

The electrodes (typically made of copper-based alloys) are wide and smooth, and they are water-cooled to maintain a low temperature. Weld nuggets, similar to those seen in spot welding, form when the electrodes exert pressure on the projections, softening and compressing them. Figure 6.12 shows detailed applications of projection welding.

(5) Percussion welding

Percussion welding (PEW) is similar to flash welding in that the weld time is extremely short, usually lasting just 1 to 10 ms. Fast heating is achieved by rapidly transferring electrical energy between the two surfaces to be connected, followed by pounding of one component against the other to form the weld. Since the heating is very localised, this method is appealing for electronic applications where the measurements are very small and neighbouring components can be heat-sensitive. The method is advantageous when it is necessary to prevent heating of the components adjacent to the joint, such as in electronic assemblies and electrical wires.

6.6 DESIGN OF WELD JOINTS, BUTT JOINTS, FILLET JOINTS

Figure 6.13 shows a butt-welded joint subjected to tensile force P. The weld's average tensile stress is given by

$$\sigma_t = \frac{P}{hl} \tag{6.1}$$

where, σ_t = tensile stress in the weld (N/mm^2)

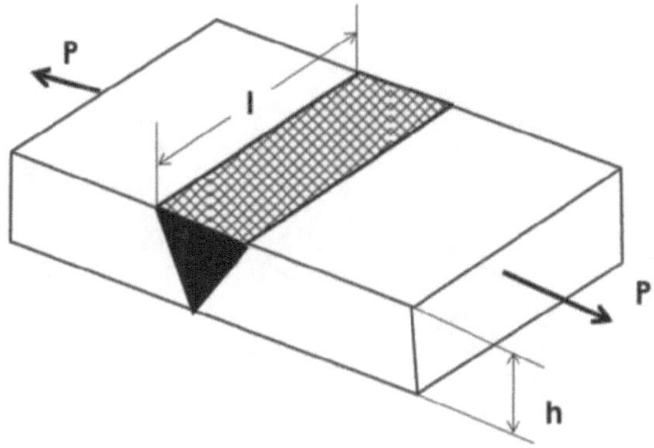

FIGURE 6.13 Butt joint.

P= tensile force on the plates (N)
h = throat of the butt weld (mm)
l = length of the weld (mm)

The weld throat does not include the bulge or reinforcement. The reinforcement is provided to compensate for weld flaws. The strength equation of a butt joint can be written by equating the weld throat h to the plate thickness t in Eq. (6.1).

$$P = \sigma_t \text{tl} \tag{6.2}$$

where, P = tensile force on plates (N)
σ_t = permissible tensile stress for the weld (N/mm²)
t = thickness of the plate (mm)

Certain standards, such as the code for unfired pressure vessels, recommend reducing the strength of a butt welded connection by a factor known as efficiency of joint.

Eq. (6.2) is adjusted and rewritten in the following manner when the strength is to be reduced:

$$P = \sigma_t \frac{t}{\eta} \tag{6.3}$$

where, η = efficiency of the welded joint (in fraction)

When properly built, a butt-welded joint has equal or more strength than plates, and there is no need to determine the stresses in the weld, or the size and length of the weld. All that is required is to match the weld material's strength to the strength of the plates.

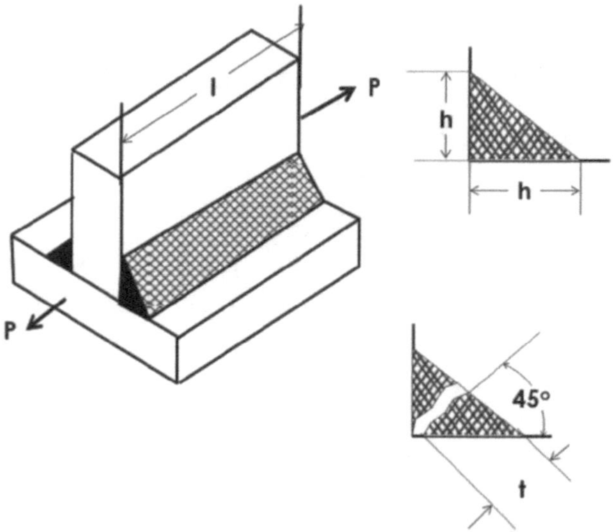

FIGURE 6.14 Parallel fillet weld.

- **Strength of parallel fillet welds**

Figure 6.14 depicts a parallel fillet weld subjected to a tensile force P. The figure shows an enlarged view of the fillet weld. The dimensions of the fillet weld are denoted by two terms: leg h and throat t. The leg length determines the size of the weld. The fillet weld's cross-section is made up of a right-angled triangle with two equal sides. A leg is the length of each of the two equal sides. In general, the leg length h equals the plate thickness. The throat is the weld's smallest cross-section, placed at 45° to the leg dimension. Therefore,

$$t = hcos\left(45^0\right)$$

or

$$t = 0.707h$$

Shear along the minimum cross-section at the throat causes the fillet weld to fail. The inclination of the plane where maximal shear stress is created in a parallel fillet weld is 45° to the leg dimension. Figure 6.14 depicts the weld's shear failure. The throat cross-sectional area is (tl) or (0.707 hl). The fillet weld shear stress is given by,

$$\tau = \frac{P}{0.707hl} \tag{6.4}$$

Rearranging the terms of Eq. (6.4), the strength equation of the parallel fillet weld can be as follows

$$P = 0.707\, hl\tau$$

where, P = tensile force on plates (N)
h = leg of the weld (mm)
l = length of the weld (mm)
τ = permissible shear stress for the weld (N/mm^2)

Usually, there are two welds of equal length on two sides of the vertical plate. In that case,

$$P = 2(0.707\, hl\tau)$$

or

$$P = 1.414\, hl\tau$$

To allow for starting and stopping of the weld run, add 15 mm to the length of each weld computed by the following equations when computing the needed length of the weld. In the case of a static load, the permitted shear stress for the fillet welds is 94 N/mm^2 according to the American Welding Society rule (AWS).

6.7 STRENGTH OF TRANSVERSE FILLET WELDS

Figure 6.15 depicts a transverse fillet weld subjected to tensile force P. Tensile stress is applied to the transverse fillet welds. The neck has the smallest cross section of the weld. As a result, tensile stress failure will occur in the throat region. The throat's cross-sectional area is (tl). The transverse fillet weld's tensile tension is given by

$$\sigma_t = \frac{P}{tl}$$

$$\sigma_t = \frac{P}{0.707hl} \tag{6.5}$$

Rearranging the terms of Eq. (6.5), the strength equation of the transverse fillet weld is written in the following form,

$$P = 0.707hl\sigma_t$$

where, σ_t = permissible tensile stress for the weld (N/mm^2)

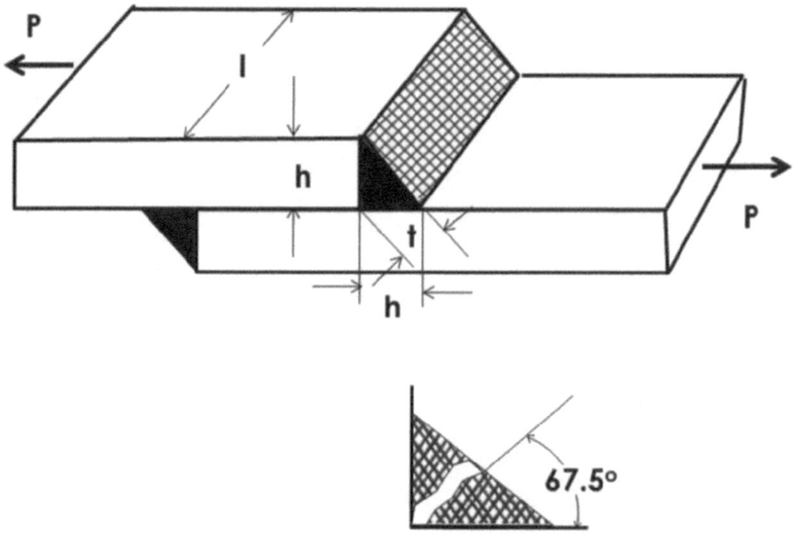

FIGURE 6.15 Failure of a fillet weld.

Usually, there are two welds of equal length on two sides of the plate as shown in Figure 6.15. In such cases,

$$P = 2(0.707hl\sigma_t)$$

or

$$P = 1.414hl\sigma_t$$

The nature of strains in the transverse fillet weld cross-section is complex. The weld is subjected to both normal and shear stress. Furthermore, the throat is subjected to a bending moment, which complicates matters further. Theoretically, the inclination of the plane where maximum shear stress is created for a transverse fillet weld is 67.5° to the leg dimension, as shown in Figure 6.15. Shear failure is frequently employed as the failure criterion in order to simplify the design of fillet welds. For any applied load direction, it is assumed that the stress in the transverse fillet weld represents shear stress on the throat area. and the equations developed for parallel fillet welds are likewise usable for transverse fillet welds under this assumption.

- **Maximum shear stress in parallel fillet welds**

A double parallel fillet weld of equal legs subjected to a force of (2P) is shown in Figure 6.16. The inclination of the plane in the weld where maximum shear stress is created, as well as the size of the maximum shear stress, must be determined.

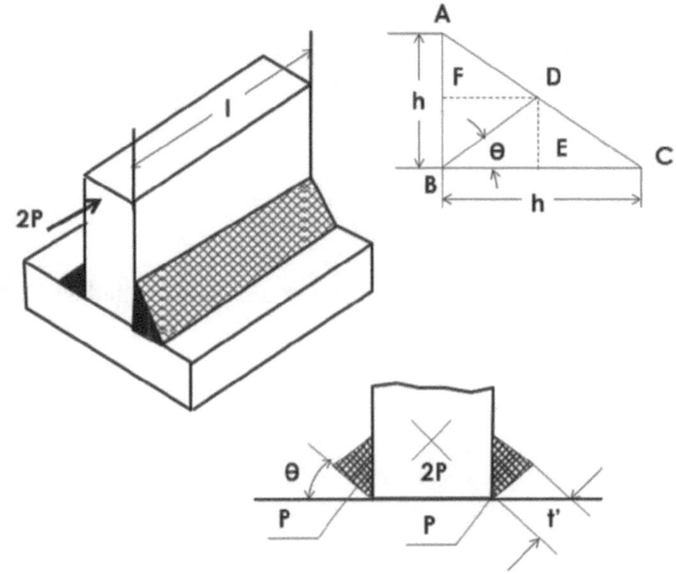

FIGURE 6.16 Free body diagram for a parallel fillet weld.

The bending effect is ignored. Figure 6.16 shows the free body diagram of forces operating on a vertical plate with two symmetrically cut welds. The symbol (cross) represents a force that is perpendicular to the plane of the paper and is directed away from the observer. The point on an inclined surface represents a force perpendicular to the plane of the paper, pointing towards the observer. The welds are cut at an angle of θ with respect to the horizontal. t' is the breadth of a plane inclined at an angle with the horizontal.

In the triangle ABC

$$AB = BC = h$$

$$< ECD = 45°$$

$$DE \perp BC$$

$$BC = BE + EC$$

$$= BE + DE \ (DE = EC)$$

$$= BD \cos\omega + BD \sin\theta$$

$$= BD\ (\cos\ \omega + \sin\ \omega)$$

or

$$h = t'(\sin\omega + \cos\omega)$$

therefore

$$t' = \frac{h}{(\sin\omega + \cos\omega)}$$

(t'l) is the area of the weld in the plane inclined at an angle with the horizontal. As a result, shear stress in this plane is given by

$$\tau = \frac{P}{t'l}$$

$$\tau = \frac{P(\sin\omega + \cos\omega)}{hl}$$

To locate the plane with the greatest shear stress, differentiate t with respect to and set the derivative to zero.

$$\frac{\partial\tau}{\partial\theta} =$$

$$\frac{P}{hl}(\cos\omega - \sin\omega) = 0$$

$$(\cos\omega - \sin\omega) = 0$$

$$cos\,\theta = sin\,\theta$$

$$tan\theta = 1$$

Therefore

$$\theta = 45^0$$

The criterion for the plane with the greatest shear stress is (= 45°). When this value of is substituted in Eq. (6.7), the maximum shear stress is given by,

$$\tau_{max} = \frac{P(\sin 45^0 + \cos 45^0)}{hl}$$

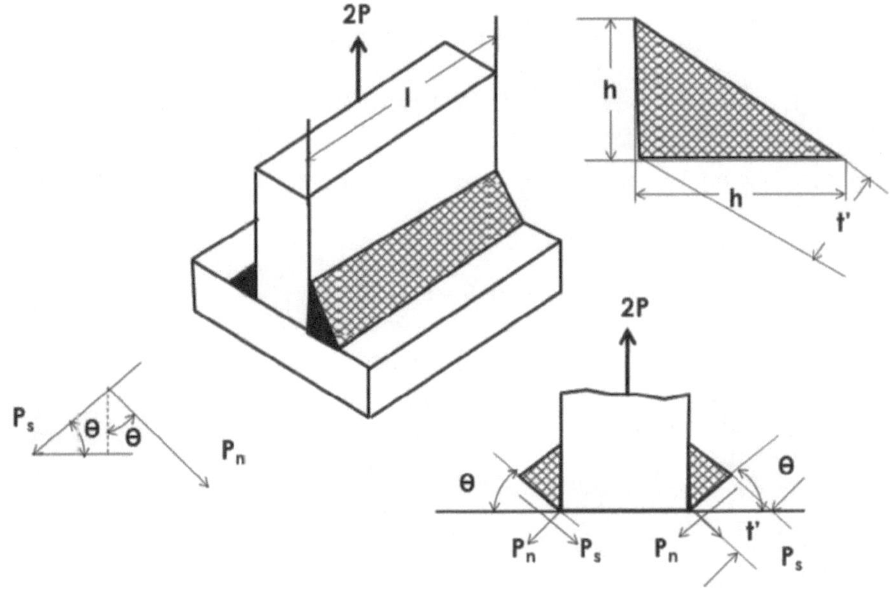

FIGURE 6.17 Double parallel fillet weld and its analysis.

$$= \frac{1.414P}{hl}$$

$$= \frac{P}{\left(\frac{1}{1.414}\right)hl}$$

$$= \frac{P}{0.707hl}$$

When (l = 1 mm) is substituted in the equation, the allowed load P_{all} per mm length of the weld is given by,

$$P_{all} = 0.707h\tau_{max} \tag{6.6}$$

- **Maximum shear stress in transverse fillet welds**

As shown in Figure 6.17, a double transverse fillet weld with equal legs is exposed to a force (2P). The inclination of the plane in the weld where the maximum shear stress is created, as well as the size of the maximum shear stress, must be determined. The bending effect is ignored.

$$2P = 2P_s \sin\theta + 2P_n \cos\theta$$

$$P = P_s \sin\theta + P_n \cos\theta$$

Because the product of P_s and P_n is vertical, their horizontal components must be equal and inverse. Therefore,

$$P_s \cos\theta = P_n \sin\theta$$

$$P_n = \frac{P_s \cos\theta}{\sin\theta}$$

$$P = P_s \sin\theta + \frac{P_s \cos\theta \cos\theta}{\sin\theta}$$

Multiplying both sides of the above equation by sin θ,

$$P\sin\theta = P_s \sin^2\theta + P_s \cos^2\theta$$

$$= P_s(\sin^2\theta + \cos^2\theta) \text{ or}$$

$$P_s = P \sin\theta$$

From Eq. (6.6), the width t′l of the plane in the weld that is inclined at an angle θ with the horizontal is given by,

$$t' = \frac{h}{(\sin\theta + \cos\theta)}$$

The area of the weld in an incline plane at an angle with the horizontal is t′l, hence the shear stress in this plane is given by

$$\tau = \frac{P_s}{t'l}$$

From the equation

$$\tau = \frac{P\sin\theta(\sin\theta + \cos\theta)}{hl} \tag{6.7}$$

To locate the plane with the greatest shear stress, differentiate with respect to and set the derivative to zero.

$$\frac{\partial \tau}{\partial \theta} = 0$$

$$\frac{P}{hl}\frac{\partial}{\partial \theta}\left[\sin\theta(\sin\theta + \cos\theta)\right] = 0$$

or

$$\frac{\partial}{\partial \theta}\left[\sin\theta(\sin\theta + \cos\theta)\right] = 0$$

$$\frac{d}{dx}(uv) = u\frac{dv}{dx} + v\frac{du}{dx}$$

$$x = 0 \; u = \sin\theta \; \mathrm{v} = (\sin\theta + \cos\theta)$$

$$\frac{du}{d\theta} = \cos\theta$$

$$\frac{dv}{d\theta} = (\cos\theta - \sin\theta)$$

Substituting

$$\frac{\partial}{\partial \theta}\left[\sin\theta(\sin\theta + \cos\theta)\right] = \sin\theta(\cos\theta - \sin\theta) + (\sin\theta + \cos\theta)\cos\theta$$

$$= 2\sin\theta\cos\theta + (\cos^2\theta - \sin^2\theta)$$

$$= \sin 2\theta + \cos 2\theta$$

$$\sin 2\theta + \cos 2\theta = 0$$

$$\sin 2\theta = -\cos 2\theta$$

$$\tan 2\theta = -1$$

$$2\theta = -1$$

$$2\theta = 135^0$$

$$\theta = 67.5^0$$

The condition for the plane with the maximum shear stress is (θ = 67.5°). Substituting the above value of θ in Eq. (6.9), the maximum shear stress is given by,

$$\tau_{max} = \frac{P\sin\left(67.5^0\right)\left(\sin 67.5^0 + \cos 67.5^0\right)}{hl}$$

$$\tau_{max} = \frac{1.21\text{P}}{hl} \tag{6.8}$$

Substituting (l = 1 mm) in Eq. (6.12), the allowable load P_{all} per mm length of transverse fillet weld is given by,

$$P_{all} = \frac{h\tau_{max}}{1.21} \tag{6.9}$$

Assume you need to calculate the allowable load per mm length of a transverse fillet weld with a permissible shear stress of 94 N/mm² and a leg dimension of 8 mm. Using Eq. (6.13),

$$\text{P}_{all} = 0.8284\text{h}\tau_{max} = 0.8284(8)(94) = 622.96\text{N}/\text{mm}^2$$

It is observed from Eqs (6.12) and (6.13) that the allowable load for a transverse fillet weld is more than that of a parallel fillet weld. Or

$$\frac{\text{P}_{all}\,fortranverseload}{\text{P}_{all}\,forparallelload} = \frac{0.8284\text{h}\tau_{max}}{0.707\text{h}\tau_{max}} = 1.17$$

Transverse fillet weld strength is 1.17 times that of parallel fillet weld strength. As noted in an earlier section, transverse fillet welds are frequently built using the same equations as parallel fillet welds. Such a design is safer and has the added benefit of being simple to calculate.

6.8 SOLVED EXAMPLES

1. A gas tank consists of a cylindrical shell of 2.8 m inner diameter. It is enclosed by hemispherical shells by means of butt welded joint as shown in Figure N6.1. The thickness of the cylindrical shell as well as the hemispherical cover is 12 mm. Determine the allowable internal pressure to which the tank may be subjected, if the permissible tensile stress in the weld is 85 N/mm^2. Assume efficiency of the welded joint as 0.85.

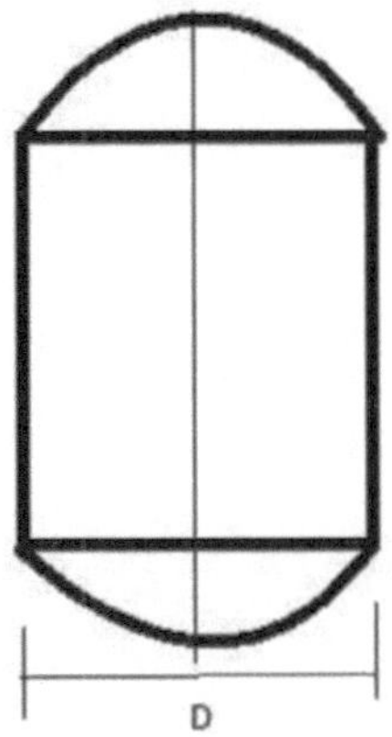

Figure N6.1

Solution:
Given For shell, D = 2.8 m, t = 12 mm
For weld, $\sigma_t = 85$ N/mm^2, $\eta = 0.85$

Step I Tensile force on plates
The length of the welded joint is equal to the circumference of the cylindrical shell.
$l = \pi D = \pi(2.8\times10^3) = 8796.45$ mm
$P = \sigma_t l\eta = (85)(12)(8796.45)(0.85)$
$= (7626.522\times10^3)$ N

Step II Allowable internal pressure
Corresponding pressure inside the tank is given by

$$p = \frac{P}{\frac{\pi}{4}D\wedge 2} = \frac{7626.522\times 103}{\frac{\pi}{4}(2.8\times10^3)\wedge 2} 1.24\ \mathrm{N/mm^2}$$

2. A steel plate, 120 mm wide and 12 mm thick, is welded to another steel plate by means of double parallel fillet welds as shown in Figure N6.2. The plates are subjected to a static tensile force of 50 kN. Determine the required length of the welds if the permissible shear stress in the weld is 94 N/mm^2.

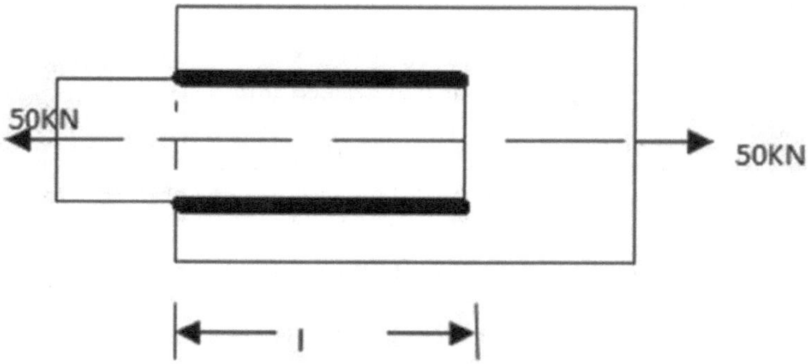

Figure N6.2

Solution:
Given P= 50 kN, τ= 94 N/mm^2, h= 12mm

Step I Length of weld
$P = 1.414hl\tau$
or $50\times10^3 = 1.414\ (12)\ l\ (94)$
$\therefore$ l=31.35 mm

Adding 15 mm of length for starting and stopping of the weld run, the length of the weld is given by,

l= 31.35+15= 46.35 mm

3. Two steel plates, 150 mm wide and 12.5 mm thick are joined together by means of double transverse fillet welds as shown in Figure N6.3. The maximum tensile stress for the plates and the welding material should not exceed 110 N/mm^2. Find the required length of the weld, if the strength of weld is equal to the strength of the plates.

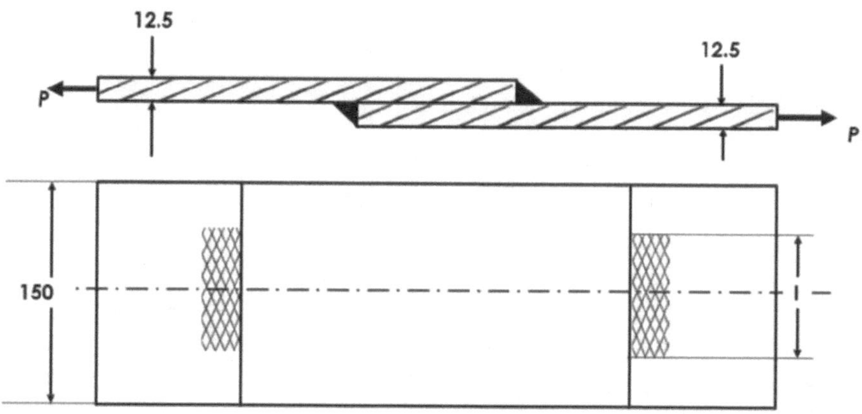

Figure N6.3

Solution:
Given For plates w = 150 mm, t = 12.5 mm
For welds h = 12.5 mm, σ_t= 110 N/mm^2

Step I Tensile force on plates
The plates are subjected to tensile stress. The maximum tensile force acting on the plates is given by,

P= (wt)σ_t= (150×12.5)(110) = 206250 N

Step II Length of the weld

P = 1.414hlσ_t
206250 = 1.414(12.5)l(110)
∴ l=106.08 mm

Adding 15 mm for starting and stopping of the weld run, the length of the weld is given by,

l= 106.08+15= 121.08 mm

4. A steel plate, 120 mm wide and 10 mm thick, is joined with another steel plate by means of single transverse and double parallel fillet welds, as shown in Figure N6.4. The strength of the welded joint should be equal to the strength of the plates to be joined. The permissible tensile and shear stresses for the weld material and the plates are 70 and 50 N/mm^2, respectively. Find the length of each parallel fillet weld. Assume the tensile force acting on the plates as static.

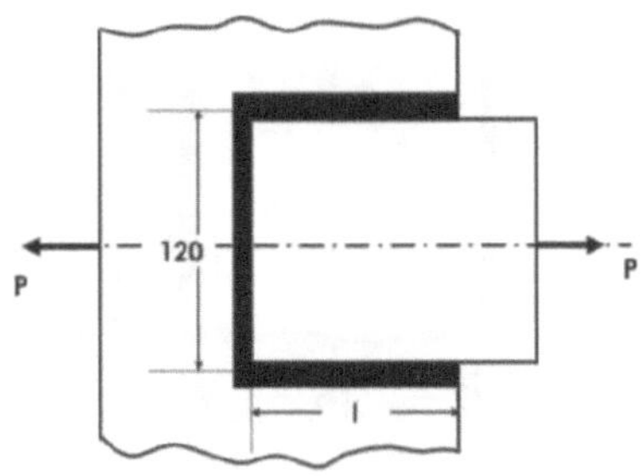

Figure N6.4

Solution:
Given For plates w = 120 mm, t = 12.5mm
For welds h = 12.5 mm, σ_t= 110 N/mm^2
τ= 50 N/mm^2

Step I Tensile strength of plate
The tensile strength of the plate is given by,
P = (w×t)σ_t = (120×10) (70) = 84000N ---------- (i)

Step II Strength of transverse and parallel fillet welds
The strength of the transverse fillet weld is denoted by P_1.

$P_1 = 0.707hl\sigma_t = 0.707\ (10)\ (120)\ (70)$
= 59388N---------- (ii)

The strength of the double parallel fillet weld is denoted by P_2.

$P_2 = 1.414hl\tau = 1.414\ (10)\ l\ (50) = 707\times l$ ---------- (iii)

Step III Length of parallel fillet weld
The strength of the welded joint is equal to the strength of the plate.
From (i), (ii), (iii),

$84000 = 59388 + 707\times l$

$\therefore l = 34.81\text{mm}$
Adding 15 mm for starting and stopping of the weld run,
$l = 34.81 + 15 = 49.81$ or 50 mm

5. Two plates are joined together by means of single transverse and double parallel fillet welds as shown in Figure N6.5. The size of the fillet weld is 5 mm and allowable shear load per mm of weld is 350 N. Find the length of each parallel fillet weld.

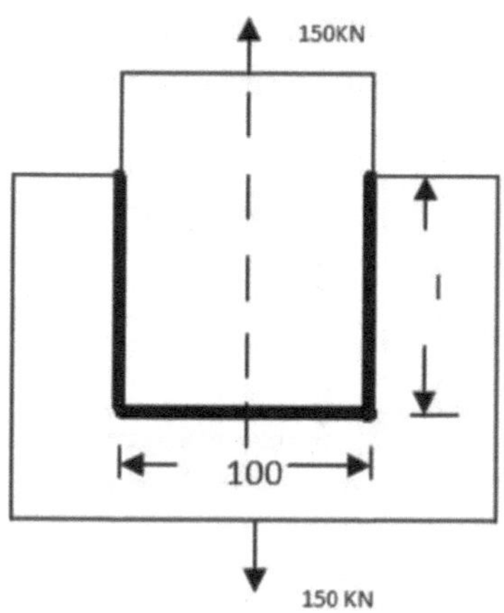

Figure N6.5

Solution:
Given P = 150 kN h = 5 mm
Allowable shear load = 350 N/mm^2

Step I Total length of weld

It is mentioned that the transverse fillet weld is designed on the basis of shear stress. In such cases, the stress in the fillet weld is considered as shear stress on the throat for any direction of applied load. With this assumption, the equations derived for the parallel fillet weld are also applicable to the transverse fillet weld.

Suppose L is the total length of welds required for the joint. Since the allowable shear load per mm length of weld is 350 N, the required length of weld is given by,

$$L = \frac{150 \times 10 \wedge 3}{350} = 428.57\text{mm} \quad \text{(i)}$$

Step II Length of parallel fillet weld
From the figure,

L= 2×l + 100 ---------- (ii)

From (i) and (ii),

2×l + 100=428.57

∴ l = 164.28 mm

Adding 15 mm for starting and stopping of the weld run,
l = 164.28+15=179.28 or 180 mm

6.9 ALLIED PROCESSES: SOLDERING, BRAZING

The joining of materials by a solid joint or the cutting of materials without the use of mechanical cutting tools, usually by means of heat, comprises the following processes: welding, brazing, thermal spraying and thermal cutting. An allied process to welding may be adhesive bonding because it also involves the joining of materials by a solid joint. However, adhesive bonding is distinctly different from welding in terms of operating conditions and equipment.

- **Soldering**

Soldering, like brazing, is a joining process in which a filler metal with a melting point (liquidus) not exceeding 450°C (840°F) is melted and distributed by capillary action between the faying surfaces of the metal parts being joined. Soldering details are identical to brazing details, and many of the heating methods are the same. Surfaces to be soldered must be pre-cleaned to remove oxides, oils and other contaminants.

The faying surfaces must be treated with an appropriate flux and heated. Solder, a filler metal, is applied to the joint and distributes itself between the closely fitted components. For certain applications, the solder is precoated onto either or both surfaces—a method known as tinning—regardless of whether the solder contains tin. Typical soldering clearances range from 0.075 to 0.125 mm (0.003–0.005 in), unless the surfaces are tinned, in which case a clearance of around 0.025 mm (0.001 in) is used.

The flux residue must be removed after solidification.

Advantages

- Low energy input in comparison to brazing and fusion welding.
- A wide range of heating methods are available.
- The joint has strong electrical and thermal conductivity.
- The capacity to create airtight and liquid-tight container seams.
- Simple to repair and rework.

Disadvantages

- Low joint strength unless mechanically strengthened.
- Possible joint cracking or melting in high-temperature operation.

Soldering applications

- Radio and television set circuitry connections
- Wiring contacts in electric connections and battery terminals
- Brass radiator tube
- Copper piping
- Brass bearings that have been halved
- Sheet metal joints, such as those used in food cans
- Roofing flashing
- Guttering
- Elements of refrigeration and plumbing machine tools.

Brazing

Brazing is a joining process in which a filler metal is melted and distributed by capillary action between the faying surfaces of the metal parts being joined. It is similar to soldering but requires higher heat and temperature.

Brazing causes no melting of the base metals; only the filler melts. In brazing, the filler metal (also known as the brazing metal) has a melting temperature (liquidus) higher than 450°C (840°F) but lower than the melting point (solidus) of the base metal(s) to be joined. Methods of brazing are described next.

Torch brazing – on previously fluxed joints, an oxyfuel gas is used. Typically a manual process, but it can be automated.

Furnace brazing – a high-volume production method that involves placing fixture parts preloaded with filler metals and, if necessary, flux in a furnace. The furnace can be a single batch or a conveyor model for continuous brazing.

Dip brazing entails immersing assembled parts in a heated chemical bath that acts as both a fluxing agent and a heat source to melt pre-applied filler material.

Induction brazing – a method of brazing that employs inductor coils to induce an alternating current into and around a pre-assembled part. The heat required to melt the filler metal is generated by the part's electrical resistance.

Advantages of brazing

- Because brazing does not melt the base metal of the joint, it allows for much tighter tolerance control and produces a clean joint without the need for secondary finishing.
- Non-metals non-similar metals (such as metalised ceramics) can be brazed together.
- Due to the uniform heating of a brazed piece, brazing produces less thermal distortion than welding.
- Complex and multi-part assemblies can be brazed at a low cost. Welded joints must occasionally be ground flush, which is expensive. As a clean joint, secondary operation is not required.
- Another one is that the brazing can be coated or clad for protection.
- Because the individual process parameters are less sensitive to variation, brazing is easily adapted to mass production and also easy to automate.

Disadvantages of brazing

- Due to the use of softer filler metals, there is a lack of joint strength compared to a welded joint.
- The strength of the brazed joint is likely to be less than that of the base metal(s), but greater than that of the filler metal.
- Brazed joints can be damaged in high-temperature environments. When brazed joints are used in an industrial setting, they must have a high level of base metal cleanliness.
- To control cleanliness, some brazing applications necessitate the use of adequate fluxing agents.
- The colour of the joint is frequently different from the colour of the base metal, which creates an aesthetic disadvantage.

Applications

(1) Pipe fitting assembly
(2) Carbide tool tips
(3) Radiators
(4) Heat exchangers
(5) Electrical components
(6) Casting repair and joining of special materials such as stainless steels.

6.10 SPECIAL WELDING PROCESSES: PLASMA ARC WELDING, METAL INERT GAS ARC WELDING (MIG), TIG WELDING

These welding machines are used in special processes for various types of welding. Special welding processes make the welding easier in special situations like a vertical

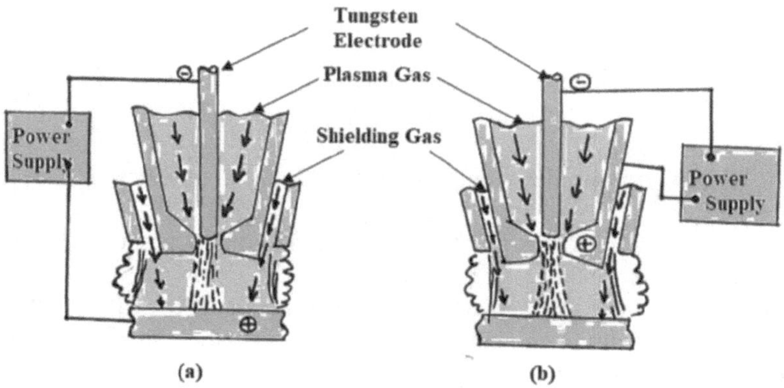

FIGURE 6.18 Plasma arc welding.

bead or the need to work well and fast. Special welding processes are explained in details.

Plasma arc welding

Plasma arc welding (PAW) is a subset of gas tungsten arc welding that involves directing a constricted plasma arc at the weld region. Plasma is the name given to hot ionised gases. When an inert gas is given enough energy, some of its electrons break free from its nucleus but continue to travel with it. After the electrons have left, the atoms are converted to a hot ionised state. It is the most common state of matter, also known as the fourth state of matter. These ionised atoms have a high heat capacity, which is used to connect two plates. This is the fundamental theory of plasma arc welding. Figure 6.18 shows details of plasma arc welding.

Non-transferred plasma arc welding

This welding procedure employs straight polarity DC current, with the tungsten electrode connected to the negative pole and the nozzle connected to the positive pole. An arc forms within the torch between the tungsten electrode and the nozzle. The ionisation of the gas within the torch will be increased as a result of this. The torch passes this ionised gas to the next step in the process. It is used to join thin sheets of metal. Transferred plasma arc welding is a form of welding that uses a plasma arc to create straight polarity DC current is also used in this operation.

The tungsten electrode is connected to the negative terminal in this operation, and the workpiece is connected to the positive terminal. An arc is formed between the tungsten electrode and the workpiece. In this method, both plasma and arc are transferred to the workpiece, increasing the process's heating capacity. It is used to join dense sheets of metal.

Working

- Instead of TIG, plasma is used to heat up the parent material in this welding. Its operation can be summarised as follows.
- The workpieces are first properly cleaned. The power source generates the arc between the tungsten electrode and the nozzle or the tungsten electrode and the workpiece.
- The tungsten electrode produces a high-intensity arc that is used to ionise gas particles and transform orifice gases into plasma. A small orifice delivers hot ionised gas to the welding plates.
- Shielding gases such as argon are delivered to the welding torch's outer nozzle through a pressure valve and a regulating valve. These gases form a shield around the welding area, protecting it from atmospheric gases such as oxygen and nitrogen.
- The plasma hits the welding plates and fuses them together. The welding torch is then pushed in the direction of the welding.
- If the welding requires filler material, it is manually fed by the welder.

Application

- This welding technique is used in the marine and aerospace industries.
- It is used to weld stainless steel or titanium pipes and tubes.
- It is mostly used in the electronic industry.
- It is used to repair tools, dies and moulds.
- It is used to weld or coat turbine blades.

Advantages

- Welding speed is high.
- A lot of energy is required for welding. It is simple to use to weld hard and thick work bits.
- The distance between the tool and the work piece has no impact on arc forming.
- Low power consumption for the same weld size.
- The PAW approach produces a more stable arc.
- It has a high-intensity arc or a high penetration rate.
- It can operate at low amperage.

Disadvantages

- Equipment costs are high.
- Noisy operation.
- Higher radiation.
- High-skilled labour is required.
- Expensive repairs.

6.10.1 Gas Metal-Arc Welding (GMAW) or Metal Inert Gas (MIG) Welding

Introduction

In gas metal-arc welding (GMAW), which was developed in the 1950s and was previously known as metal inert gas (MIG) welding, the weld region is protected by an essentially inert atmosphere of argon, helium, carbon dioxide or other gas mixtures. A wire-feed drive motor automatically feeds consumable bare wire through a nozzle into the weld arc. In addition to inert shielding gases, deoxidisers are normally present in the electrode metal itself to prevent oxidation of the molten-weld puddle.

Working

First, a high-voltage current is converted into a low-voltage DC current supply with a high current. This current flows through the welding electrode. As an electrode, a consumable wire is used. The electrode is connected to the negative terminal, and the work piece is connected to the positive terminal. Due to the power supply, a fine strong arc will form between the electrode and the work object. This arc produces heat, which melts the electrode and the base metal. The base metal is typically used to make electrodes in order to achieve a uniform joint. Figure 6.19 shows details of metal inert gas welding.

The shielding gases provide adequate protection for this arc. These gases shield the weld from other reactive gases that might weaken the welding joint. The electrode moves constantly around the welding area to ensure a proper weld joint. The angle of the travel direction should be held between 10–15^0. The angle for fillet joints should be 45^0.

Advantages

- Consumable electrodes are simple to feed.
- No necessity for a filler rod.
- Welding is a simple process.
- An inert gas shield automatically protects the weld.

Disadvantages

- Improper welding will cause solid impurities to float on top of the liquid weld.
- If the weld is not properly handled, it will become porous.
- MIG welding exposes welders to potentially toxic gases.
- Precautions must be taken to prevent the development of less ductile welds.
- Before welding, workpieces and electrodes should be kept clean.

Applications

- Automobile part repairs, such as arm, axle beam, axle housing, and so on.
- Pipe joint preparation.
- Reinforcing of worn-out railroad track surfaces.
- Automotive industry, for straight and circular welds.

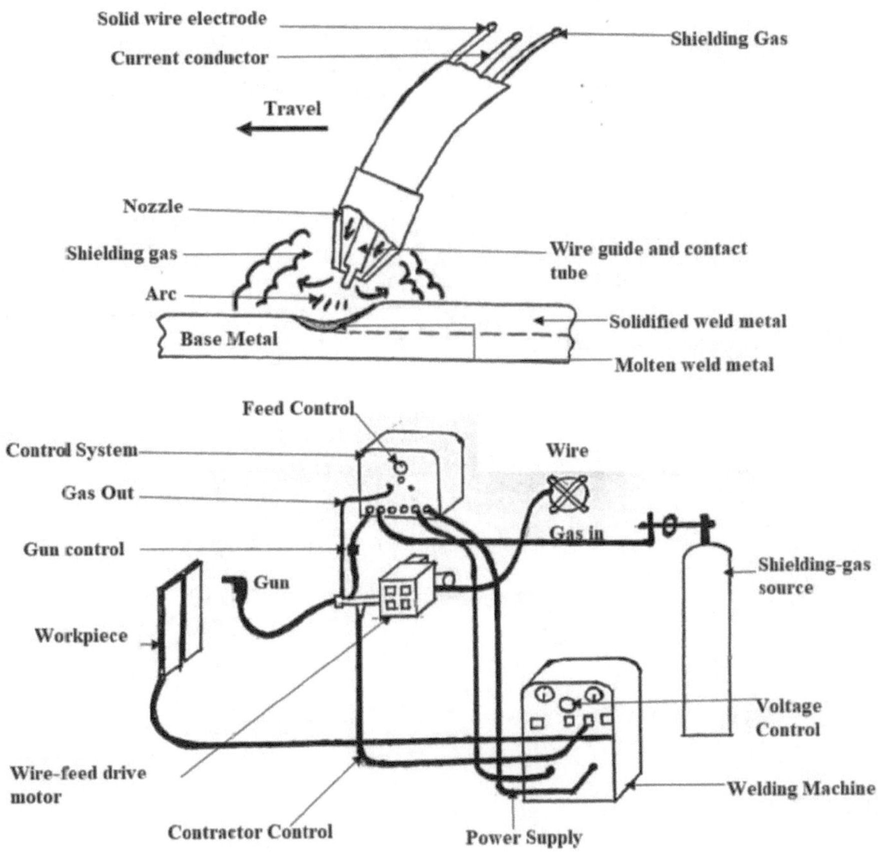

FIGURE 6.19 Metal inert gas welding.

6.10.2 Tungsten Inert Gas (TIG) Welding

Introduction

Filler metal is supplied by a filler wire in gas tungsten-arc welding (GTAW), formerly known as TIG (for "tungsten inert gas") welding. Figure 6.20 shows TIG welding process. Since the tungsten electrode is not consumed in this process, a steady and stable arc distance at a constant current level is preserved. The filler metals are close to the metals to be welded, and no flux is used. Usually, the shielding gas is argon or helium (or a mixture of the two).

Process

First, the power source supplies a low-voltage high-current supply to the welding electrode or tungsten electrode. Generally, the electrode is connected to the negative terminal of the power source, and the workpiece is connected to the positive terminal. The current

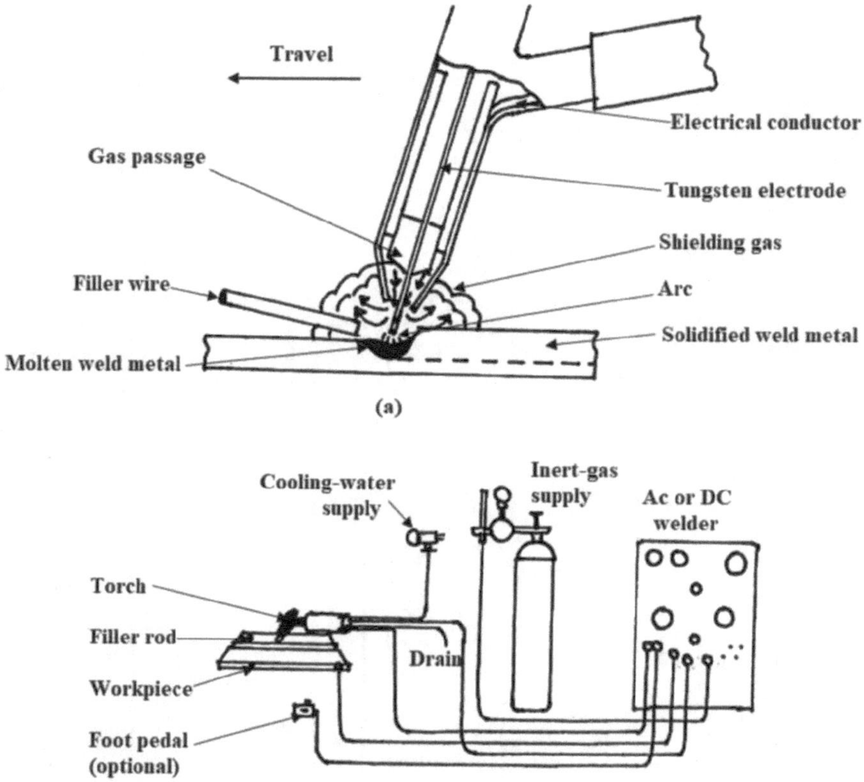

FIGURE 6.20 TIG (tungsten inert gas) welding.

is produced by a spark between a tungsten electrode and the workpiece. Tungsten is a non-consumable electrode that produces a very intense arc. This arc produces heat, which melts the base metals and forms the welding joint. Shielded gases such as argon and helium are supplied to the welding torch through a pressure valve and a regulating valve. These gases form a barrier that prevents oxygen and other reactive gases from entering the weld region. These gases also produce plasma, which increases the heat power of the electric arc and hence the ability to weld. When welding a thin material, no filler metal is required; however, when welding thick material, some filler metal is used in the form of rods that are manually fed into the welding zone by the welder.

Advantages

- TIG welding produces a stronger joint than shield arc welding, and the joint is more corrosion-resistant and ductile.
- A wide range of joint designs can be created.
- It does not necessitate flux.
- It is easy to automate.
- This welding is ideal for thin walls.

- It has a decent surface finish and there is very little metal splatter or weld sparks to harm the surface.
- Because of the non-consumable electrode, a flawless joint can be formed.
- Greater control over welding parameters as compared to other types of welding.
- As a power source, both alternating current and direct current can be used.

Disadvantages

- The weldable metal thickness is limited to around 5 mm.
- It necessitates highly skilled labour.
- In comparison to arc welding, the initial or setup cost is high.
- Welding speed is very low.

Applications

- Aerospace industry
- Space ship manufacturing
- Welding small-diameter tubing for thin-walled tubing
- Welds for different sizes of piping
- Hardware and dies for repair
- Affixing thin sheet material
- Used in the production of chemical plants and boilers.

6.11 AUTOMOTIVE APPLICATIONS OF WELDING

The automotive industry has concentrated on producing lighter, stronger and more fuel-efficient vehicles using new and improved alternative materials. This has necessitated the use of the most efficient welding techniques, which has resulted in lighter weld joints without the addition of additional material. Resistance spot welding (RSW), resistance seam welding (RSEW), metal inert gas (MIG) welding, tungsten inert gas (TIG) welding, laser beam welding (LBW), friction welding (FW) and plasma arc welding (PAW) are the most commonly used welding methods for automotive applications.

- **Resistance spot welding**

On average, a car's conventional steel body contains 4500 spot weld joints. Resistance spot welding has long been the primary joining method used in the automotive industry. The heat generated by the resistance of work pieces to the flow of current and the application of pressure creates the joint in this method. The weld is restricted to the overlapped work pieces and thus is not continuous.

- **Resistance seam welding**

The joint is produced progressively along the length of the weld in this type of resistance welding. In sheet metals, this results in a continuous and leak-tight joint. The weld can be created using overlapping or continuous workpieces. This welding process is used in the automotive industry to create leak-proof fuel tanks.

- **Friction welding**

The joint is formed in solid-state welding by applying pressure without significantly melting any of the work parts. Friction welding (FW) is a type of solid-state welding in which heat is generated by mechanically induced sliding motion between the welded parts. Under pressure, the weld parts are held together. In most cases, frictional heat is produced by rotating one part against the other. When a certain temperature is reached, the rotational motion is stopped, and pressure is applied to weld the parts together. FW is used in the automotive industry to make a variety of components such as half shafts, axle cases, steering columns, hydraulic cylinders, piston rods and engine valves, among others.

- **Laser beam welding**

Because of its distinct advantages, the use of laser technology for welding high-volume automotive components has grown in popularity. The main advantages are increased flexibility, increased productivity and significant savings on maintenance and energy costs while producing a strong weld. The heat generated when a focused laser beam impinges on the joint is used in the LBW process. Metal sheets with thicknesses ranging from 0.2 to 6 mm can be laser welded with ease. The majority of automotive industries use cross-flow CO_2 laser systems with power ratings ranging from 3 to 5 kW.

- **Magnetic pulse welding (MPW)**

Recently, there has been an increase in demand for lighter and more fuel-efficient vehicles. Lighter components are being developed by automotive manufacturers. This contributes to making existing cars more fuel efficient while also meeting the requirements of alternative fuel-powered vehicles such as fuel-cell powered cars and hybrid gas/electric vehicles. The use of lighter materials such as aluminium, as well as the development of new manufacturing processes that use less steel in welds, can help to achieve the aforementioned goal. The additional material deposited on the weld joint, as in conventional welding, adds to the weight of the welded component. MPW is a technology developed by Dana Corporation in the United States for bonding aluminium and steel (dissimilar metals) without the use of additional metal at the weld joint. Using precision-machined die cavities, pre-shaped aluminium and steel tubular stock are subjected to high pressure in this method. This is known as hydro-forming, and it results in more precise fitting of structural components that require very little fill material in the subsequent welding steps.

The hydro-formed components are then loosely assembled. A rapidly switching magnetic field produced by an inductor (either internally or externally) causes one of the metallic components to form quickly and impact the other stationary metal part with enough velocity and force to form a weld. High-end machines are required to produce proper welds in complex geometric designs. The MPW process increases manufacturing productivity by reducing production steps, materials, equipment and labour costs. The vehicle frame welded with the MPW process is claimed to be two-thirds lighter, resulting in an 8–10% increase in fuel efficiency. This reduces air pollution even more. Reduced energy consumption and shielded gases also help to reduce air pollution. Frames, side rails, cradles, stampings, space frames and bumper reinforcements are more efficiently produced using the MPW process.

With the commercialisation of the MPW process, it is now possible to weld dissimilar metals most effectively (bimetallic welding). This aids in the development of new geometries for automobile transmissions and undercarriage systems, which use various combinations of lighter materials to improve fuel economy and cost savings.

6.12 WELDING PROCEDURE FOR AUTOMOTIVE MANUFACTURING

Welding procedures are the guidelines used to perform a weld. They are designed to provide a record of the welding variables used and the inspection results obtained during the procedure qualification test. They can also provide the instructions for the welder to use in production in order to produce acceptable welds.

1. Basic information

- The basic information about the welding procedure
- Name of the company and the person who developed it
- Specification number and date
- Welding process and type
- Most WPS will be backed on a PQR (Procedure Qualification Record)
- Number and last revision.

Company Name ______________ By: ______________
Welding Procedure Specification No. ______ Date: ______ Supporting PQR No.(s) ______
Revision No. ______________ Date ______________
Welding Process(es) ______ SMAW ______ Type (s) ______ Manual ______

2. Joints

Details of joint design, including root spacing and backing (if the joint needs it and backing material). This information should feature a graphic representation of the joint to make it easier for the welder.

JOINTS (QW-402)

Joint Design: See permitted figures (right) or per constr. dwg

Root Spacing: ______

Backing (Yes) Yes (No) Yes

Backing Material (Type) Material or Weld Metal

(refer to both backing and retainers)

☐ Metal ☐ Nonfusing Metal

☐ Nonmetalic ☐ Other

Details

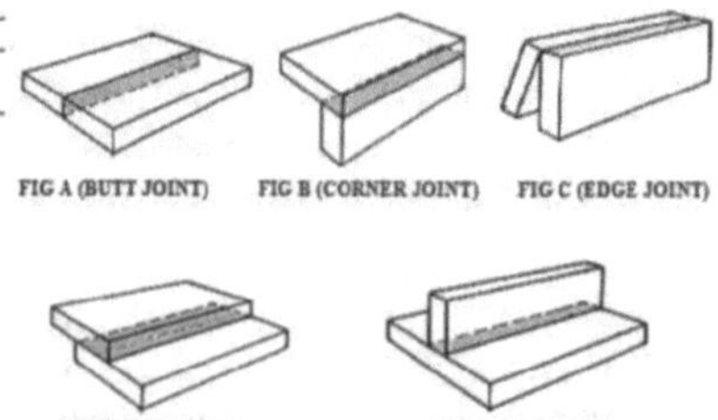

FIG A (BUTT JOINT) FIG B (CORNER JOINT) FIG C (EDGE JOINT)

FIG D (LAP JOINT) FIG E (TEE JOINT)

Sketches, Production Drawings, Weld Symbols or Written Description should show the general arrangement of the parts to be welded. Where applicable, the root spacing and the details of weld groove may be specified.

(At the option of the Mfgr., sketches may be attached to illustrate joint design, weld layers and bead sequence, e.g., for notch toughness procedures, for multiple process procedures etc.)

☐ Other (Describe) ______

☐ Figure A ☐ Figure B ☐ Figure C ☐ Figure D ☐ Figure E

3. Base and filler material

- Information about the base and filler metals that are going to be used in the welding procedure
- Its needs to specify the weld type (fillet or groove) and the thickness range of the base metal.

*BASE METALS (QW-403)

P-No. ______ Group No. ______ to P-No. ______ Group No. ______

OR

Specification type and grade ______
to Specification type and grade ______

OR

Chem. Analysis and Mech. Prop. ______
to Chem. Analysis and Mech. Prop. ______

Thickness Range:
Base Metal: Groove ______ Fillet ______
Other ______

Maximum Pass Thickness <=1/2 inch (13 mm) (Yes) ☐ (No) ☐

*FILLER METALS (QW-404)

Spec. No. (SFA)				
AWS No. (Class)				
F-No.				
A-No.				
Size of Filler Metals				
Filler Metal Product Form				
Supplemental Filler Metal				
Weld Metal				
Thickness Range:				
Groove				
Fillet				
Electrode-Flux (Class)				
Flux Trade Name				
Consumable Insert				
Other				

4. Additional details

Pre- and post-weld heat treatment required, gas used and the positions in which the procedure will be performed. The welding progression should be specified if it is required.

POSITIONS (QW-405)		POST WELD HEAT TREATMENT (QW-407)	
Position(s) of Groove		Temperature Range	
Welding Progression:	Up Down	Time Range	
Position(s) of Fillet			

PREHEAT (QW-406)		GAS (QW-408) Percent Composition	Gas(es)	(Mixture)	Flow Rate
Preheat Temp. Min.		Shielding			
Interpass Temp. Max.		Trailing			
Preheat Maintenance	(Continuous or special heating where applicable should be recorded)	Backing			

5. Electrical characteristics

Information about each weld pass including the following parameters:

- Process
- Current type and polarity
- Filler metal classification and diameter
- Volts range
- Amps range
- Travel speed range
- Wire feed speed range
- Energy of power range
- Other electrical specifications.

Techniques

- String or weave bead orifice, nozzle or gas cup size
- Method of back gouging
- Electrode spacing
- Multiple or single electrodes
- Peening
- Initial and interpass cleaning
- Contact tube to work distance
- Oscillation
- Multiple or single pass
- Other.

POSITIONS (QW-405)		POST WELD HEAT TREATMENT (QW-407)	
Position(s) of Groove		Temperature Range	
Welding Progression:	Up Down	Time Range	
Position(s) of Fillet			

PREHEAT (QW-406)		GAS (QW-408) Percent Composition	Gas(es)	(Mixture)	Flow Rate
Preheat Temp. Min.		Shielding			
Interpass Temp. Max.		Trailing			
Preheat Maintenance	(Continuous or special heating where applicable should be recorded)	Backing			

ELECTRICAL CHARACTERISTICS (QW-409)

Current AC or DC ______ Polarity ______

Amps (Range) ______ Volts (Range) ______

(Amperage and voltage range should be recorded for each electrode size, position, and thickness, etc. This information may be listed in a tabular form similar to that shown to the right or below.)

Tungsten Electrode Size and Type ______
(Pure Tungsten, 2% Thoriated, etc.)

Mode of Metal Transfer for GMAW ______
(Spray arc, short circuiting arc, etc.)

Electrode Wire feed speed range ______

TECHNIQUE (QW-410)

String or Weave Bead: String or Weave

Orifice or Gas Cup Size ______

Initial and Interpass Cleaning (Brushing, Grinding, etc.): Brushing, grinding, chipping

Method of Back Gouging: Grinding

Oscillation ______

Contact Tube to Work Distance ______

Multiple or Single Pass (per side): Multiple

Multiple or Single Electrodes: Multiple

Travel Speed (Range) ______

Peening ______

Other: No Single Pass To Exceed 1/2" In Thickness Pulsed Power Supply not permitted

Flux is not to be added to the face of the joint

Weld Layer(s) or Pass (es)	Process	Filler Metal		Current		Voltage Range	Travel Speed Range	Other (e.g., Remarks, Comments, Hot Wire Addition, Technique, Torch Angle, Etc.)
		Classification	Diameter.	Type and Polarity	Amperage Range			

6.13 SUMMARY

Welding is the method of fusing two metals that can be identical or dissimilar. It joins various metals/alloys with or without pressure and with or without the use of a filler metal. Welded structures are put together using five different types of joints: butt, lap, corner, T and edge joints. Welding processes are basically classified into two types: fusion welding and solid-state welding. The most commonly used welding methods for automotive applications include resistance spot welding (RSW), resistance seam welding (RSEW), metal inert gas (MIG) welding, tungsten inert gas (TIG) welding, laser beam welding (LBW), friction welding (FW) and plasma arc welding (PAW).Welding procedures are the guidelines used to perform a weld. They are designed to provide a record of the welding variables used and the inspection results obtained during the procedure qualification test.

6.14 UNSOLVED EXAMPLES

1. A plate, 75 mm wide and 10 mm thick, is joined with another steel plate by means of single transverse and double parallel fillet welds, as shown in Figure N6.6. The joint is subjected to a maximum tensile force of 65 kN. The permissible tensile and shear stresses in the weld material are 80 and 50 N/mm^2, respectively. Determine the required length of each parallel fillet weld. (Answer: 46.93 mm)

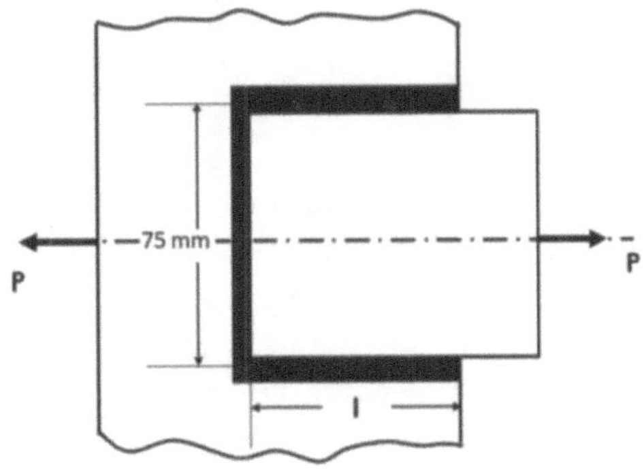

Figure N6.6

2. A steel plate, 100 mm wide and 10 mm thick, is joined with another steel plate by means of single transverse and double parallel fillet welds, as shown in Figure N6.7. The strength of the welded joint should be equal to the strength of the plates to be joined. The permissible tensile and shear stresses for the weld material and the plates are 80 and 60 N/mm^2, respectively. Find the length of each parallel fillet weld. Assume the tensile force acting on the plates as static. (Answer: 42.62 mm)

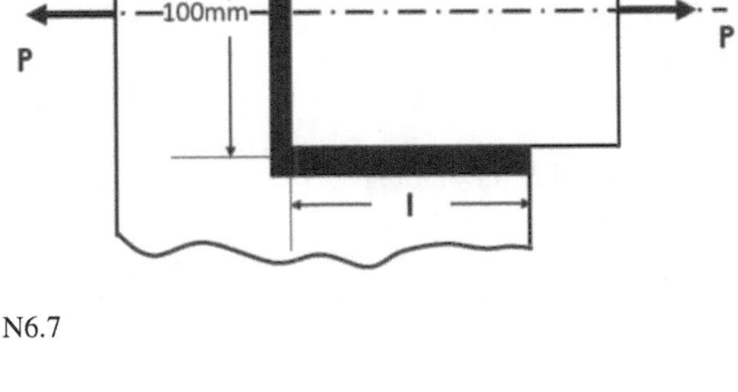

Figure N6.7

3. Two plates are joined together by means of single transverse and double parallel fillet welds as shown in Figure N6.8. The size of the fillet weld is 5 mm and allowable shear load per mm of weld is 300 N. Find the length of each parallel fillet weld. (Answer: 215 mm)

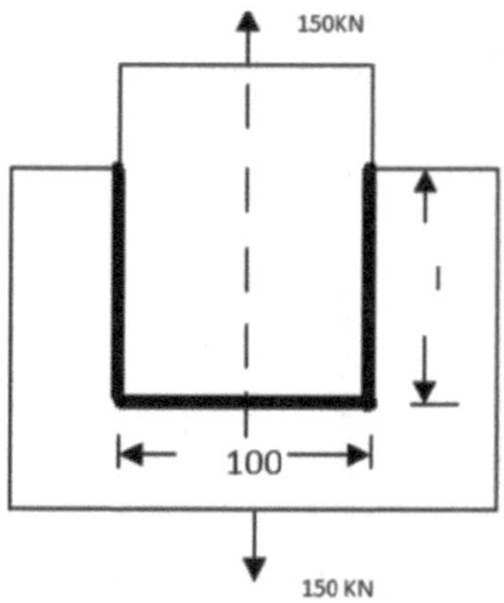

Figure N6.8

6.15 REVIEW QUESTIONS

1. Define welding, weld ability, fusion welding, solid welding.
2. List the arc welding equipment.
3. State the advantages of welding.
4. State the limitations of the welding process.
5. State the types of arc welding process.
6. State the types of resistance welding.
7. Describe the gas welding process with advantages, disadvantages and limitations.
8. Describe with a neat sketch metal inert gas arc welding (MIG) with advantages, disadvantages and limitations.
9. Describe with a neat sketch tungsten inert gas arc welding (TIG) with advantages, disadvantages and limitations.
10. Describe with a neat sketch spot welding processes with advantages, disadvantages and limitations.
11. Classify the welding process.
12. Sketch and describe welding flames.
13. Describe with a neat sketch plasma arc welding with advantages, disadvantages and limitations.
14. Differentiate TIG and MIG welding.
15. Differentiate soldering and brazing.
16. Justify plasma welding as a better option than other welding processes.
17. The use of welding processes is increasing daily in automobile manufacturing, justify this statement.

7 Material Removal Processes

LEARNING OBJECTIVES

- To understand the fundamentals and principles of material removal processes.
- To get details of mechanics of metal cutting, important parameters, cutting forces and their analysis.
- To study the cutting tools their reconditioning, failures advanced tool material and their important aspects.
- To get details of cutting fluids their composition, types and use as per application requirements.

7.1 INTRODUCTION

The material removal processes are a class of shaping operations that eliminate extra material from a beginning workpiece, leaving just the desired final shape. Different material removal methods using mechanical, electrical, laser or chemical means are available to achieve the desired shape and surface qualities. The most common metal shaping process is mechanical material removal or metal machining, also known as metal cutting; in this process, a wedge-shaped single-point or multipoint cutting tool is used to remove surplus metal (chip) from a workpiece such that the remaining metal has the desired component shape.

The term "machining" refers to any process in which material is gradually removed from a workpiece, such as metal cutting with single-point or multi-point tools (tools with geometrically defined cutting edges) and grinding with abrasive wheels that contain a large number of micro-cutting edges that are randomly shaped and oriented (i.e., tools with geometrically undefined cutting edges). Figure 7.1 shows classification of material removal process.

7.2 MECHANICS OF METAL CUTTING

The workpiece is tightly gripped in a machine tool vice, clamps, chuck or collet. A wedge-shaped tool is set to a specific depth of cut and is forced to travel in the direction as shown in Figure 7.2. All classic machining procedures necessitate the use of a cutting tool with a basic wedge shape at the cutting edge. If the tool is used, it

DOI: 10.1201/9781003367321-7

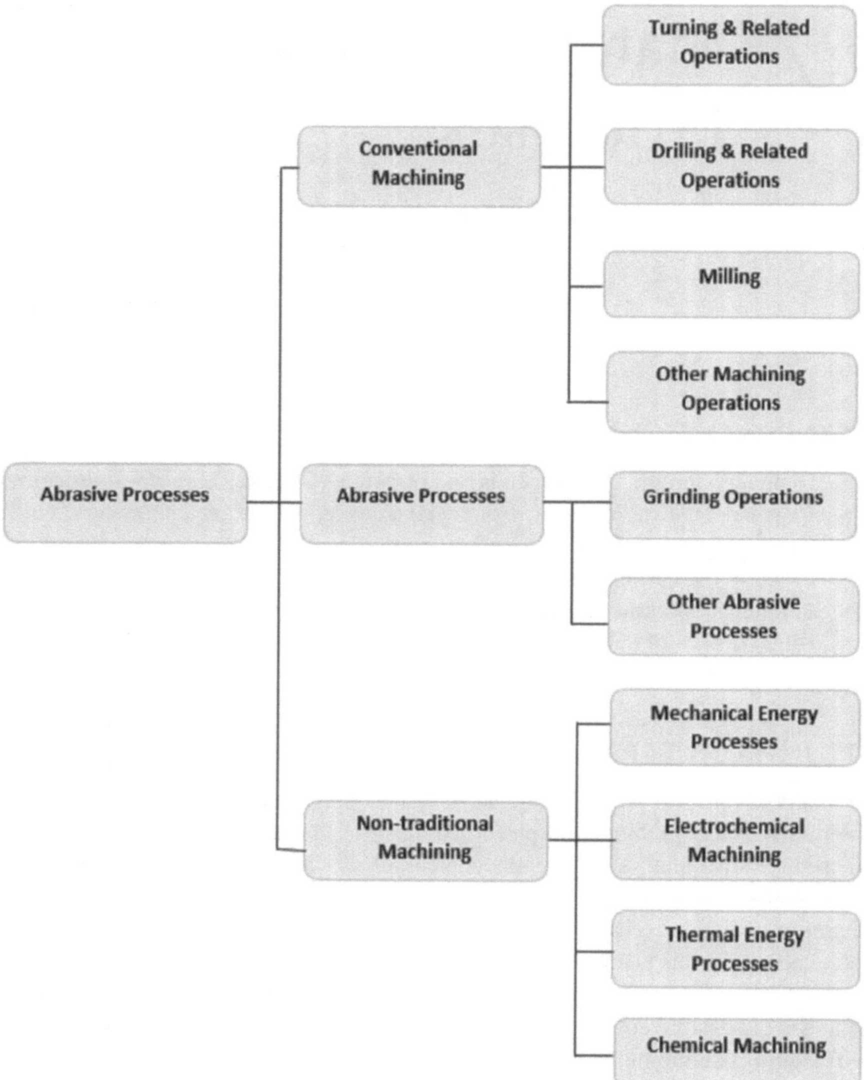

FIGURE 7.1 Classification of the material removal process.

will cut or shear the metal provided that it is hard than the metal. The tool is suitably shaped so that its edge can cut the metal effectively. The tool is strong enough to withstand cutting pressures while also being sharp enough to cut the metal, and as long as the tool moves relative to the material or vice versa, a cutting action is enabled. The majority of metal cutting is done using high-speed steel or carbide tools.

When cutting metal, the tool does not slip through it like a jack knife through wood, nor does it split it like an axe splits a log. In reality, the metal is driven off the workpiece by being squeezed, sheared off, and sliding down the cutting tool's face.

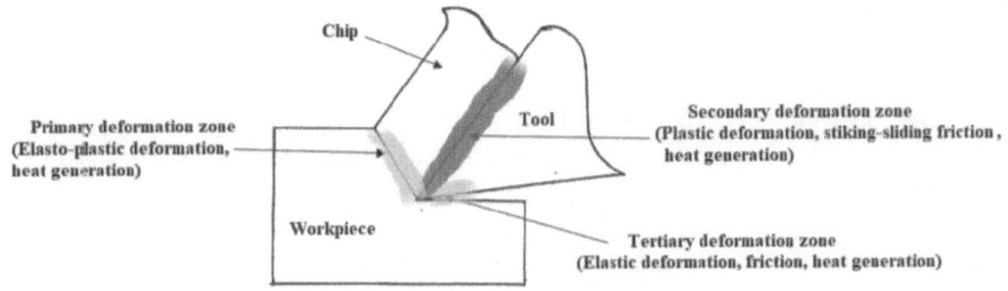

FIGURE 7.2 Mechanism of metal cutting.

The following describes how a cutting tool slices metal. All metals in solid form have a distinct crystalline structure, which is commonly referred to as grain structure. The size of the grain or crystals varies from very fine to very coarse, depending on the metal and heat treatment. The cutting tool advances in the workpiece once more. Extensive forces are applied to the crystals in front of the tool face. These crystals, in turn, impose identical pressures on the crystals ahead of them, along the direction of the cutter's cut or force. As the tool advances, the material at the sheared spot is sheared by the tool's cutting edge or pulled loose by the action of the bending chip that is being created. As the tool progresses, maximum stress is exerted along the sheared line, also known as the shear plane. This plane is roughly perpendicular to the tool's cutting face. A shear zone exists on both sides of the shear plane; when the force of the tool exceeds the strength of the material at the shear plane, the crystalline grain structure ruptures or slips, generating the metal chip. The chip separates from the workpiece material and advances up the tool face. Furthermore, when the metal is sheared, the crystals elongate, with the direction of elongation opposite that of shear. After leaving the shearing plane, the circles that represent the crystals in the uncut metal stretch into ellipses.

7.2.1 Mechanism of Chip Formation

Machining is a semi-finishing or finishing method used to impart required or defined dimensions, form correctness and surface quality to enable the product to:

- Fulfil its core functional requirements
- Provide better or improved performance
- Have a long service life.

The shape of the chips is an important machining index since it reveals, directly or indirectly:

- The nature and behaviour of the work material while it is being machined
- Requirements for specific energy in machining work (amount of energy required to remove unit volume of work material)
- The nature and scope of interaction at chip–tool interfaces.

The following factors heavily influence the shape of machined chips:

- Workplace material
- The material and geometry of the cutting tool
- Cutting velocity and feed rates, as well as, to a lesser extent, depth of cut
- The machining environment or cutting fluid that affects temperature and friction at the chip–tool and work–tool interfaces
- Understanding the fundamental mechanism(s) of chip production aids in understanding chip characteristics and creating acceptable chip shapes.

Types of chips

The chips that are formed during metal cutting operations can be classified into four types:

1. Discontinuous or segmental chips
2. Continuous chips
3. Continuous chips with built-up edge.
4. Non-homogeneous chips.

Continuous chips

Continuous chips, as the name implies, have a continuous segment. This chip is formed during high-speed cutting of ductile materials such as aluminium, mild steel, copper and so on. The shape is the result of continuous plastic deformation of the material caused by the use of a tool. Figure 7.3 shows continuous chip formation process.

These chips have the same thickness all the way down the length, and they have a good surface quality.

The best circumstances for creating continuous chips are as follows.

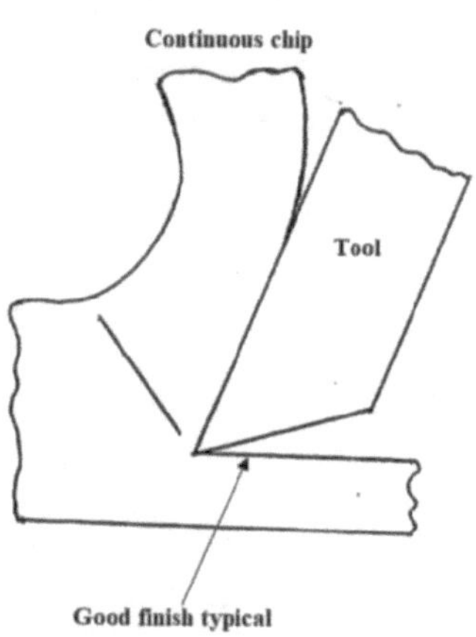

FIGURE 7.3 Continuous chip.

- The workpiece should be ductile.
- A large rack angle is required.
- Friction between the work item and the tool should be kept to a minimum.
- The cutting speed should be high.
- The depth of cut should be kept to a minimum.
- Correct use of coolant and lubricant.
- The tool's coefficient of friction should be low.

Because of the following advantages, continuous chips are the most preferred type of chip.

- They produce a high surface polish while cutting ductile materials.
- Continuous chips form when friction is minimal, minimising friction loss.
- Tool life is extended due to less friction.
- Power usage is minimal.

7.2.2 Discontinuous Chips or Segmental Chips

As the name implies, this chip is segmented. It is formed while machining brittle materials such as cast iron, brass and so on at a slow cutting speed. Chips are cut into little segments during the cutting process. Figure 7.4 shows details of discontinuous ship or segmental chip formation process. This is generated when cutting at a slow speed with a small rack angle. When the friction between the tool and the workpiece is high, these chips occur in ductile material. Discontinuous chips in ductile material cause poor surface quality and slow machine operation. It is a suitable form of machining brittle material chips. The favourable circumstances for producing this type of chip are as follows:

- The work should be brittle in nature
- Slow cutting speed

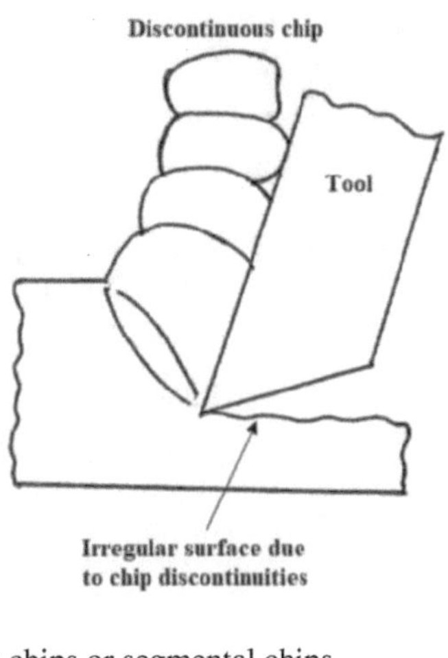

FIGURE 7.4 Discontinuous chips or segmental chips.

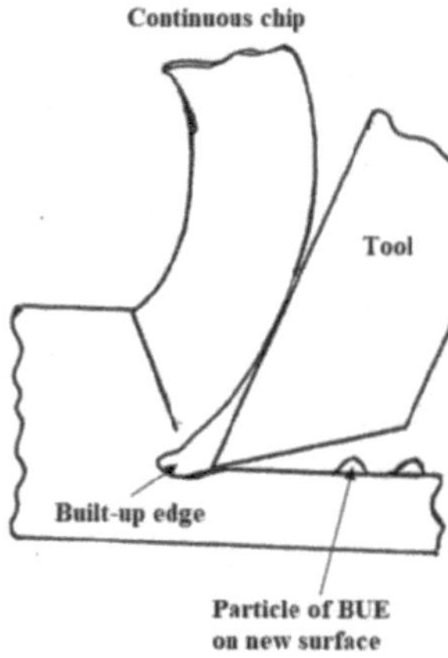

FIGURE 7.5 Continuous chip with built up edge.

- Tool has a small rack angle
- The depth of cut should be large.

7.2.3 Continuous Chips with Built Up Edge

This type of chip is similar to continuous chips except that a built edge is formed at the face of the tool. Figure 7.5 shows details of continuous chip with built up edge formation process. This is formed during ductile metal machining with considerable friction between the tool and workpiece. This chip does not have the same smoothness as continuous chips. The built up edge is formed as a result of the high temperature between the tool and the work item. The high temperature is caused by the strong friction force between the tool and the workpiece. The common factors a promoting built up edge are as follows:

- Ductile metal cutting
- High friction force at the tool's face
- A high temperature existing between the tool and the work item
- Inadequate coolant and lubrication.

A comparison of continuous chip, discontinuous chip and continuous chip with built up edge is as follows:

Type of chip	Material type	Rack angle	Depth of cut	Cutting speed
Continuous chips	Ductile	High	Small	Large/medium
Discontinuous chips	Brittle, ductile but hard	Medium	High	Low
Continuous chips with built edge	Ductile	Low/medium	Medium	Medium

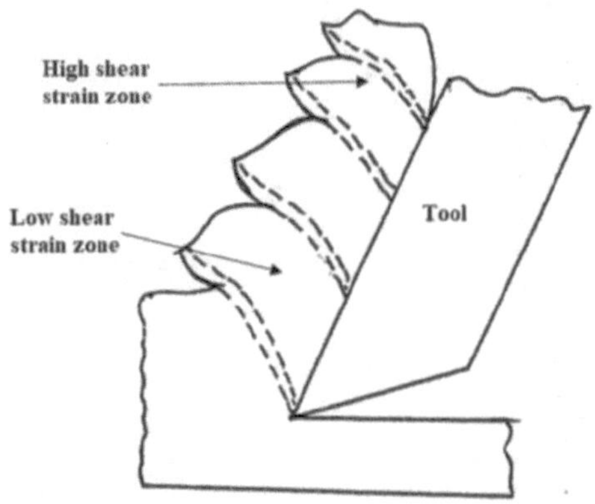

FIGURE 7.6 Serrated chips.

7.2.4 Serrated Chips

(The term shear-localised is also used for this fourth type of chip. Figure 7.6 shows details of serrated chip formation process.)

These chips are semi-continuous in the sense that they have a sawtooth look caused by a cyclical chip generation of alternating high shear strain and low shear strain. When manufactured at greater cutting speeds, this fourth form of chip is most closely connected with certain difficult-to-machine metals such as titanium alloys, nickel-base super alloys and austenitic stainless steels. When common work metals (e.g., steels) are cut at high speed, this phenomenon can be observed.

7.2.5 Types of Metal Cutting Processes

The metal cutting process is divided into two types, as follows:

- Orthogonal cutting process (two-dimensional cutting) – The cutting edge or face of the tool is 90^0 to the line of action or path of the tool in the orthogonal cutting process
- Oblique cutting process (three-dimensional cutting) – The tool's cutting edge or face is inclined at an angle less than 90^0 to the tool's line of action or route. Figure 7.7 shows details of orthogonal cutting and oblique cutting process.

7.3 CUTTING TOOLS

Cutting tools can be classed in a variety of ways, including being classified into two groups based on the number of cutting points on the tool:

1. Single-point cutting tool
2. Multipoint cutting tool

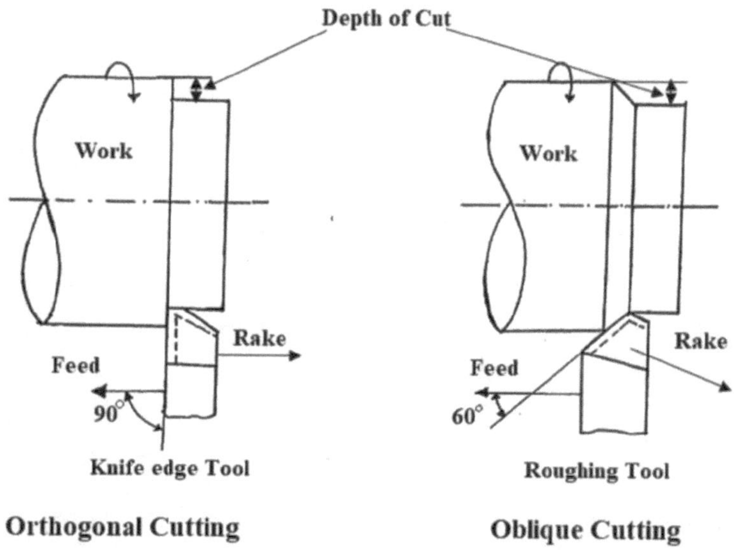

FIGURE 7.7 Orthogonal cutting, oblique cutting.

There is only one cutting point or edge on a single-point cutting tool. Single-point tools are those used for turning, boring, shaping or planing operations, such as those used on lathes, boring machines, shapers and planers. A multi-point tool is one that has two or more cutting points (for example, tools used on drilling machines, milling machines, broaching machines and so on). A multi-point tool is essentially a collection of single-point tools. It is classed as follows based on the construction of the cutting tool:

1. Solid tools
2. Tipped cutting tools

The solid cutting tools are made entirely of the same material, whereas, in a tipped cutting tool, an insert of cutting tool material is brazed or held mechanically to the shank of another material.

- **Single-point cutting tools**

A single-point cutting tool is a tool that helps to perform several operations (such as turning, facing, and producing a flat surface) on lathe, shaper and planer machines.

- **Single-point tipped cutting tools**

Carbides, ceramics, cast alloys, diamond, CBN and UCON are used as tips or inserts that are either brazed into a prepared seat machined on a strong steel tool shank or clamped to the shank. Indexable inserts or throwaway tips are the second type of tips or inserts.

- **Tools with a brazed tip**

Tool material tips or inserts with suitable forms are brazed in a steel shank. When the tip or insert becomes worn, it is resharpened using specific grinding wheels, or it can be replaced or indexed. The tool wheel must be withdrawn from the machine for resharpening, which necessitates a resetting procedure. The fundamental disadvantage of a brazed tip is that due to the difference in coefficient of expansion of tip material and tool shank material, the brazing must be done with extreme caution.

- **Multi-point cutting tools**

A multi-point cutting tool contains more than two main cutting edges that simultaneously engage in cutting action in a pass. Sometime, cutters with two cutting edges (more than one) are also considered multi-point cutting tools (instead of considering it as a double-point cutter). The number of cutting edges present in a multi-point cutter may vary from three to a few hundred. Since the cutting edge appears at the intersection of the rake surface and flank surface, a set of rake surfaces and flank surfaces also exists for each cutting edge.

- **Milling cutters**

Milling cutters are multi-point cylindrical cutting tools with cutting teeth spread evenly around the perimeter. The manner of providing relief on the tools is the most appropriate way to define milling cutters. Sharpening a narrow area behind the cutting edge produces profile-relieved cutters. When the cutting edge becomes dull, this narrow area is resharpened by grinding; relieved cutters have a curved relief behind the cutting edge and are sharpened by grinding the tool face. Because the relief angle is fixed during the manufacturing process, there is more freedom in altering relief angles in profile-relieved cutters. However, because the relief is not changed during resharpening, this kind is more suited for cutters with complicated shapes/profiles.

Milling cutters can also be classified according to how they are attached, for example, arbor type, shank type or spindle-mounted type. The majority of milling cutters are solid HSS, although they are also available with tipped teeth or disposable tips of other tool materials.

Milling cutters are classified into two types based on this:

1. Profile-relieved cutters
2. Form-relieved cutters.

- **Broach tool**

A broach is a multi-point cutting tool that consists of a bar with a surface that contains a succession of cutting teeth or edges that gradually grow in size from the beginning or entering through the back end. Broaches can be used to machine internal or external surfaces. Surfaces can be flat, round or of any complicated shape. Broaching is the process of pushing or pulling a broach over or through the surface of a workpiece. Each tool tooth scrapes a tiny slice from the surface. Broaching of the interior surface is referred to as "internal or hole broaching," and broaching of the exterior surface is referred to as "surface broaching."

- **Drill**

Drilling is the technique of creating a round hole from a solid material. The tool (drill) is turned and fed into the material along its axis rather than the workpiece. Drills can be classified in a variety of ways, including material, number and type of flutes, drill size, shank type (straight or taper), and cutting point geometry, among others.

The fluted twist drill, on the other hand, is the most popular type of drill. It is made from a round bar of tool material and includes three major parts: the tip, the body and the shank. The shank of the drill is used to hold and rotate it. The cutting elements are contained in the point, whereas the body controls the drill during operation. The drill's body has two helical grooves called "flutes" carved into it. The flutes form the cutting surface and also aid in chip removal from the drilled hole. The two cutting edges are straight and separated by the drill's web thickness, which is supplied to reinforce the drill structure.

- **Reamers**

A reamer is a rotary cutting tool that is generally cylindrical in shape and is used to enlarge and finish holes to precise dimensions to a previously made hole. It is a multi-edge cutting instrument with cutting edges on its circumference. A reamer is made up of three major parts: a fluted section, a neck and a shank.

7.3.1 Cutting Tool Geometry

Figure 7.8 shows details of single point cutting tool geometry.

- **Tool geometry**: A single-point cutting tool's many angles serve vital tasks in machining operations.
- **Shank**: This is the portion of the tool bit having a rectangular cross section which is not ground to form cutting edges.
- **Face**: The portion of the cutting tool against which the chip slides upward.
- **Flank**: A cutting-flank tool's surface that faces the workpiece.
- **Heel**: A single point tool's heel is the lowest section of the side-cutting edges.

Back rake angle

This is the angle measured in a perpendicular plane via the side cutting edge between the tool's face and a line parallel with the tool's base. If the slope face is downward toward the nose, it is a negative back rake angle; if it is upward toward the nose, it is a positive back rake angle. This angle helps in removing the chips from the workpiece.

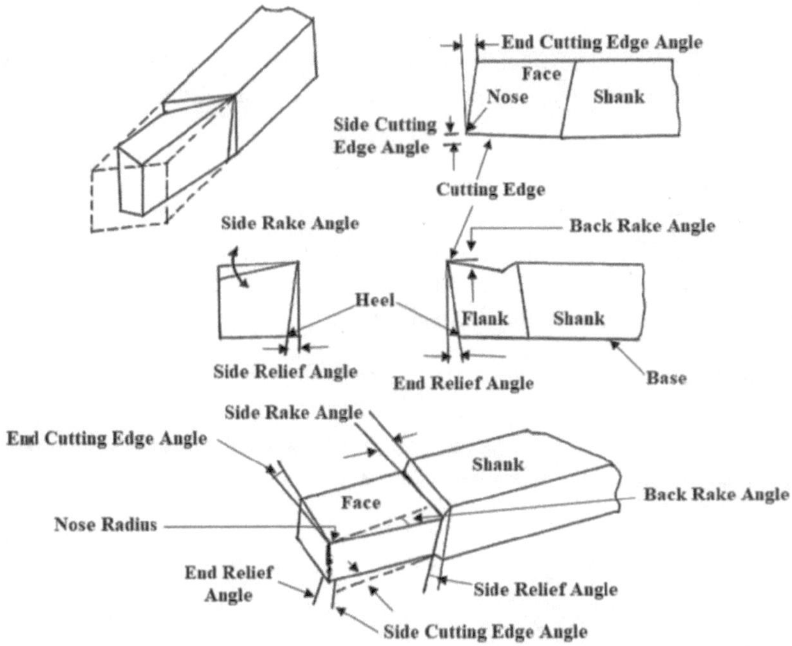

FIGURE 7.8 Single-point cutting tool geometry.

Side rake angle

This is the angle at which the tool's face is tilted sideways, and it determines the thickness of the tool behind the cutting edge. It is placed on the tool to provide clearance between the workpiece and the tool, preventing the workpiece from rubbing on the tool's end flake. It is placed on the tool to provide clearance between the workpiece and the tool, preventing the workpiece from rubbing on the tool's end flake. It is the angle of the tool that permits it to cut without rubbing on the workpiece. It is defined as the angle measured at right angles to the flank between the section of the end flank immediately below the cutting edge and a line perpendicular to the base of the tool. Extra end clearance, also known as end clearance angle, is sometimes given on the tool. This is the secondary angle that lies right beneath the end relief angle.

Side relief angle

This is the angle that prevents interference while the tool is inserted into the material. It is the angle measured at right angles to the side between the area of the side flank immediately below the side edge and a line perpendicular to the base of the tool. It is built into the tool to offer relief between the tool's flank and the workpiece surface. Extra side clearance, also known as side clearance angle, is sometimes supplied on the tool. It is the secondary angle that is right beneath the side relief angle.

End cutting edge angle

This is the angle formed between the tool's end cutting edge and a line perpendicular to the tool's shank. It creates space between the cutting edge of the tool and the work item.

Side cutting edge angle

This is the angle formed by the straight cutting edge on the tool's side and the side of the shank. It is sometimes referred to as the lead angle. It is in responsible for directing the chip away from the finished surface.

Nose radius

This is the nose point that connects the side and end cutting edges. It has a tiny radius, which is responsible for the workpiece's surface finish.

7.3.2 Tool Signature

Tool signature or tool nomenclature is a convenient approach to define tool angles by using a standardised abbreviated format. It denotes the angles used by a tool during a cut. It provides the tool's active angles normal to the cutting edge. As long as the tool shank is attached at right angles to the workpiece axis, this will always be true. The seven elements that comprise the signature of a single point cutting tool can be stated in the following order:

Tool signature 0-7-6-8-15-16-0.8

- Back rake angle (0°)
- Side rake angle (7°)
- End relief angle (6°)
- Side relief angle (8°)
- End cutting edge angle (15°)
- Side cutting edge angle (16°)
- Nose radius (0.8 mm).

Tool signature (designation) under ASA (American Standards Association) System is given in the order: αb-αs-θe-θs-Ce-Cs-R αb = back rake angle; αs = side rake angle; θe = end relief angle; θs = side relief angle; Ce = end cutting edge angle; Cs = side cutting edge angle; R = nose radius.

7.4 CUTTING TOOLS MATERIALS

A cutting tool is a device that is used to remove undesirable material from a workpiece. The material properties of a cutting tool include the following:

- The material should be harder than the workpiece so that it can penetrate it, and it should have hot hardness, or the ability to maintain hardness at high temperatures.
- For a superior surface polish and reduced wear, the coefficient of friction at the tool chip interface should be low.
- The material should be wear-resistant to prevent cutting tool surface wear and tear.
- It should be chemically stable so that it does not react with the workpiece and chemically inert so that no oxidation occurs and thus no scales or pits form on the surface.
- The material must be strong and durable enough to endure shocks and vibrations.
- The thermal conductivity should be high so that heat created during the machining process can be dissipated, extending the life of the cutting tool.

The following materials are commonly used in cutting tools:

1. **Carbon steel**: Carbon steels with carbon percentages as high as 1.5% are employed as tool materials, however they cannot withstand very high temperatures and so operate at modest cutting speeds.
2. **High-speed steel (HSS)**: This is a special alloy steel made by alloying tungsten, chromium, vanadium, cobalt and molybdenum with steel. In comparison to carbon steel, HSS has higher hot hardness, wear resistance, and a three to four times faster cutting speed. The following are the compositions of the most regularly used HSS.
   (a) 18-4-1 HSS, which is composed of 18% tungsten, 4% chromium and 1% vanadium, with a carbon concentration of 0.6–0.7%. If the vanadium content is 2%, the result is 18-4-2 HSS.
   (b) Cobalt high-speed steel: This is also known as super high-speed steel. Cobalt is added in amounts ranging from 2% to 15%. The most frequent composition is tungsten at 20%, chromium at 4%, vanadium at 2% and cobalt at 12%.
   (c) Molybdenum high-speed steel: This steel comprises 6% tungsten, 6% molybdenum, 4% chromium and 2% vanadium.
3. **Cemented carbide**: This is essentially carbon that has been cemented together by a binder. It is a powder metallurgy product with cobalt as the primary binder. The main components are tungsten carbide (82%), titanium carbide (10%) and cobalt (8%). These materials have high hardness and wear resistance, and their cutting speed is six times that of high-speed steel (HSS).
4. **Ceramics**: Ceramics are made up primarily of aluminium oxide (Al_2O_3) and silicon nitride (Si_3N_4). Ceramic cutting tools are hard and have high hot hardness, and they do not react with the workpiece. They can be utilised at higher temperatures and at four times the cutting speed of cemented carbide. They have a low thermal conductivity.
5. **Diamond**: This is the hardest known substance; with a cutting speed 15 times that of high-speed tools.

6. **Cubic boron nitride (CBN)**: This is the second hardest substance after diamond and a more cost-effective alternative. High temperature and pressure are used to bond cubic boron crystals with a ceramic or metal binder to generate a polycrystalline structure with nitride particles. It is a good cutting tool material due to its exceptional high hot hardness at temperatures up to 2000°C.

7.5 ADVANCED CUTTING TOOL MATERIALS

Advanced tool materials include cermet, ceramic, cubic boron nitride (CBN) and diamond. The most advanced materials are harder than common tool materials, such as carbide, and they have a range of properties and applications.

7.5.1 Coated Carbides

Although coated carbides have been around since the late 1960s, they did not attain their full potential until the mid-1970s. The early coated carbides were little more than normal carbide grades that had been coated. As manufacturers developed experience in creating coated carbides, they realised that the coating was only as good as the base carbide beneath it (known as the substrate).

Coating materials of several varieties are utilised, each for a specialised application. When using coated carbides, it is critical to follow the do's and don'ts.

The most common coating materials are: titanium carbide, titanium nitride, ceramic coating, diamond coating and titanium carbo-nitride. These materials provide benefits such as:

- Reduced abrasion, adhesion, and diffusion wear
- Reduced friction and BUE generation
- Heat resistance, as well as reduced thermal cracking and plastic deformation
- Decrease in cutting forces and power consumption
- Increase in tool life (by 200–500%) for the same Vc or increase in Vc (by 50–150%) for the same tool life
- Enhancements to product quality
- Efficient and effective machining of a diverse range of work materials.

7.5.2 Cermets

Cermets are ceramic–metal alloys. They are made up of two phases: ceramic and a metal. Composite tools made of Al_2O_3 and TiC but without a metallic binder have two ceramic phases but no free metallic phase and hence are not cermets. Tools containing 15–30 w/o TiC and the remainder Al_2O_3 are less brittle than Al_2O_3 tools but also less refractory. These are formed by hot pressing the mixture and sintering it. Such tools are most useful in high-speed machining of hard cast irons (HB = 350–600 kg/mm^2) and relatively hard steels under limited mechanical shock circumstances. In contrast to sintered tungsten carbides, these cermets have the following characteristics.

The grains are composed of TiCN (instead of WC) and Ni or Ni-Co and Fe as a binder (in place of Co).

They also are harder, more chemically stable and thus more resistant to wear. They are more brittle and less resistant to heat shock.

The weight percentage of binder metal used ranges from 10% to 20%. Unlike coated carbide inserts, the cutting edge sharpness is retained. Modern TiCN-based cermets with bevelled or slightly rounded cutting edges are ideal for high-speed finishing and semi-finishing of steels and stainless steels, but not for jerky interrupted machining and machining of aluminium and related materials. Cermet characteristics and performance are continually being improved through research and development.

7.5.3 Coronite

As previously stated, the characteristics and performance of HSS tools may be significantly enhanced by refining the microstructure, powder metallurgical process and surface coating. Recently, a novel tool material known as coronite was invented for the production of tools such as small and medium-sized drills and milling cutters, which were previously mostly manufactured of HSS. Coronite is essentially formed by mixing HSS for strength and toughness with tungsten carbides for heat and wear resistance. The matrix is equally disseminated with microfine TiCN particles. The coronite-based tool, unlike a solid carbide tool, is composed of three layers:

1. The HSS or spring steel core in the centre
2. A coronate coating with a thickness of around 15% of the tool diameter
3. A thin (2–5 μm) TiCN PVD coating.

Such tools not only increase productivity but also improve product quality.

In terms of cutting forces, tool life and surface finish, coronite tools manufactured by hot extrusion followed by PVD-coating of TiN or TiCN surpassed HSS tools.

7.5.4 High-Performance Ceramics (HPC)

Significant gains in strength and toughness, and hence overall performance of ceramic tools, have become possible during the last few years through a variety of ways, including the addition of TiO_2 and MgO to increase the sinterability, microstructure, strength and toughness of Al_2O_3 ceramics to some extent. Transformation toughening is done with the addition of an appropriate amount of partially or totally stabilised zirconia in Al_2O_3 powder.

Isostatic and hot isostatic pressing (hipping) are effective but costly methods. Silicon nitride ceramic consists of (1) plain nitride ceramics, (2) SIALON, (3) whisker toughened. While alumina is toughened by (1) zirconia, (2) SiC whiskers, (3) metal (silver), etc.

7.5.5 Plain Nitride Ceramics Tools

With the development of modern manufacturing technology, traditional carbide cutting tools have been unable to meet the production needs of new and difficult-to-make materials. Silicon nitride ceramic cutting tools are favoured for their high hardness, strength, good fracture toughness, high-temperature oxidation resistance and thermal shock resistance, which are widely used in hard cutting, high-speed cutting and so on. It offers greater bending strength, toughness and conductivity, greater resistance to fracturing caused by mechanical and thermal shocks, and is better suited for rough and interrupted cutting of various materials.

7.5.6 SIALON Tools

Silicon nitride and SiAlONs are materials possessing a unique combination of fracture toughness and hardness. A composite ceramic tool is formed by hot pressing and sintering an adequate mixture of Al_2O_3 and Si_3N_4 powders. These tools are hard, strong, and resistant to wear.

7.5.7 Nitrite Toughened

These tools are very expensive, but they are ideal for high-volume machining of a variety of soft and hard materials, even when cutting is interrupted.

7.5.8 Zirconia Toughened Alumina (ZTA)

Zirconia toughened alumina (ZTA) provides mechanical strength and wear resistance and is used as the ideal intermediate solution between zirconia and alumina. Zirconia toughened alumina is manufactured using stress-induced transformation of fine tetragonal zirconia particles. It possesses the following properties:

- High hardness
- High hot hardness
- High-speed machining application
- Expensive.

7.5.9 Cubic Boron Nitride (CBN)

Due to similarities in the layer-type crystal structure of hexagonal boron nitride and graphite, Wentdorf decided to look into the prospect of a high-temperature, high-pressure stable cubic form of boron nitride similar to diamond but not found in nature. Early experiments with hexagonal boron nitride powder and the then-familiar diamond-forming catalysts did not yield a cubic form, even at pressures as high as 100 kbar and temperatures as high as 2000°C (Wentdorf, 1960). The alkali metals, alkaline earth metals, their nitrides, antimony, tin and lead, or combinations of all of these, were discovered to be excellent catalyst solvents for CBN production.

In the presence of a catalyst, larger CBN crystallites (300 m or greater) of economic value were discovered to form.

When utilised as a catalyst at 80 kbar and 1700–1900°C, a mixture of magnesium nitride and sodium metal produced massive crystals. Because the size of cubic boron nitride particles increases with time, it is desirable to maintain the reaction conditions between 3 and 5 minutes, even though the reaction period is only approximately 0.5 minute. In 1967, CBN grits were commercially marketed.

At high machining speeds, it remains inert and retains high hardness and fracture toughness. It performs admirably when grinding materials with high hardness and strength. Because of its extreme hardness, toughness, chemical and thermal stability, and wear resistance, CBN cutting tool inserts have been developed for a high material removal rate (MRR) as well as precision machining, imparting outstanding surface integrity to the products. Such one-of-a-kind tools are effectively and advantageously used in machining a wide range of work materials, including high-carbon and alloy steels, non-ferrous metals and alloys, exotic metals such as Ni-hard, Inconel, Nimonic, and others, as well as many non-metallic materials that are difficult to machine with conventional tools. It is extremely stable at temperatures as high as 1400^{0}C. When machining grey cast iron, the operative speed range for CBN is 300–400 m/min. Other materials' speed ranges are as follows:

- Hard cast iron (more than 400 BHN): 80–300 m/min
- Super alloys (more than 35 RC): 80–140 m/min
- Hardened steels (more than 45 RC): 100–300 m/min.

Aside from speed, the most critical aspect influencing the performance of CBN inserts is the cutting edge preparation. It is preferable to utilise CBN. Tools with polished or chamfered edge preparation, particularly for interrupted cut CBN tools, like ceramics, are only accessible as indexable inserts. Its main drawback is its expense.

7.5.10 Diamond

Diamond crystals, whether natural or synthetic, are utilised as cutting tool tips/edges. Natural single crystal is employed for numerous applications due to its great hardness and sharp edges, particularly when high accuracy and precision are required.

- Single-point cutting tool tips and small drills for high-speed machining of non-ferrous metals, ceramics, polymers, composites and other materials, as well as effective machining of difficult-to-machine materials
- Drill bits used in mining, oil exploration, and so forth
- Cutting and drilling tool for glassware, stones, ceramics, FRPs and other materials
- Extrusion and drawing dies for wire
- Super-abrasive wheels for fine grinding.

Natural diamond's limited supply, increasing demand, high cost and ease of cleavage have necessitated a more dependable source of diamond. This resulted in the invention and production of artificial diamond grits via an ultra-high temperature and pressure synthesis technique, allowing for large-scale production of diamond with some control over size, shape and friability of the diamond grits as desired for diverse purposes.

7.5.11 Polycrystalline Diamond (PCD)

Polycrystalline diamond (PCD) tools are made up of a layer of fine grain size, randomly oriented diamond particles sintered with a suitable binder (typically cobalt) and then metallurgically attached to a suitable substrate, such as cemented carbide or Si_3N_4 inserts. PCD has great wear resistance, a sharp edge, little friction in the cut, strong fracture strength and good thermal conductivity. These characteristics contribute to the extended life of PCD tooling in conventional and high-speed machining of soft, non-ferrous materials (aluminium, magnesium, copper, and so on), advanced composites and metal–matrix composites, superalloys, and non-metallic materials. PCD is especially well suited for abrasive materials (such as drilling and reaming metal matrix composites), where it outlasts carbides by 100 times.

Because of the high solubility of diamond (carbon) in these materials at extreme temperatures, PCD is not commonly suggested for ferrous metals. However, under certain conditions, they can be used to machine certain of these materials; for example, light cuts in grey cast iron are successfully created. The main advantages of such a PCD tool are increased toughness due to a finer microstructure with variable grain orientation and less cleavage. However, such a one-of-a-kind PCD has some drawbacks, including:

- It is an expensive tool.
- The presence of cobalt as a binder reduces wear resistance and thermal stability.
- Complex tool forms, such as built-in chip breakers, are not possible.
- Size constraints, particularly when producing tools with very small diameters.
- Diamond-coated tools have almost completely addressed the above-mentioned limitations of polycrystalline diamond tools.

7.5.12 Diamond-Coated Carbide Tools

Since the advent of low-pressure synthesis of diamond from the gaseous phase, there has been an ongoing endeavour to utilise thin-film diamond in the cutting tool industry. These are typically employed as thin (50 µm) or thick (> 200 µm) diamond films produced using CVD for cutting tools, dies, wear surfaces and even abrasives for abrasive jet machining (AJM) and grinding. A thin layer is put directly on the tool surface. A thick film (> 500 m) is developed on a suitable substrate and brazed to the actual tool substrate, and the original substrate is removed by dissolving or another method. Thick film diamond is used to make inserts, drills, reamers, end mills and routers. Coating has become more popular than single diamond crystal and PCD for a variety of reasons, including:

- Binder-free, higher hardness, greater resistance to heat and wear than PCD, and qualities similar to natural diamond
- Extremely pure, dense, and free of single-crystal cleavage
- Allows for a broader range of tool size and shape, and can be deposited on any shape of tool, including rotary tools
- Relatively less costly.

7.5.13 Carbon with Diamond-Like Properties (DLC)

DLC is a non-crystalline (amorphous) type of carbon that was discovered in 1973. It can only be created at low pressures and temperatures lower than those required to produce crystalline carbon. These dense carbon deposits are created with the use of an energy-assisted method, such as radio frequency plasma or a low-energy ion beam. These amorphous diamond-like formations are thought to be connected to quenching caused by a cold substrate. This is especially important for coating HSS tools, where the surface temperature must be lower than the tool's transition temperature. Diamond-like features of DLC films include great hardness, minimal friction, high heat conductivity and undistorted image transmission. DLC films are an effective scratch-resistant coating for eyeglass and other optical lenses.

7.5.14 Carbonado

This is a polycrystalline aggregate of extremely fine diamond particles, graphite and other impurities.

This black substance is found naturally in Brazil, where diamond particles are sintered together to form enormous polycrystalline lumps. It has the same hardness as a single crystal diamond but is far less fragile. A crack that forms in one particle only spreads until it reaches the next particle with a different crystal orientation.

7.5.15 Extra-Large Synthetic Diamond

Sumitomo Electric Industries, Inc. has produced single crystals of diamond up to 1.2 carats in size utilising the thermal gradient method with a "belt" device since 1985. Doping with nitrogen results in a single yellow crystal, whereas doping with aluminium results in a colourless or light-blue single crystal. In the HPHT apparatus, a seed is put with a carbon source and a solvent. The temperature is then raised to 1450°C at 50,000 atm.

Diamond and CBN hybrid tools, as well as nano-cutting tools that have been modified for high-speed machining of hard and strong steels and related materials, are possible and viable. Diamond tools are exceptionally useful for machining stones, slates, glass, ceramics, composites, FRPs and nonferrous metals, particularly those that are sticky and BUE formers like pure aluminium and its alloys. CBN and diamond tools are also commonly employed for ultra-precision, micro- and nano-machining.

Microstructural enhancement of machining tools, such as nano-structuring and hybridisation, is a relatively new topic. Diamond is extremely hard and has both static and dynamic strength. Chemical wear is caused by diamond graphitisation followed by carbon diffusion to the material being machined. Chemical wear in diamond tools can be decreased by using CBN as a partial replacement for diamond.

7.6 TOOL COATING PROCESS

The coating on a cutting tool has a significant impact on the mechanical and tribological properties of the tool, as well as the end-result of the product. A cutting tool can be made of a variety of materials, but cemented tungsten carbide is the most commonly used material in the industry today because its properties meet the requirements of manufacturers. To improve the performance of cutting tools, various coatings have been applied to the tools, including single coatings, multi-layered coatings, nanocomposites and superlattices. The key reasons for utilising coating tools are to increase wear resistance, oxidation resistance, friction reduction, resistance to metal fatigue and resistance to thermal shock.

When cutting tools are properly coated and perform as intended, the end-user benefits from higher cutting data, longer tool life, and the option of dry machining.

7.6.1 Coating Techniques for Cutting Tools

Cutting tools are coated using two basic processes: CVD (chemical vapour deposition) and PVD (physical vapour deposition). Each strategy has its own set of pros and downsides.

1. Chemical vapour deposition

For many years, CVD coating was the original and most widely utilised coating process. The CVD process includes heating the substrate in a chemical reactor and then exposing it to a gas stream. On the hot substrate surface, the gases decompose and form a coating layer. In general, temperatures about 1,000^{0}C are required for the CVD process.

A common coating produces TiN (titanium nitride) + HCl by combining the three gases $TiCL_4$ (titanium tetrachloride), H_2 (hydrogen) and N_2 (nitrogen) (hydrogen chloride). The HCl produced as a by-product of the operation must be disposed of in accordance with strict environmental standards.

The CVD process has several advantages, including good layer adhesion and consistent layer distribution.

The downsides of the CVD process are high temperatures that harm the substrate, a limited number of appropriate coating materials because the coating material is fed in a gaseous form and long cycle periods.

2. Physical vapour deposition

PVD coating is a modern tool coating technology that is gaining popularity in the industry. The PVD process involves moving laminate material in a vacuum from a source to the substrate via a transport space.

The laminate material is vaporised by utilising either thermal or electrical energy from the power source, allowing the vaporised material to adhere to the substrate.

The PVD technique has the benefit of a wide range of appropriate coating materials and relatively moderate operating temperatures, around 450°C, allowing for the coating of sharp cutting edges.

The downsides are that coating internal surfaces is difficult (coating requires a line of sight from the laminate material to the substrate) and the substrate's surface requirements are substantially greater.

- **Methods of primary coating**

PVD coats various substrates using two basic technologies: the arc method (arc discharge) and the sputtering method (cathodic sputtering). Both approaches have one additional advantage in that the coating chambers are relatively simple to build. The arc process entails an electrical power source (similar to a lightning bolt) striking the laminate material and converting it from a solid to a liquid to a gaseous phase. The arc process entails an electrical power source (similar to a lightning bolt) striking the laminate material and converting it from a solid to a liquid to a gaseous phase. This technique has the advantage of having high layer rates (in relation to sputtering). However, because the laminate material exists in all three phases (solid, liquid and gas), droplets (minor liquid particles) can be formed during the process. These droplets never reach the gaseous state. The sputtering process employs a thermal energy source, which instantly converts the solid laminate to a gaseous state. Because the substance skips the liquid phase, no droplets form. Lower layer rates (in comparison to arc) result in longer cycle periods.

7.6.2 Coatings for Hard Materials

Most hard materials (coating is one of them) are made up of a metal and a metalloid. TiN (titanium nitride), TiCN (titanium carbo nitride), TiAlN (titanium aluminium nitride), AlTiN (aluminium titanium nitride) and AlCrN (aluminium chromium nitride) are some common cutting tool coatings. The periodic table of elements displays the metals and metalloids that are prospective coating possibilities. During the coating process, the smaller metalloid – in the case of TiN, nitrogen or N – lodges itself in the metal titanium's lattice vacancies (Ti). When transitioning to TiCN, a portion of the nitrogen (N) is replaced by carbon (C). The metals and metalloids necessary for the additional sample coatings may be determined using the same approach. One of the benefits of the PVD technique is that almost any metal can be coated because the metal is solid in the PVD chamber (CVD introduces in a gaseous phase). Of course, not all metals are ideal, but they can be used.

- **Structures of coating layers**

Layer structures have changed and improved significantly during the years of developing coatings. The coating technique typically employs five different layer structures. As the name implies, the monolayer structure consists of only one layer of covering. Tall columns of coating can be seen when examining the structure under a microscope. This is simple to apply, but it is also susceptible to cracking and damage. Consider a ball striking a number of columns. Columns will begin to collapse, and the break will easily penetrate all the way to the substrate. The multilayer structure is made up of numerous separate monolayer structures stacked on top of one another. Damascene steel is a historical example of this sort of structure, which combines the qualities of many elements to create a robust and hard surface. Nano-layer structures are similar to multilayer structures, but much smaller; the layers are at the atomic level of thickness. Nano-composite coatings employ technology similar to that of carbide cutting instruments. The nanostructure combines the toughness of the binder phase (cobalt with carbide as an example) with the hardness of the nano-composite coating. The gradient structure gains its capabilities by beginning soft and elastic at its core and then becoming hard and wear-resistant near the surface.

7.7 TOOL FAILURE

The term tool failure may be defined as "the condition of the tool when it starts giving unsatisfactory performance while cutting." Every cutting tool is subjected to wear while machining. A smooth, safe and cost-effective tool in machining can be achieved by:

- Preventing the cutting tools from failing prematurely and catastrophically
- Reducing the rate of wear on the tool in order to extend its life.

To achieve the aforementioned goals, one must first understand why and how cutting instruments fail. Cutting tools typically fail as a result of:

(i) Mechanical failure as a result of high stresses and shocks. Such tool failures are random and catastrophic in character, and thus exceedingly harmful.

(ii) Rapid dulling caused by plastic deformation as a result of high loads and temperatures. This form of failure also occurs quickly and is extremely harmful and unwelcome.

(iii) The cutting tool's flanks and rake surface gradually wear out. The first two causes of tool failure are extremely hazardous not only to the tool but also to the task and machine tool. As a result, these types of tool failure must be avoided by selecting appropriate tool materials and shape based on the work material and cutting condition.

(iv) However, failure due to gradual wear, which is unavoidable, cannot be avoided; it can only be slowed in order to extend the tool's service life. The cutting tool is removed as soon as it fails or, if possible, just before it completely fails. To do so, one must recognise that the instrument has failed or will fail soon.

One or more of the following situations indicate that the tool has failed or is likely to fail:

(a) In research and development laboratories
- Complete tool or tool tip breakage
- Significant fracture at the cutting edge
- A significant increase in cutting forces and/or vibration
- The average wear (flank or crater) meets the desired limit

(b) In the machining industry
- Excessive (in excess of the limit) current or power consumption
- Abnormal sound and/or excessive vibration (chatter)
- Complete tool failure
- Dimensional deviance that exceeds tolerance
- Fast deterioration of surface finish
- Poor chip formation.

7.8 CUTTING FORCES AND POWER CONSUMPTION

Cutting force is the resistance of the material against the intrusion of the cutting tool. The force directions and amplitudes differ in different cutting processes such as turning, milling, drilling, etc. performed in manufacturing machines. Power is the product of cutting force and cutting speed. The power required in a cutting operation is equal to the cutting force multiplied by the cutting speed.

Manufacturers as the owners of the machine production will save money and gain a healthier sustainability achievement if their power demands are minimised. However, most of them do not perceive power efficiency as a major priority to gain higher profits. In fact, it is estimated that two-thirds of the electrical power used by the machining process is utilised to operate motors and drives for cutting tools. Developing power efficiency of production processes involves information on the power demand as a means of the machining and cutting process itself. One of the processes generally required in the production is the turning process.

The importance of cutting force determination and prediction is well-known, being crucial for process optimisation in terms of tool performance, tool design, tool life and optimal parameter determination

7.8.1 Chip Thickness Ratio (or Cutting Ratio)

In practice, the thickness of the chip created is more than the actual depth of cut. The reason for this is that a chip moves upwards at a slower rate than the cut velocity. The shear plane angle (($\varnothing$)) has a direct effect on chip flow velocity; the smaller this angle, the slower the chip flow velocity and the thicker the chip.

$$\text{Chip thickness ratio (r)} = \frac{t}{t_c} \tag{7.1}$$

Further, the reverse of "r" is called chip reduction ratio or coefficient (K).

$$\text{Chip reduction ratio (K)} = \frac{1}{r} = \frac{t_c}{t} \tag{7.2}$$

Because "t_c" is always greater than "t,", the chip thickness ratio (r) is always smaller than 1. The greater the value of "r," the better the machining operation. Because orthogonal cutting is being examined, the breadth of the chip equals the width of the cut. Assuming that the volume of chip created equals the volume of metal cut, and that the breadth and specific gravity of metal are the same in both situations,

$$t.1 = t_c.l_c$$

where l = length of chip before cutting = ∏D/rev
l_c = length of chip

or
$$r = \frac{t}{t_c} = \frac{l_c}{l} \quad (7.3)$$

$$= \frac{Length\ of\ chip\ out}{Length\ of\ chip\ before\ cutting}$$

$$\left(or\, uncut\, chip\, \frac{length}{rev}\right)$$

Then,
$$K = \frac{1}{r} = \frac{l}{l_c} = \frac{t_c}{t} \quad (7.4)$$

Chip thickness ratio (or cutting ratio) r may also be defined as the ratio of chip velocity (V_c) to the cutting speed (V).
In Figure 7.9, the ∅ = shear angle and α = tool rake angle
Depth of cut = t = ML sin ∅
And chip thickness = t_c = ML cos (∅− ∝)
Then, chip thickness ratio (r):

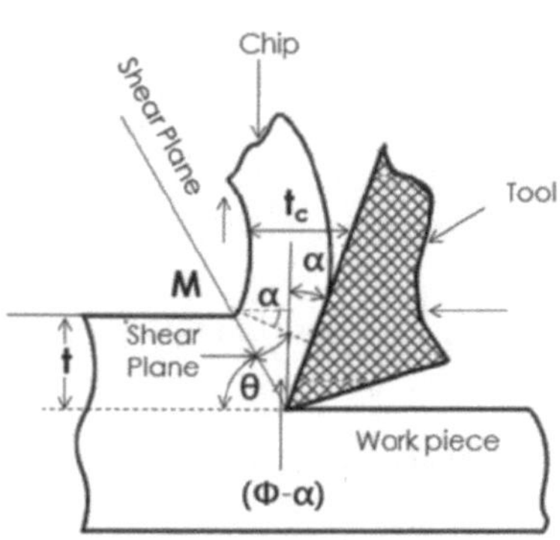

FIGURE 7.9 Shear plane, shear angle and rake angle.

$$r=\frac{t}{t_c}=\frac{\text{ML} \sin\varnothing}{ML\cos(\varnothing-\propto)}=\frac{\sin\varnothing}{cos\varnothing cos\propto + sin\varnothing sin\propto}=\frac{sin\varnothing}{\cos(\varnothing-\propto)} \tag{7.5}$$

(Dividing the numerator and denominator by $\sin\varnothing$)

$$=\frac{1}{cot\varnothing cos\propto+\sin\propto}$$

Or $r(\cot\varnothing\cos\propto+\sin\propto)=1$

Or $\cot\varnothing \cot\propto=\frac{1-rsin\propto}{r}$

Or $$\tan\varnothing\frac{rcos\propto}{1-rsin\propto} \tag{7.6}$$

Hence,

$$\text{Shear angle}(\varnothing)=\tan^{-1}\left[\frac{rcos\propto}{1-rsin\propto}\right] \tag{7.7}$$

7.8.2 Velocity Relationship in Orthogonal Cutting

The following three velocities are involved in orthogonal cutting:

V = cutting velocity or velocity of tool relative to work
V_c = velocity of chip flow or velocity of chip flow relative to tool
V_s = velocity of shear or velocity of displacement of the chip along the shear plane relative to work

Cutting velocity (V) and rake angle (α) are both known. The following method is used to determine V_c and V_s, as well as the relationship between the three velocities. Refer to Figure 7.9 for a velocity diagram in which:

$$\vec{V}=\overrightarrow{V_c}+\overrightarrow{V_s}$$

By applying the sine rule,

$$\frac{V_c}{sin\varnothing}=\frac{V_s}{\sin\left[(90^0-\varnothing)+(\varnothing-\propto)\right]}=\frac{V}{\sin[(180-(90-\varnothing+\varnothing-\propto+\varnothing)]}$$

Or

$$\frac{V_c}{sin\varnothing} = \frac{V_s}{cos \propto} = \frac{V}{\cos(\varnothing - \propto)}$$

Hence,

$$\text{Velocity of chip flow } (V_c) = \frac{V.sin\varnothing}{cos(\varnothing - \propto)} \tag{7.8}$$

And

$$\text{Velocity of shear } (V_s) = \frac{V.cos \propto}{cos(\varnothing - \propto)} \tag{7.9}$$

$$\text{Since } r = \frac{sin\varnothing}{\cos(\varnothing - \propto)}, \text{ then}$$

Velocity of chip flow (V_c) = cutting velocity (V) × chip thickness ratio (r),

$$V_c = V.r \tag{7.10}$$

7.8.3 Forces Acting on the Chip in Orthogonal Cutting (Merchant's Analysis)

Merchant established a relationship between various forces acting on the chip during orthogonal metal cutting, but only under the following conditions:

- (i) The cutting velocity is always constant.
- (ii) The cutting edge of the tool stays sharp throughout the cutting process, with no contact between the workpiece and the tool flank.
- (iii) Chip does not flow in the other direction.
- (iv) Only continuous chips are manufactured.
- (v) There is no built-up edge.
- (vi) The chip's inertia force is not taken into account.
- (vii) The width of the tool is greater than the width of the cut.
- (viii) The chip's behaviour is similar to that of a free body in a stable equilibrium due to the influence of two resultant forces that are equal, opposite and collinear.

The aforementioned assumptions were eventually adjusted due to a number of faults and practical challenges. In orthogonal cutting, the forces acting on the chip are caused by the cutting force (R) delivered through the tool. With reference to Figure 7.10, these forces are described below.

F_s = Shear force or metal resistance to shear during chip formation. It acts along shear plane.
F_c = Backing up or compressive normal force exerted by workpiece on the chip. It acts normal to shear plane.
N = Force exerted by tool on the chip. It acts normal to the tool face.
F = Frictional force (or μN) or resistance of the tool against the chip flow. It acts along the tool face. Here μ is a kinetic coefficient of friction between the tool face and chip and F/N = tanβ, where β is the angle of friction.

The chip's free-body diagram is depicted in Figure 7.10. Forces F_s and F_c have their resulting force R, whereas forces F and N do not. R and R′ are the resulting forces, which are equal in magnitude, opposite in direction and collinear. As a result, the chip can be viewed as an autonomous body held in mechanical equilibrium by the action of two equal and opposing forces R′ exerted on the chip by the workpiece and the tool.

As previously stated, the resultant force R comprises two components, F_c and F_s. Let the resultant force R be further resolved into two further component forces (FH) and (FV), as shown in Figure 7.11 and explained below.

FH denotes the horizontal component of the resultant force (R). It is referred to as cutting force (F_t) or tool tangential force on workpiece. FV denotes the vertical component of the resultant force (R). It is referred to as axial feed force (F_f or F_a) or thrust force acting in the opposite direction of feed. F_t = FH will be regarded as the cutting force in the following discussion, while F_f = FV = F_a will be considered as the feed force.

For the convenience of further relationships between various forces, the two triangles of forces of the free-body diagram of the chip in above have been considered together in Figure 7.12, called the Merchant circle diagram. For the sake of simplicity, the cutting forces are plotted at the tool point instead of their actual point of application and a composite cutting force circle is obtained wherein the diameter of the circle is R (note that R = R'). From this diagram, various force relationships can be obtained.

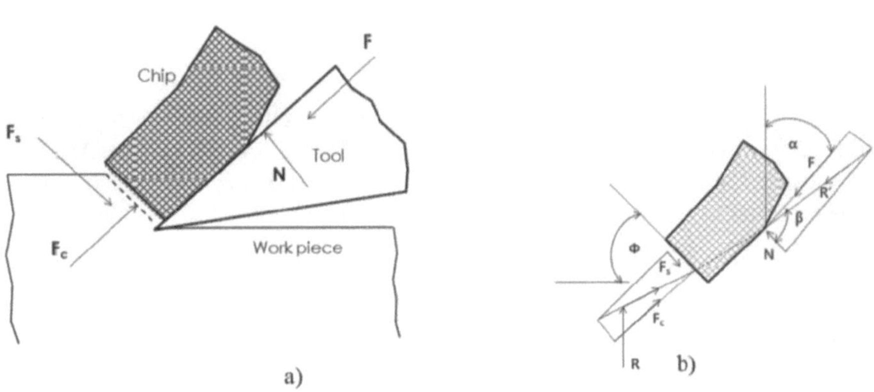

FIGURE 7.10 Forces acting on the chip in orthogonal cutting.

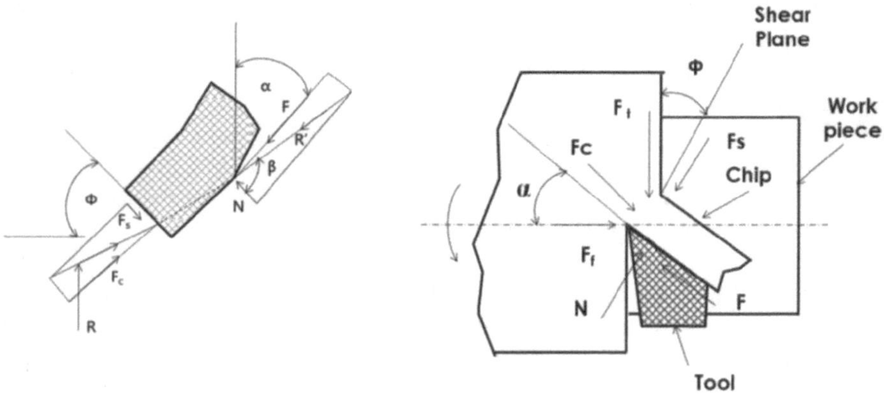

FIGURE 7.11 Free body diagram and force system.

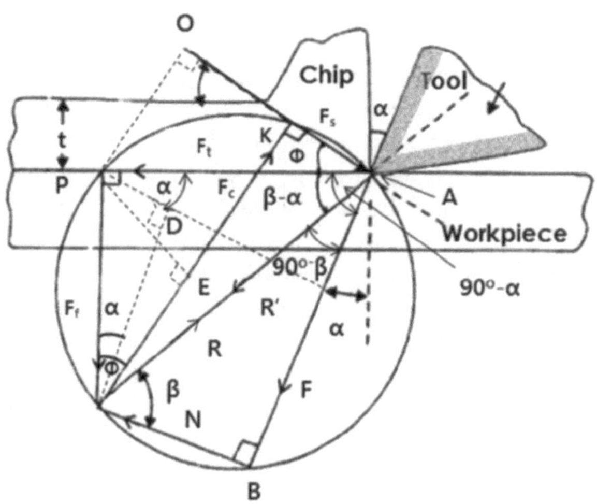

FIGURE 7.12 Merchant's circle.

Figure 7.13 shows velocity relationship in orthogonal cutting. The cutting force (F_t) and feed force (F_f or F_a) can be found with the help of a force dynamometer. When laid out as in Figure 7.12, resultant R can be found easily. Knowing the rake angle (α) of the tool, forces F and N can be determined. Shear angle (ϕ) can be found as

$$\tan \varnothing = \frac{\cos \propto}{K - \sin \propto} \tag{7.11}$$

where $\propto$ = rake angle
K = chip reduction coefficient

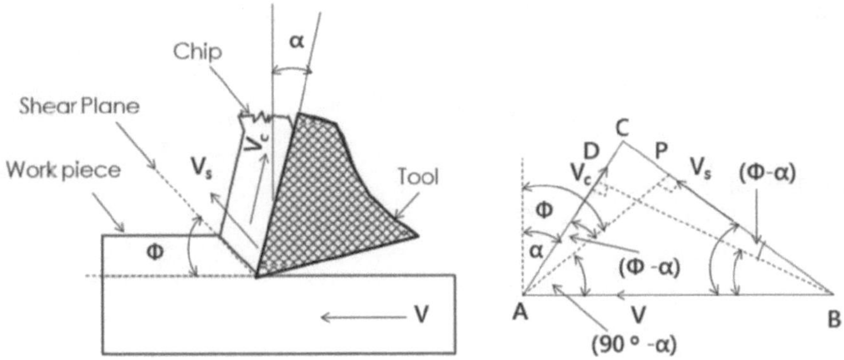

FIGURE 7.13 Velocity relationship in orthogonal cutting.

$$= \frac{Chipthickness(t_c)}{Uncutthickness(orfeedinturning)(t)}$$

Knowing the above, all other component forces on the chip may be determined from the geometry of Figure 7.12 wherein

F_t = cutting force

F_f = feed force

F_c = Compresssive or Normal force on shear plane

F_s = Shear force on shear plane

F = frictional force along the a rake angle of tool

N = Normal force at the rake face of tool

Now, F = AQ + QB = AQ + DC (as QB = DC)

$$\text{Or} \quad F = F_t \sin\propto + F_f \cos\propto \tag{7.12}$$

Then, $N = PQ - PD = F_t cos \propto - F_f \sin \propto$

$$\text{or} \quad N = F_t cos \propto - F_f \sin \propto \tag{7.13}$$

Further, $F_s = AO - OK = AO - PE = F_t cos\varnothing - F_f \sin\varnothing$

or $$F_s = F_t cos\varnothing - F_f \sin\varnothing \tag{7.14}$$

Then, $F_c = CK = CE + EK$
$= CE + PO$ (as $EK = PO$)

or $$F_c = F_f cos\varnothing + F_t \sin\varnothing \tag{7.15}$$

and $$F_t = R\cos(\beta - \propto) \tag{7.16}$$

$$Ff = R\sin(\beta - \propto) \tag{7.17}$$

$$\text{Also}, F_s = R\cos(\beta - \propto + \varnothing) \tag{7.18}$$

and $$\frac{F_t}{F_s} = \frac{Rcos(\beta - \propto)}{Rcos(\beta - \propto + \varnothing)} = \frac{cos(\beta - \alpha)}{cos(\beta - \alpha + \varnothing)}$$

or $$F_t = F_s\left[\frac{cos(\beta - \alpha)}{cos(\beta - \alpha + \varnothing}\right] \tag{7.19}$$

and $$\frac{F}{N} = \frac{F_t sin\alpha + F_f cos\alpha}{F_t cos\alpha - F_f sin\alpha}$$

or $$\frac{F}{N} = \frac{F_f + F_t tan\alpha}{F_t - F_f tan\alpha} = \tan\beta \tag{7.20}$$

or $\frac{F}{N} = \tan\beta = \mu$ (kinetic coefficient of friction)
where β angle of friction $= \tan^{-1}\mu$

Also, $\tan PAC = \tan(\beta - \alpha) = \frac{CP}{AP} = \frac{F_f}{F_t}$

OR $$\frac{F_f}{F_t} = \tan(\beta - \alpha) \tag{7.21}$$

Merchant developed a relationship between shear angle (ϕ), angle of friction (β) and tool rake angle (α) as follows:

$$2\varnothing + \beta - \alpha = C \tag{7.22}$$

7.8.4 Stress and Strain on the Chip

Figure 7.14 shows geometry of chip formation in orthogonal cutting process. Chips are produced due to the plastic deformation of the metal; they experience stress and strain. Two forces, F_c and F_s (perpendicular to each other), act at the shear plane, wherein:

A = cross-sectional area of uncut chip = bt

where b = width of cut and t = depth of cut or uncut chip thickness

A_s = shear plane area or
$A = A_s \sin \varnothing$

- **Mean normal stress (σ)**

$$\sigma = \frac{F_c}{A_s} = \frac{F_c}{A / sin\varnothing} = \frac{F_c sin\varnothing}{A} \tag{7.23}$$

Putting the values of $F_c = F_f cos\varnothing + F_t \sin \varnothing$

$$\text{Mean normal stress}(\sigma) = \frac{\left(F_f cos\varnothing + F_t sin\varnothing\right) sin\varnothing}{A} \tag{7.24}$$

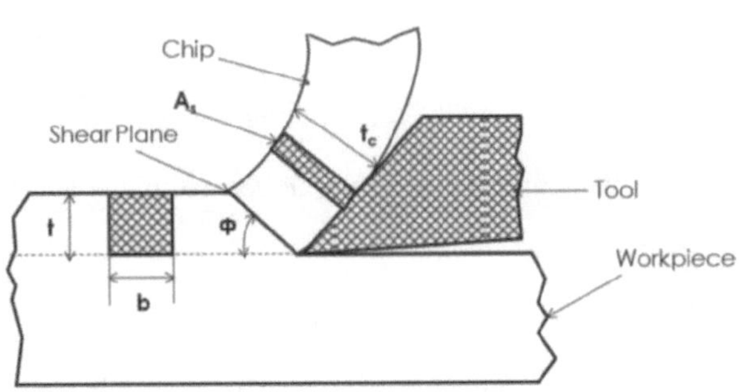

FIGURE 7.14 Geometry of chip formation in orthogonal cutting.

Mean shear stress (τ)

$$\tau = \frac{F_s}{A_s} = \frac{F_s sin\varnothing}{A}$$

Or

$$F_s = \frac{\tau.b.t}{sin\varnothing} \quad (7.25)$$

It can be shown that

$$\tau = \frac{F_t cos\varnothing - F_f sin\varnothing}{A_s} = \frac{\left(F_t cos\varnothing - F_f sin\varnothing\right) sin\varnothing}{A} \quad (7.26)$$

Shear strain (ε)

$$\varepsilon = \cot\varnothing + \tan\left(\varnothing - \alpha\right) \quad (7.27)$$

$$\text{And also } \varepsilon = \frac{cos\alpha}{sin\varnothing.cos\left(\varnothing - \alpha\right)} \quad (7.28)$$

7.8.5 Forces of a Single-Point Tool

The workpiece metal provides resistance to the cutting tool during metal cutting. The cutting force applied by the tool overcomes this resistance. This force's work in cutting is spent shearing the chip from the workpiece, deforming the chip and overcoming the chip's friction on the tool face. The cutting force is determined by the workpiece material, feed, depth of cut, cutting speed, tool angles and lubrication or coolant utilised. Figure 7.15 depicts the forces acting on a single-point turning tool during oblique or conventional cutting. F_a = axial feed force or thrust force acting in horizontal plane parallel to the axis of work but in the direction opposite to the feed.

F_r = radial force acting in horizontal plane along a radius of work, i.e. along the axis of tool.

F_t = cutting force or tangential force acting in vertical plane and tangential to the work surface.

F_t has the greatest magnitude of the three forces listed above, whereas F_r has the smallest. For turning operations, F_a ranges between 0.3 and 0.6 F_t and F_r ranges between 0.2 and 0.4 F_t. The components F_t and F_a are easily calculated using an appropriate force dynamometer. The resultant force (R) is calculated as follows:

$$R = \sqrt{F_a^2 + F_t^2 + F_r^2} \quad (7.29)$$

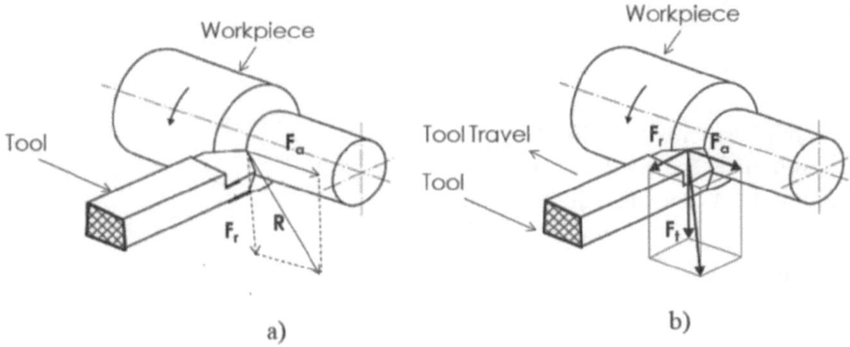

FIGURE 7.15 Forces acting on a cutting tool: (a) orthogonal cutting and (b) oblique cutting.

In orthogonal cutting, only two forces (F_a and F_t) come into play and F_r is zero (Figure 7.15a). The resultant force (R) is as follows wherein F_a and F_t are axial (feed) force and cutting force (or tangential force), respectively

$$R = \sqrt{F_a^2 + F_t^2} \tag{7.30}$$

$$\text{Torque } (T) = \frac{F_t.D}{2 \ X \ 1000} (\text{Nm}) \tag{7.31}$$

where D = diameter of work, mm, F_t = cutting force, N
Heat produced (H)
Heat produced = work done in metal cutting

$$H = \frac{F_t X V}{60 X 1000}, kN \frac{n}{s} orkWorkJ / s \tag{7.32}$$

where V = cutting speed, m/min, F_t = cutting force, N. Heat produced is also equal to the following where F_t is in kgf and V is m/min

$$H = \frac{F_t.V}{427}, \ kcal/min \tag{7.33}$$

Power required

$$P = \frac{F_t.V}{60 \ X \ 1000 \ X \eta}, \ kW \tag{7.34}$$

where F_t = cutting force, N
V = cutting speed, m/min
η = efficiency (say 80 to 90%)

Metal removal rate (MRR) = V.b.t, cm^3/min (7.35)

where

V = cutting velocity, cm/min
b = width of cut (cm) or feed rate, cm/rev.
t = depth of cut or uncut chip thickness, cm

$$\text{MPR max} = \frac{\textit{Max. power available at machine spindle } (kW)}{\textit{Power required}\left(\dfrac{\frac{kW}{cm^3}}{\min}\right)},\ cm^3/\min \quad (7.36)$$

7.8.6 Popular Theories on the Mechanics of Metal Cutting

Several relationships have already been developed for shear angle (ϕ), friction angle (β) and rake angle (α). Several researchers have put in a lot of effort to build a realistic relationship between α, ϕ and β and have developed several hypotheses with subtle differences in their assumptions and outcomes. Among the more notable theories are: (i) Earnst–Merchant theory, (ii) Merchant theory, (iii) Stabler theory and (iv) Lee and Shaffer's theory. The two theories listed below are most popular among metal cutting engineers.

Earnst–Merchant hypothesis is a theory proposed by Earnst and Merchant

This theory is founded on the idea of lowest energy consumption, which states that when cutting, the metal shear should occur in the direction with the least amount of energy required for shearing. Furthermore, the behaviour of machined metal is similar to that of an ideal plastic. Also, shear stress is highest and constant in the shear plane, and it is independent of shear angle. They devised the following relationship:

$$\phi = \frac{\pi}{4} - \frac{\beta}{2} + \frac{\alpha}{2} = \frac{\pi}{4} - \frac{1}{2}(\beta - \alpha) \quad (7.37)$$

Lee and Shaffer's theory

The process of orthogonal cutting has been explored in this theory using the theory of plasticity for an ideal hard plastic material. The following assumptions have been made:

(i) The workpiece metal in front of the cutting tool behaves as if it were a perfect plastic material.
(ii) Metal deformation occurs on a single shear plane.
(iii) The chip is not hardened.
(iv) At the shear plane, the chip separates from the parent metal of work.

$$\phi = \frac{\pi}{4} + \alpha - \beta = 45^0 + \alpha - \beta$$

Or $$\phi + \beta - \alpha = 45^0 \tag{7.38}$$

This was further modified as:

$$\phi = \frac{\pi}{4} + \alpha + \theta - \beta \tag{7.39}$$

where θ covers the changes in different parameters because of the formation of built-up edge

7.8.7 Tool Life

Tool life is defined as the time span between two subsequent grindings or resharpenings of the tool during which the tool performs satisfactorily. Thus, tool life is essentially a tool's functional life. Tool life can be used to evaluate the performance of a tool material, evaluate the machinability of a workpiece material and determine the cutting conditions. The following are some examples of how tool life is expressed: (i) the time interval in minutes between two successive tool grindings; (ii) the number of components machined in the two successive grindings; and (iii) the amount of metal removed between two grindings. Volume of metal removed per minute (V_m):

$$V_m = \pi D.t.f.N,\ \text{mm}^3/\text{min} \tag{7.40}$$

where D = diameter of workpiece, mm
t = depth of cut, mm
f = feed, mm/rev
N = number of revolutions of job per minute

If T is the time for tool failure in minutes, then total volume of metal removed up to tool failure

$$= \pi D.t.f.N,T\ \text{mm}^3 \tag{7.41}$$

We know that

$$\text{Cutting speed} = \text{V}\frac{\pi DN}{1000}, \text{m / min}$$

Or $$\pi DN = 1000\ \text{V}$$

Substituting

$$\text{Tool life } (T_L) = 1000V.f.\ T.\ t\ (mm^3) \tag{7.42}$$

7.8.9 Factors Affecting Tool Life

The following are some of the factors that influence tool life:

(i) The cutting speed has the biggest impact on tool life. The shorter the tool life, the faster the cutting speed. The reduction in tool life associated with an increase in cutting speed is parabolic. According to F.W. Taylor's studies, the relationship between cutting speed and tool life is as follows:

$$V.\ T^n = C \tag{7.43}$$

where V = cutting speed, m/min

T = tool life, min
n = an exponent (also called tool life index) depending largely on tool material and
n = 0.1 to 0.15 for an HSS tool, 0.2 to 0.5 for a cemented carbide tool, 0.6 to 1.0 for carbide tools.
C = machining constant which is equal to cutting speed (m/min) that will give a tool a life of 1 minute

Normally, the feed rate is regulated between 0.2 and 0.8 mm/rev. The depth of cut for rough turning is kept at 2–5 mm, and it is kept at 0.5–1 mm for finishing.

(ii) The influences of feed and depth of cut on tool life are comparable. Increases in feed or depth of cut have less of an impact on tool life than increases in cutting speed.

(iii) Tool geometry, specifically tool angles, affects tool performance and life. Rake angle increases tool life (to a point) by reducing cutting force and heat generation. A rake angle that is too large, on the other hand, weakens the cutting edge. The optimal rake angle is between –5 and +10°, with the minus sign denoting negative rake, which results in a stronger tool since cemented carbide and ceramic tools have negative rake and can operate at higher cutting speeds. Relief angles or clearance angles are used on cutting tools to prevent the tool flank from rubbing against the machined surface of the work. As a result, these angles limit heat generation, which enhances tool life. Large relief angles, on the other hand, result in a less effective instrument. Up to a point, increasing the cutting edge angles (both front and end) at the same time improves tool life. The nose radius adds to a more polished surface and a stronger tool.

(iv) Tool material: At all speeds, an optimal tool material eliminates most material from the task. There is, however, no such thing as an ideal tool material. As a result, the higher the hot hardness and toughness of a tool material, the longer its tool life.

(v) The work material and its microstructure have a considerable impact on tool performance; for example, free graphite and ferrite in cast iron and steel contribute to softness. The increase in cutting temperature and power usage is related to the workpiece material's hardness. The harder the work material, the more the tool wears and the shorter the tool life. The surface of the work material also has an impact on the machining operation and tool life (scaly or smooth).

(vi) Tool life is influenced by the manner of cutting, whether continuous or intermittent. In intermittent cutting, the tool is subjected to repetitive impact loading. For increasing tool life, continuous cutting is better.

(vii) The stiffnesses of the machine tool and the work are essential for preventing unwanted vibrations during cutting.

(viii) The proper type and quantity of cutting fluids contribute to extended tool life by keeping the tool's cutting edge cool and lubricated.

7.9 COOLANT TECHNOLOGY

Cutting fluids have been widely used in machining to achieve the following results:

- Reduce friction and wear, thereby prolonging tool life and improving workpiece surface finish
- Cool the cutting zone, thereby increasing tool life and lowering workpiece temperature and thermal distortion
- Minimise the utilisation of force and energy
- Flush the chips away from the cutting zone to keep them from interfering with the cutting operation, particularly during drilling and tapping
- Prevent corrosion induced by the environment on the machined surface.

Depending on the type of machining process, the cutting fluid used may be a coolant, a lubricant or both. A multitude of factors influence cutting fluid efficacy, including the kind of machining operation, tool and workpiece materials, cutting speed and application method.

Cutting fluids are available in a variety of forms.

Cutting fluid uses in machining can be divided into four categories:

1. Flooding: Flow rates for single-point tools typically range from 10 L/min (3 gal/min) to 225 L/min (60 gal/min) per cutter for multiple-tooth cutters, as in milling. Fluid pressures ranging from 700 to 14,000 kPa (100 to 2000 psi) are used to flush away chips created in some activities, such as drilling and milling, to avoid interfering with the operation.
2. Mist: This form of cooling, similar to utilising an aerosol can, delivers fluid to inaccessible places and improves the visibility of the workpiece being machined (compared with flood cooling). It is particularly effective with

water-based fluids at air pressures ranging from 70 to 600 kPa (10 to 80 psi). It does, however, have a limited cooling capability. Mist application needs venting to avoid the machine operator and people around from inhaling airborne fluid particles.

3. High-pressure system: As the speed and power of current computer-controlled machine tools have increased, heat generation in machining has become a serious problem. The use of high-pressure refrigerated coolant systems to accelerate the rate of heat evacuation from the cutting zone is very beneficial. High pressures are also employed to supply the cutting fluid through specially engineered nozzles, which aim a forceful jet of fluid at the zone, particularly into the clearance or relief face of the tool. The pressures used, which are typically between 5.5 and 35 MPa (800 and 5000 psi), operate as a chip breaker in instances where the chips created would otherwise be long and continuous, interfering with the cutting operation. To avoid damage to the workpiece surface from the impact from any particles present in the high-pressure jet, the contaminant size in the coolant should not exceed the proper size. Also ongoing fluid filtering is required to preserve quality.
4. Through the cutting tool system: The difficulty of delivering fluids into the cutting zone and flushing away chips has been used to describe the severity of certain machining procedures. For a more effective application, tiny passageways in cutting tools and tool holders can be created through which high-pressure cutting fluids can be applied.

This approach has two applications.

(a) Gun drilling with a long, tiny hole through the drill's body, and boring bars have a long hole through the shank (toolholder) into which an insert is secured. Similar concepts have been developed for cutting tools and inserts, as well as for distributing cutting fluids through the machine tool's spindle.

Cutting fluid effects: Considerations such as the consequences of a cutting fluid on the environment should also be considered, Machine tools and workpiece materials, biology concerns and the surroundings should be taken into account. When choosing a cutting fluid, consider whether the machined component will be subjected to stress and detrimental consequences, which could result in stress-corrosion cracking. This is especially true for cutting fluids containing sulphur and chlorine compounds. Cutting fluids containing sulphur, for example, should not be utilised with nickel-based alloys.

(b) Chlorine-containing fluids should not be utilised with titanium: Cutting fluids can also have a negative impact on machine tool components, thus their compatibility with various metallic and non-metallic materials in the machine tool must be examined.

Machined parts should be cleaned and washed to remove any cutting-fluid residue. This procedure can be time-consuming and expensive. As a result, for simplicity of cleaning and filtering, water-based, low-viscosity fluids are becoming popular. Because the machine-tool operator is frequently in close proximity to cutting fluids, the health impacts of fluid contact on the operator should be a main consideration.

Cutting fluid mist, fumes, smoke and odours can cause serious skin reactions and respiratory difficulties, particularly when utilising fluids with chemical elements such as sulphur, chlorine, phosphorus, hydrocarbons, biocides and different additives. Much progress has been achieved in assuring the safe use of cutting fluids in manufacturing plants, including the reduction or elimination of their use by implementing the most recent trends in dry or near-dry machining techniques. Cutting fluids (as well as other metalworking fluids used in manufacturing activities) can undergo chemical changes over time if they are used regularly. These changes could be the result of environmental factors or pollution from a variety of sources, such as metal chips, fine particles created during machining, and tramp oil (from leaks in hydraulic systems, oils on sliding members of machines and lubricating systems for the machine tools). The modifications involve the growth of microorganisms (bacteria, moulds and yeast), which becomes an environmental danger and also has an adverse effect on the properties and effectiveness of the cutting fluids. For clarifying used cutting fluids, several procedures (such as settling, skimming, centrifuging and filtration) are available. Recycling entails treating fluids with various additives, agents, biocides and deodorisers, as well as treating water (for water-based fluids). These fluids must be disposed of in accordance with national, state and local rules and regulations.

Cutting fluids have been used extensively in machining operations to achieve the following results:

- Reduce friction and wear, thus improving tool life and the surface finish of the workpiece
- Cool the cutting zone, thus improving tool life and reducing the temperature and thermal distortion of the workpiece
- Reduce forces and energy consumption
- Flush away the chips from the cutting zone, thus preventing the chips from interfering with the cutting process, particularly in operations such as drilling and tapping
- Protect the machined surface from environmental corrosion.

Depending on the type of machining operation, the cutting fluid needed may be a coolant, a lubricant or both. The effectiveness of cutting fluids depends on a number of factors, such as the type of machining operation, tool and workpiece materials, cutting speed and the method of application. The need for a cutting fluid depends on the severity of the particular machining operation, which may be defined as the level of temperatures and forces encountered and the ability of the tool materials to withstand them, the tendency for built-up edge formation, the ease with which chips produced can be removed from the cutting zone and how effectively the fluids can be applied to the proper region at the tool–chip interface.

7.9.1 Types of Cutting Fluids

Briefly, four general types of cutting fluids are commonly used in machining operations:

1. Oils (also called straight oils), including mineral, animal, vegetable, compounded and synthetic oils, typically are used for low-speed operations where temperature rise is not significant.
2. Emulsions (also called soluble oils), a mixture of oil and water and additives, generally are used for high-speed operations because the temperature rise is significant. The presence of water makes emulsions highly effective coolants. The presence of oil reduces or eliminates the tendency of water to cause oxidation.
3. Semi-synthetics are chemical emulsions containing a little mineral oil, diluted in water, and with additives that reduce the size of oil particles, making them more effective.
4. Synthetics are chemicals with additives, diluted in water, and containing no oil. Because of the complex interactions among the cutting fluid, the workpiece materials, temperature, time and cutting-process variables, the application of fluids cannot be generalised.

Machined parts should be cleaned and washed to remove any cutting-fluid residue. This procedure can be time-consuming and expensive. As a result, for simplicity of cleaning and filtering, water-based, low-viscosity fluids are becoming popular. Because the machine-tool operator is frequently in close proximity to cutting fluids, the health impacts of fluid contact on the operator should be a main consideration.

Cutting fluid mist, fumes, smoke and odours can cause serious skin reactions and respiratory difficulties, particularly when utilising fluids with chemical elements such as sulphur, chlorine, phosphorus, hydrocarbons, biocides and different additives. Much progress has been achieved in assuring the safe use of cutting fluids in manufacturing plants, including the reduction or elimination of their use by implementing the most recent trends in dry or near-dry machining techniques. Cutting fluids (as well as other metalworking fluids used in manufacturing activities) can undergo chemical changes over time if they are used regularly. These changes could be the result of environmental factors or pollution from a variety of sources, such as metal chips, fine particles created during machining and tramp oil (from leaks in hydraulic systems, oils on sliding members of machines, and lubricating systems for the machine tools). The modifications involve the growth of microorganisms (bacteria, moulds and yeast), which becomes an environmental danger and also has an adverse effect on the properties and effectiveness of the cutting fluids. For clarifying used cutting fluids, several procedures (such as settling, skimming, centrifuging and filtration) are available. Recycling entails treating fluids with various additives, agents, biocides and deodorisers, as well as treating water (for water-based fluids). These fluids must be disposed of in accordance with national, state and local rules and regulations.

7.10 RECONDITIONING OF THE TOOLS

High-performance cutting tools can boost efficiency and production, but they can also deplete tooling budgets. When these tools get worn or damaged, they must often be reground and reconditioned in order to be cost-justified. An effective reconditioning programme minimises tooling expenses by extending the life of the tool as much as possible.

While high-performance cutting tools and modern machine tools are expensive, they are required for performing high-speed machining, hard milling, thread milling and helical pocketing – operations that were unusual not long ago. These machining processes necessitate the use of cutting tools manufactured to stringent standards and fine tolerances, which adds to the complexity of the reconditioning process.

Reconditioning has become so intricate that it is unusual to find a shop that does not send its tools to be reconditioned. This is due to the fact that reconditioning high-performance cutting equipment takes more than a standard tool and cutter grinder and a pot of hot wax. Monoset grinders and vitrified wheels have been replaced by multi-axis CNC tool grinders and super-abrasive wheels.

Complex tool geometries, multilayer coatings and difficult tool production procedures are frequently kept under wraps. Only the tool manufacturer can accurately recondition its tools in some situations, but most major toolmakers offer reconditioning services that return equipment to OEM specifications and come with a performance guarantee.

A cost-effective reconditioning program must be carefully designed and actively handled. The first step toward developing a successful reconditioning program is proper tool application during the machining process.

Major toolmakers often provide engineering support and aid in the development of efficient machining techniques. Engaging toolmakers and learning how to use their tools correctly helps to prevent damage caused by misuse. Damaged flutes and chipped corners can swiftly lead to catastrophic tool failure and the scrapping of the machined product. Furthermore, because regrinding broken tools is more difficult, they cost more to recondition than worn tools.

Tools that have been refurbished should perform just as well as new tools. If performance suffers after reconditioning, it may be time to look for a new vendor. Even a significant reduction in reconditioning costs per tool can be outweighed by the cost of lost production caused by a badly reconditioned tool. Shop employees frequently avoid using refurbished tools that perform poorly, which defeats the objective of reconditioning.

The following instructions should be followed when reconditioning a tool.

1. Sharpen the tool to the original geometric parameters. Restoring the original tool geometry will assist the tool in producing consistent results in future uses. Tool resharpening with computer numerical control (CNC) grinding machines has made it easier to restore a tool's original geometry.

2. Fine-grind the cutting edges and surfaces. Rough finishes left by poor and abusive regrinding impair resharpened tool performance. Tops of ridges left by rough grinding on coated tools will break away during early tool use, leaving uncoated and unprotected surfaces that will cause premature tool failure.
3. Remove any remaining burrs from the resharpened cutting edges. Premature failure can occur if a tool with a burr is coated because the burr will break away in the first cut, leaving an uncoated surface exposed to wear.
4. Avoid resharpening procedures that overheat and burn or melt (referred to as glazed over) the tool surfaces, since this will cause coating adhesion issues. Similar issues arise when instruments are polished or wire-brushed.

Each recoating costs around one-fifth of the cost of purchasing a new tool. Depending on the quantity of pieces being machined, recoating can reduce tooling costs per workpiece by 20–30%.

7.11 SPECIAL CASE STUDY: DEVELOPMENT OF A HYBRID TOOL FOR BORING, REAMING AND CHAMFERING OPERATION

7.11.1 Introduction

Grey cast iron casting (GCIC) materials are widely used, particularly in automotive industries. However, the high cost of processing these materials limits the use of their improved mechanical properties. Tool life is one of the most important factors in machining operations of such materials and it is mainly affected by cutting conditions including the cutting speed, depth of cut, insert material and cooling environment, along with the length and diameter of the tool body. In addition, modern industry is moving towards automating the manufacturing processes. Therefore, tool life monitoring is important to achieve an efficient manufacturing process. The gear housing is a mechanical housing that surrounds the mechanical components of a gear box. It provides mechanical support for the moving components, mechanical protection from the outside world for those internal components, and a fluid-tight container to hold the lubricant which can wash the components.

One of the leading automobile manufacturers was using a CNC workstation for machining of gear housing which is an important element in tractor assembly. The processing of the component was done on a five-axis CNC machine with capacity to accommodate 60 tools in a tool magazine. The processing of components of gear housing requires three operations to be performed, i.e. boring, reaming and chamfering. These tools were stored in a tool magazine and operations were performed sequentially by removing the concerned article from the tool magazine. The machine cycle time which consists of time to load the machine, the actual machining or machine time, and the unloading time of the component, was too high. Hence productivity was hampered with more time required during machining and replacing the tool. The need for reducing the cycle time and tooling was observed for this workstation.

7.11.2 Development of the Tool

This rigorous study was undertaken by the researcher Dr. Rajendra Kadu under his supervisors Dr. G. K. Awari (author), Dr. J. P. Modak and the industry-academia mentor Dr. Prashant Kadu. In this study, a tool wear prediction model during the boring machining operation of grey cast iron was studied. This was based on the monitoring of tool performance in controlled machining tests with measurements of tool life, surface finish, bore size variation, cutting time and load on spindle in terms of % current under different combinations of cutting parameters (cutting speed, depth of cut, tool nose radius, length and diameter of tool, tool material, and coolant pressure and concentration). The influence of cutting parameters on the tool life was studied experimentally by performing more than 120 cutting tests. A prediction model was then developed to predict tool wear. The basic steps used in generating the model adopted in the development of the prediction model were: collection of data; analysis, pre-processing and feature extraction of the data; design of the prediction model; training of the model; and finally testing the model to validate the results and its ability to predict tool wear.

The geometry of three tools used separately in an operation was analysed and it was decided to combine the three operations together to reduce the cycle time. The geometry for a hybrid tool which can combine these operations was designed and manufactured. Practical runs were conducted on a hybrid tool where three operations, i.e. boring, reaming and chamfering, were combined in a single tool.

In machining, boring is the process of enlarging a hole that has already been drilled (or cast) by means of a single-point cutting tool. Reaming is a finishing operation of high-precision holes performed with a multi-edge tool. High surface finish, superb hole quality and close dimensional tolerance are achieved at high penetration rates and small depths of cut, and chamfering is the operation of bevelling the extreme end of the workpiece. The form tool used for taper turning may be used for this purpose. Chamfering is an essential operation after thread cutting so that the nut may pass freely on the threaded workpiece. The evolution of boring machining operation properties using different parameters is a complex phenomenon. There are many factors (like cutting speed, depth of cut, insert material and cooling environment along with length and diameter of the tool body) affecting the performance of a cast iron boring machining operation resulting in poor tool life.

This case study presents an experimental investigation and sequential classical experimentation technique used to perform experiments for various independent parameters. An attempt of the mini–max principle has been made to optimise the range-bound process parameters for minimising cutting time and surface finish during cast iron boring machining operation. The test results proved that cutting time and surface finish were significantly influenced by changing important four parameters. The process parameters grouped these terms and suggested the effective guidelines to the manufacturer for improving tool life by changing any one or all from the available process parameters. The boring process is carried out on a horizontal machine or vertical machine, and the automatic boring process is carried out by CNC control. The boring process is illustrated in Figures 7.16 and 7.17.

FIGURE 7.16 Boring operation.

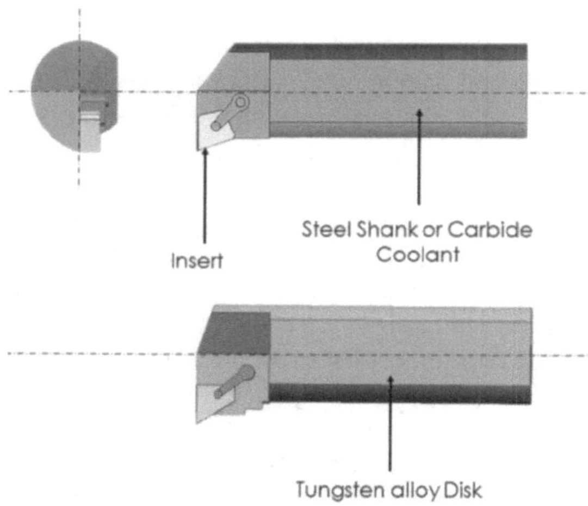

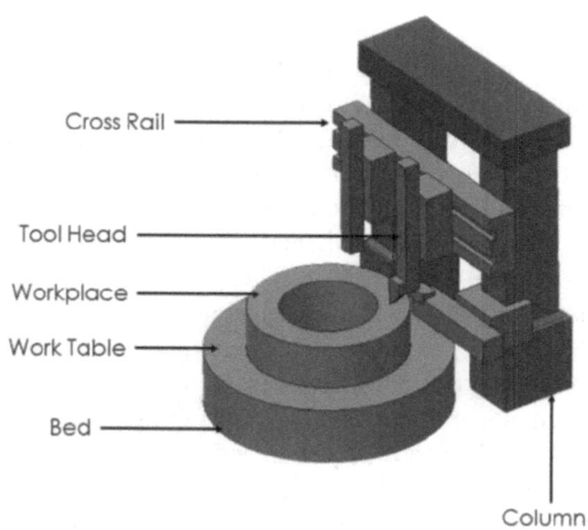

FIGURE 7.17 Boring process on a vertical CNC machine.

FIGURE 7.18 Hybrid tool.

The inventors of this tool, Dr. Rajendra S. Kadu, Dr. G. K. Awari (author), Dr. J. P. Modak and Dr. Prashant S. Kadu, finally obtained a patent titled "Hybrid Tool concept for Boring, Reaming and Chamfering in a Single Tool" for this product which was filed for an esteemed automobile manufacturer in 2013. This innovative tool helped to increase tool life by 25 times compared with the earlier boring process. It has also contributed to reducing the operation cycle time and cost of operation. Figure 7.18 depicts the innovative tool developed and the combined operations.

7.11.3 Conclusion

This case study provides an overview of the implementation of boring, reaming and chamfering operations in an industry. Dedicated research and development efforts led to the birth of an innovative tool combining three machine operations together. Due to this, the tool machine cycle time was reduced and it resulted in reducing the cost of operation and the productivity of the machine has been enhanced.

7.12 SOLVED EXAMPLES ON TOOL LIFE, MACHINING FORCES

1. Determine percentage change in cutting speed required to give 50% reduction in tool life (i.e., to reduce tool life to 1/5 of its previous value). Take n = 0.2.

Given: Tool life in first case = 1 Tool life in second case = 1/5 exponent n = 0.2
To find: Percentage change in cutting speed.
Formula used: Taylor's tool life equation $VT^n = C$,
where V= cutting speed (m/min)

T= tool life (min)
C = machining constant

Procedure: Using the Taylor's tool life equation

$$VT^n = C$$

$$V_1T_1^n = V_2T_2^n$$

$$\frac{V_2}{V_1} = \left(\frac{T_1}{T_2}\right)^n = \left(\frac{1}{1/5}\right)^{0.2} = 1.38$$

$$\frac{V_2}{1} = 1.38 \ldots\ldots\ldots\ldots\ldots as\, V_1 = 1$$

$$V_2 = 1.38$$

$$\%\,\text{increase in}\, V_2 = (1.38 - 1) * 100\%$$

$$= 38\%$$

Result: The percentage increase in cutting speed is 38% by having a half tool life.

2. Find the optimum tool life and cutting speed at optimum tool life, i.e., optimum cutting speed for machining medium-carbon steel, the conditions are:
Cost of operating machine = Rs 0.3 per min
Total cost of tool change = Rs 8
Cutting speed = 40 m/min
Tool life = 50 min
Exponent index n = 0.2

Given: Cost of operating machine = C_1 = Rs 0.3 per min
Total cost of tool change = C_2 = Rs 8
Cutting speed = V = 40 m/min
Tool life = T = 50 min
Exponent index = n = 0.2

To find: (i) optimum tool life (T_{QpT}), (ii) optimum cutting speed (V_{opT})

Formula used:

1) $T_{opt} = \left(\frac{1}{n} - 1\right)\frac{c_2}{c_1}$
2) $VT^n = V_{QPT} \times T_{OPT}^n$

Procedures:

(i) Finding the optimum tool life T_{OPT}

$$T_{opt} = \left(\frac{1}{n} - 1\right).\frac{c_2}{c_1}$$

$$\Rightarrow T_{opt} = \left(\frac{1}{0.2} - 1\right).\frac{8}{0.3} = 106\,\text{min}$$

(ii) Finding the optimum cutting speed (V_{OPT})

$$VT^n = V_{opt} T_{opt}{}^n$$

$$\Rightarrow 40(50)^{0.2} = V_{opt}\left(106\right)^{0.2}$$

$$\Rightarrow \frac{40(50)^{0.2}}{\left(106\right)^{0.2}} = V_{opt}$$

$$\Rightarrow V_{opt} = 31.41 m/min$$

Results:

(i) Optimum tool life, T_{OPT} = 106 min
(ii) Optimum cutting speed, V_{OPT} = 34.41 m/min

3. Find the tool life equation for a tool life of 80 min obtained at a cutting speed of 30 m/min and 8 min at 60 m/min.

Given: Tool life T_1 = 80 min, cutting speed V_1 = 30 m/min

Procedure: Applying $V_1T_1{}^n = V_2T_2{}^n$
30(80f = 60 $(8)^n$
=> n = 0.3
Hence, the tool life equation becomes $VT^{0.3} = C$.

Result:
Tool life equation $VT^{0.3} = C$.

4. Determine the shear angle for a single-point cutting tool with 12^0 rake angle used to machine a steel workpiece. The depth of cut, i.e., uncut thickness, is 0.81 mm. The chip thickness under orthogonal machining condition is 1.8 mm.

Given: α =12c, t = 0.81 mm, t_c = 1.8mm

Procedure: Shear angle, $\tan\varphi = [r\cos\alpha/1 - r\sin\alpha]$ .... (7.i)
Chip thickness ratio, $r = t/t_c$
= 0.81/1.8
= 0.45

From equation (7.i), $\tan\varphi = [0.45\cos 12^0/1 - 0.45\sin 12^0]$
$\varphi = \tan^{-1}(0.486) = 25.91^0 \approx 26^0$.

5. Determine the values of shear angle and shear strain in an orthogonal machining operation:
Uncut thickness = 0.5 mm, cutting speed = 20 m/min, rake angel = 15^0
Width of cut = 5 mm, chip thickness = 0.7 mm, thrust force = 200 N, cutting force = 1200 N

Given: t = 0.5mm, V = 20 m/min, $\alpha = 15^0$, w= 5mm, t_c = 0.7 mm, F_t = 200 N, F_c = 1200 N

Procedure: We know, from the merchant's theory
Chip thickness ratio, $r = t/t_c = 0.5/0.7 = 0.714$
For shear angle, $\tan\varphi = [r\cos\alpha/1 - r\sin\alpha]$
Substitute the values, we get
$\tan\varphi = [0.714\cos 15^0/1 - 0.714\sin 15^0] = 0.689/0.815 = 0.845$
$\varphi = \tan^{-1}(0.845) = 40.2^0$
Shear strain, $s = \cot\varphi + \tan(\varphi - \alpha)$
$s = \cot(40.2^0) + \tan(40.2^0 - 15^0)$
$= \cot 40.2^0 + \tan 25.2 = 1.183 + 0.470 = 1.65$

6. Find the closet value of ϕ in a single-point turning tool, the side rake angle and orthogonal rake angle are equal. ϕ is the principal cutting edge angle and its range is $0^0<\phi<90^0$. The chip flows in the orthogonal plane.

Procedure: Interconversion between ASA (American Standards Association) system and ORS (Orthogonal Rake System) $\tan\alpha_s = \sin\varphi\tan\alpha - \cos\varphi\tan i$ where α_s = side rake angle, α = orthogonal rake angle, φ = principle cutting edge angle = $0^\circ<\phi<90^0$, i = inclination angle (i = 0 for ORS) $\alpha_s = \alpha$
(Given)
$\tan\alpha_s = \sin\varphi\tan\alpha - \cos\varphi\tan(0^\circ)$ $\tan\alpha_s = \sin\varphi\tan\alpha$
$\tan\alpha_s/\tan\alpha = \sin\varphi$ $\varphi = \sin^{-1}(1) = 90^0$.

7.13 SUMMARY

The term "machining" refers to any process in which material is gradually removed from a workpiece, such as metal cutting with single-point or multi-point tools (tools with geometrically defined cutting edges) and grinding with abrasive wheels that contain a large number of micro-cutting edges that are randomly shaped and oriented (i.e., tools with geometrically undefined cutting edges). Discontinuous or segmental chips, continuous chips, continuous chips with built-up edge and non-homogeneous chips are the important types of chips. The coating on a cutting tool has a significant impact on the mechanical and tribological properties of the tool, as well as the end result of the product. Carbon steel, high-speed steel, cemented carbide, ceramics, diamond and cubic boron nitride (CBN) are the commonly used cutting tool materials. Cutting tools typically fail as a result of high stresses and shocks during machining, high load and temperature. High-performance cutting tools can boost efficiency and production, but they can also deplete tooling budgets. When these tools get worn or damaged, they must often be reground and reconditioned in order to be cost-justified. Cutting fluids (as well as other metal-working fluids used in manufacturing activities) can undergo chemical changes over time if they are used regularly.

7.14 UNSOLVED EXAMPLES

1. Calculate the following: (1) expected chip thickness; (2) shearing stress on the shear plane; (3) machining power; (4) specific cutting energy for a shaper tool, making an orthogonal cut, that has a 10° rake angle. The depth of cut to = 0.6 mm, the width of cut = 3 mm. The cutting speed V = 40 m/min. A two-component dynamometer is used to determine the main cutting force (P_s = 3600 N), and the thrust force (P_t = 2400 N). A high-speed photograph shows a shear plane angle $\phi = 20°$.
2. Calculate: chip thickness, shear plane angle, resultant cutting force, machining power, specific cutting energy for An orthogonal cut with 3.0 mm depth made at a speed of 45 m/min and a feed rate of 0.25 mm/rev, with a high-speed steel tool having a 15° rake angle. The chip thickness ratio is found to be 0.58, the main cutting force is 1000 N and the thrust force is 280 N.
3. Calculate the main cutting force component for the following turning operation. Material: mild steel spec. cutting energy = 3500 N/mm^2 initial dia. of work = 80 mm. final dia. of work = 74 mm; feed rate = 0.4 mm/rev. Calculate the machining power if the spindle speed n = 710 rpm.
4. Determine the tool life with a cutting speed of 80 m/min (n = 0.09)? The machining cast iron with a high-speed steel tool life of 50 minutes was observed with a cutting speed of 100 m/min.
5. What is the percentage increase in tool life when the cutting speed is halved in a machining experiment, where tool life was found to vary with the cutting speed in the following manner.

Cutting speed (m/min)	Tool life (minutes)
60	81
90	36

6. Calculate the percentage increase in tool life when the cutting speed is reduced by 50% (assume $n = 0.5$ and $C = 400$). Use the Taylor's equation, $V.T^n = C$

7.15 REVIEW QUESTIONS

(1) Give a detailed classification of the material removal process.
(2) List the various types of coolants used in the machining process.
(3) Describe various types of chips produced in the machining process.
(4) Justify the importance of the machining process in manufacturing processes.
(5) Describe various tool failures.
(6) List various tool materials.
(7) Describe the tool coating process.
(8) Describe the role of coolant in the machining process.
(9) Describe single point cutting tool geometry.
(10) Give detail classification of different types of cutting tools used in machining processes.

8 Plastic Processing in the Automotive Industry

LEARNING OBJECTIVES

- To understand the fundamentals and principles of plastic processing.
- To get details of tools and equipments used in the plastic processing industry.
- To understand the details of extrusion process their classifications and other details.
- To understand the spectrum of plastic processing industry in vehicular applications.

8.1 INTRODUCTION

Plastics can be shaped into a wide range of items, including moulded components, extruded segments, films and sheets, electrical wire insulation coatings, automotive body parts and textile fibres. Furthermore, plastics are often the main ingredient of other materials such as paints and varnishes, adhesives and various polymer matrix composites. Plastics' economic and technical significance can be attributed to a number of factors, including: plastics can be moulded into intricate component geometries by moulding, typically with no additional processing needed. They work well with net shape processing. Plastics have an appealing set of properties for many engineering applications where strength is not an issue.

They have low density in comparison to metals and ceramics, high strength-to-weight ratios for some (but not all) plastics, low thermal and electrical conductivity, and high corrosion resistance. Plastics are cost-competitive with metals on a volumetric basis. This is commonly valid since the temperatures used to deal with these materials are much lower than those used with metals.

Since some plastics are translucent and/or clear, they can compete with glass in certain applications. Plastics are normally used in composite materials.

8.2 EXTRUSION OF PLASTICS

Extrusion is a compression process in which material is subjected to forces flowing through an orifice of the die to produce a long continuous product with a cross-sectional form defined by the orifice's shape.

DOI: 10.1201/9781003367321-8

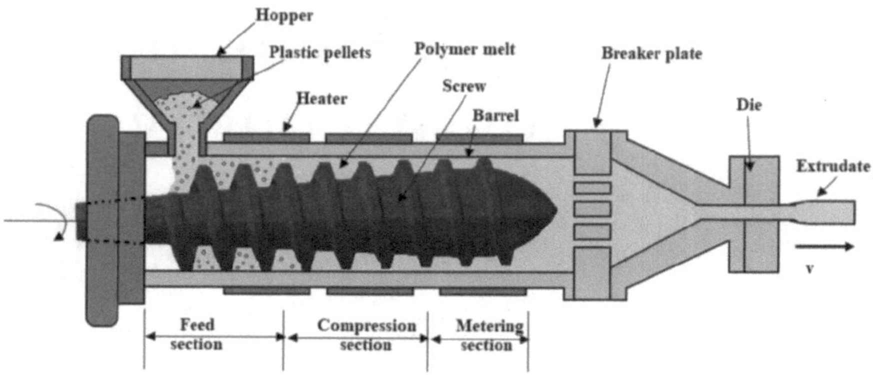

FIGURE 8.1 Extrusion of plastics.

Extrusion is a high-volume production technique. Using heat, the plastic material gets melted and by passing through a die it can be extruded into the required shapes. Inside a barrel, a cylindrical spinning screw pushes molten plastic material through a die. The extruded material takes the form of the die's cross-section. Figure 8.1 shows the extrusion of plastics.

- **Process**

Plastic material in the form of granules or pellets is gravity-fed into the barrel from a top-mounted hopper, and additives such as ultraviolet inhibitors and colorants (liquid or pellet form) may be mixed in the hopper. The plastic material reaches the feed throat and makes contact with the spinning screw. The plastic beads are propelled forward into the barrel by the spinning screw. The heating elements are used to heat the barrel up to the melting point of the plastic. The heating elements are used in such a way that the temperature of the barrel steadily rises from the back to the front. A rotating screw is divided into three sections: the feed section, the compression section and the metering section. As the plastic beads are forced through the barrel in the feed portion, they eventually melt. In the compression portion, the plastic material is fully melted.

A thermostat is used to keep the inside temperature of the barrel constant. Overheating of plastics should be avoided because it can lead to material property deterioration. During the process, a cooling fan or water cooling device is used to keep the barrel temperature stable. At the front of the barrel, the molten plastic exits the screw and passes through a screen pack to eliminate any pollutants. A breaker plate reinforces the screens. Back pressure in the barrel is also generated by the breaker plate assembly. The back pressure ensures that the molten plastic material is uniformly melted and mixed into the barrel. Molten plastic reaches the die after passing through the breaker plate. The die shapes the plastic product to the desired shape. Unwanted stresses in the plastic product will result from an uneven flow of molten plastic. These stresses can cause warping after molten plastic has solidified.

Since plastics are excellent thermal insulators, they are extremely difficult to cool rapidly. The plastic product is cooled by being dragged through a series of cooling rolls.

8.2.1 Types of Extrusion Process

Depending upon the specific applications, the extrusion process can be classified into six different types:

(a) Sheet/film extrusion

The molten plastic material is extruded through a flat die in this extrusion process. The cooling rolls determine the thickness and surface texture of the sheet/film. Sheet thicknesses ranging from 0.2 to 15 mm are possible.

It is possible to produce a thin flat sheet or film of plastic material. Polystyrene plastic is commonly used as a raw material for the sheet extrusion process.

(b) Blown film extrusion

The die in the blown film process resembles a vertical cylinder having a circular profile. Figure 8.2 shows the blown film extrusion process. A pair of nip rollers pulls the molten plastic upwards from the die. In order to inflate the tube, compressed air is used. An air ring is mounted around the die. The aim of the air ring is to keep

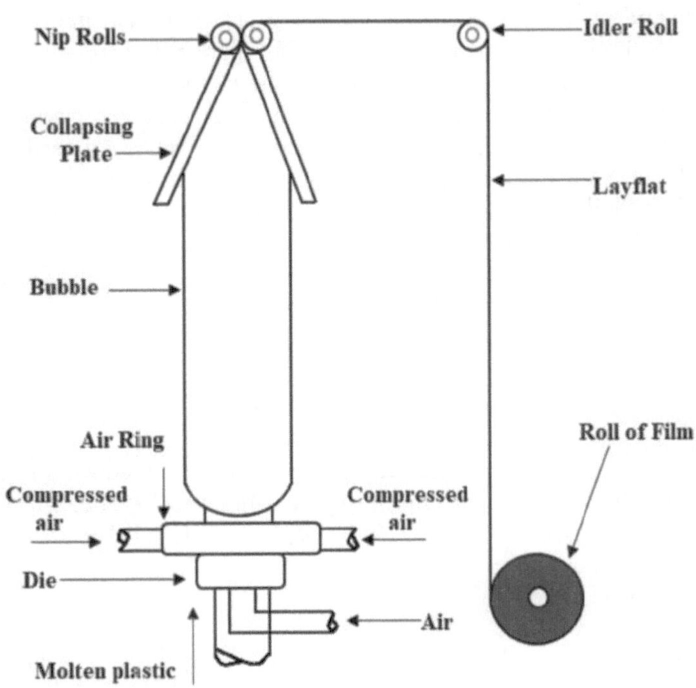

FIGURE 8.2 Blown film extrusions.

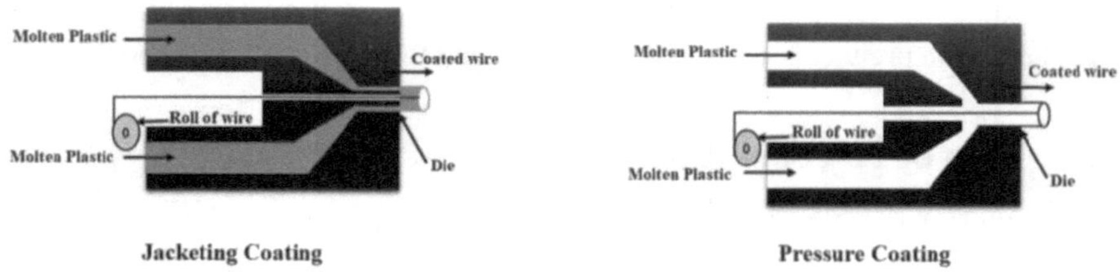

FIGURE 8.3 Over jacketing extrusion.

the film cool as it travels upwards. There is an air inlet in the centre of the die from which compressed air can be pushed into the centre of the circular profile, forming a bubble. The extruded circular cross-section can be increased by two to three times the diameter of the die. The collapsing plate is used to collapse the bubbles. The nip rolls flatten the bubble, creating a double layer of film known as lay flat. The film's wall thickness can be adjusted by varying the speed of the nip rollers. The lay flat can be spooled as a roll or cut into the desired shapes. The bottom side of the lay flat is sealed with heat and cut across further up to form an opening, allowing it to be used to make a plastic bag. The diameter of the die will range from 1 to 300 centimetres. Polyurethane plastic is commonly used in this operation.

(c) Over jacketing extrusion

This is also known as the wire coating technique. A bare wire is pulled through the centre of a die during this process. As shown in Figure 8.3, there are two forms of extrusion tooling used for coating over a wire: pressure tooling and jacketing tooling. When intimate contact or adhesion between the wire and the coating is needed, pressure tooling is used. Jacketing tooling is used if adhesion is not needed. The wire is retracted within the die for pressure tooling, where it comes into contact with the molten plastic at a much higher pressure. In the case of jacketing tooling, the wire will stretch and molten plastic will form a cover over the wire after die. The bare wire is fed through the die without coming into direct contact with the molten plastic before it exits the die. The primary distinction between jacketing and pressure tooling is the location of the wire in relation to the die.

(d) Tubing extrusion

In this method, molten plastic is extruded through a die, and hollow cross-sections are created by inserting a mandrel into the die. A tube with several holes can also be created for particular applications by inserting a variety of mandrels in the centre of the die.

(e) Co-extrusion

Co-extrusion is the method of extruding several layers of material at the same time. It is used to add one or more layers on top of the base material to achieve specific properties such as ultraviolet absorption, grip, matte surface and energy reflection, whereas base material is better suited for other applications such as impact resistance and structural efficiency. It can be used in any of the following processes: blown film, over jacketing, tubing and sheet/film extrusion. Two or more extruders are used in this process to deliver materials, which are then merged into a single die that extrudes the materials in the desired form. The speed and size of the individual extruders supplying the materials influence the layer thickness.

(f) Extrusion coating

Extrusion coating is used to apply an additional layer to an existing roll stock of paper, foil or film.

For example, polyethylene coating is used to increase the water resistance of paper. Extrusion coating is used in liquid packaging, photographic paper, envelopes, sack lining for fertilizer packaging and medical packaging. Polyethylene and polypropylene are commonly used.

Materials used

Polyethylene, polypropylene (PP), acrylic, polyvinyl chloride (PVC), acetal, nylon (polyamides), polystyrene, acrylonitrile butadiene styrene (ABS) and polycarbonate are some of the plastic materials that can be used in the extrusion method.

Applications

Extrusion is used to make rods, plates, and tubes, wire and cable coating, hose liners, hose mandrels, filaments, sheet, multilayer film, medical packaging and food packaging, among others.

Advantages

- High production volumes
- In comparison to other moulding processes, it is relatively inexpensive
- Design adaptability
- Short lead times
- Wire may be coated to obtain the desired properties
- It is possible to manufacture a continuous component.

Disadvantages

- Pieces are of limited complexity
- Only a uniform cross-section can be made.

8.3 INJECTION MOULDING – PLUNGER AND SCREW MACHINES

Injection moulding is the most widely used moulding process for thermoplastics. It is based on the ability of thermoplastic materials to be softened by heat and to harden when cooled. The process thus consists essentially of softening the material in a heated cylinder and injecting it under pressure into the mould cavity, where it hardens by cooling. Each step is carried out in a separate zone of the same apparatus in the cyclic operation. Most of the plastic articles are produced by the injection moulding method. Thermoplastic can be used as a raw material for this process.

Working principle of injection moulding

The thermoplastic polymer is heated to a plastic state during this process. The hot plastic in its plastic state is forced to flow under high pressure into a prepared mould cavity, where it solidifies. The finished product is the solidified plastic, which is separated from the mould cavity. The injection moulding process is extremely quick, with product cycle times ranging from 10 to 30 seconds. Since a single mould may have multiple mould cavities, it is possible to mould more than one part in a single cycle period.

Equipment for injection moulding

Injection moulding equipment is similar to metal die casting equipment. This equipment is called an injection moulding machine. It consists of two important parts:

(i) Molten plastic injection unit
(ii) Mould clamping unit.

1. Molten plastic injection unit

This is made up of a barrel with a feed hopper at one end, which supplies the plastic pellets. A screw similar to an extruder screw is located within the barrel. This screw can be rotated to mix the prepared material in the barrel and also serves as a ram for injecting molten plastic into the mould. A non-return valve near the screw stops molten plastic from moving backward down the screw threads. The screw retracts to its original location without interrupting the flow of molten plastics. The injection unit of the moulding machine performs three tasks:

(a) Causes the plastics to melt (polymer)
(b) Renders the molten plastics homogeneous
(c) The plastics are then correctly injected into the mould cavity.

2. Mould clamping unit

This is concerned with mould activity. It serves three purposes:

(a) It keeps the mould's two halves in good harmony with respect to each other.
(b) It applies clamping force to hold the mould closed during the injection process. The injection force is resisted by the clamping force.
(c) It opens and closes the mould at the correct times during the moulding period.

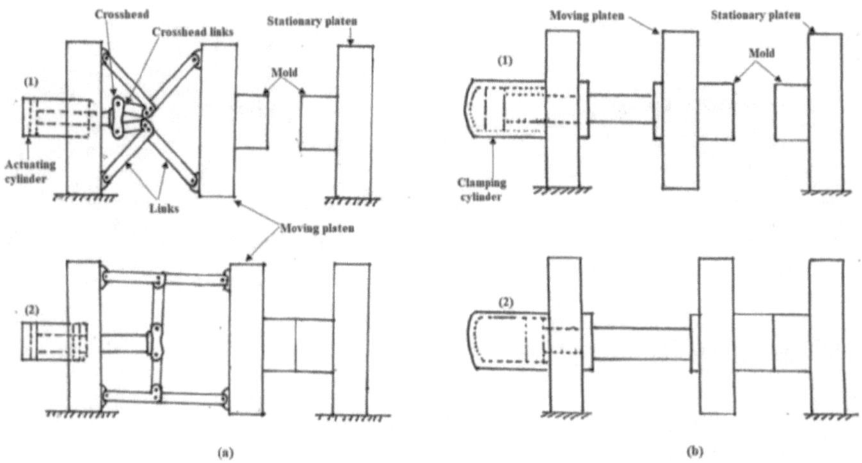

Two clamping designs: (a) one possible toggle clamp design: (1) open and (2) closed; and (b) hydraulic clamping: (1) open, and (2) closed.

FIGURE 8.4 Mould clamping unit.

The clamping unit is made up of two platens, one fixed and one movable, as well as a mechanism for translating the latter. There are three types of clamping designs: toggle, hydraulic and hydro-mechanical. Figure 8.4 depicts a typical clamping process with its cycle-by-cycle operation. There are two clamping designs: (a) one possible toggle clamp design: (1) open and (2) closed; and (b) hydraulic clamping: (1) open and (2) closed.

Working of an injection moulding machine

The working of injection moulding is similar to extrusion process. Moulding material/raw material is poured into the hopper by a feeding device. After that, moulding material goes down under the action of gravity into the cylinder (barrel). A circumferential heater which is located on the barrel is used to melt the material. When a powder form of moulding material goes down into the barrel from the hopper it starts melting and a hydraulic ram or rotating screw pushes the material forward into the mould by applying pressure. Molten plastic material is injected into a closed mould attached on the other side of barrel; in this a split mould is used. Moulding material goes forward continuously by the rotating screw. Pressure is applied by the hydraulic system. Injection pressure is generally 100–150 MPa. After injection, pressure is applied for some time or locked at the same position with some force.

After the whole process is completed, the parts are cooled sufficiently. Then the mould is opened and some ejectors are used for proper removal of the part without causing damage. After removing the part, the mould is closed again. This process is very fast and automatically repeated. Here complex shape parts can be easily manufactured. the production capacity of injection moulding is 12,000–16,000 parts per cycle.

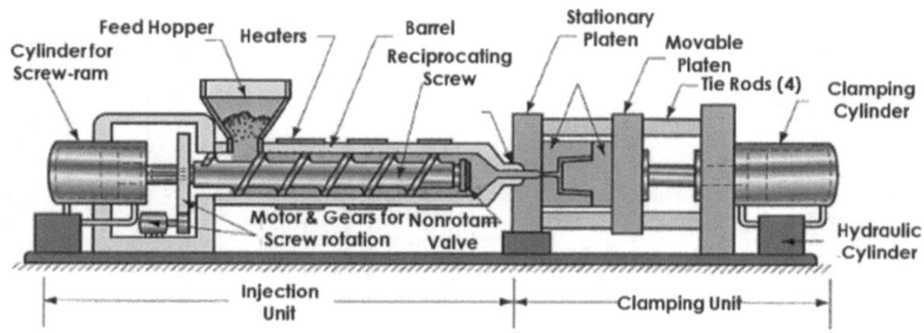

FIGURE 8.5 Injection moulding machine.

8.3.1 Screw Type Injection Moulding Machine

Figure 8.5 shows details of the screw type injection moulding machine. The operation of an injection moulding machine, in detail, cycle by cycle, is as follows:

- The first action is to open the mould, and the machine is then ready to begin a new moulding.
- When the mould is closed and clamped, the first part of the moulding period begins. The raw material, plastic or rubber, is fed into the barrel through a hopper, where it is heated and brought to the proper temperature and viscosity.
- The ram injects it under pressure into the mould cavity.
- The injection process is the second stage of the moulding cycle when the screw (ram) moves towards the mould cavity.
- As the injected plastic comes into contact with the cold surface of the mould cavity, it coalesces and starts to solidify.
- Ram pressure is maintained to pack molten plastic into the cavity, compensating for plastic contraction during solidification.
- The screw is rotated and retracted back with the non-return valve open to allow fresh polymer to flow into the forward portion of the barrel during the third stage of the moulding cycle.
- The polymer in the mould cavity has fully solidified by the end of the third part. The fourth stage of the moulding cycle involves opening the mould and removing the component.

8.3.2 Plunger Type Injection Moulding

Figure 8.6 shows details of a plunger type injection moulding machine. When the plunger is removed, granular material (the plastic resin) falls from the hopper into the barrel. After that, the plunger forces the material into the heating region, where it is heated and softened (plasticized or plasticated). Rapid heating occurs as a result of the polymer expanding into a thin film around a torpedo. The still molten polymer that has been replaced by this new material is forced forward through the nozzle, which is in close contact with the mould. The molten polymer flows through the

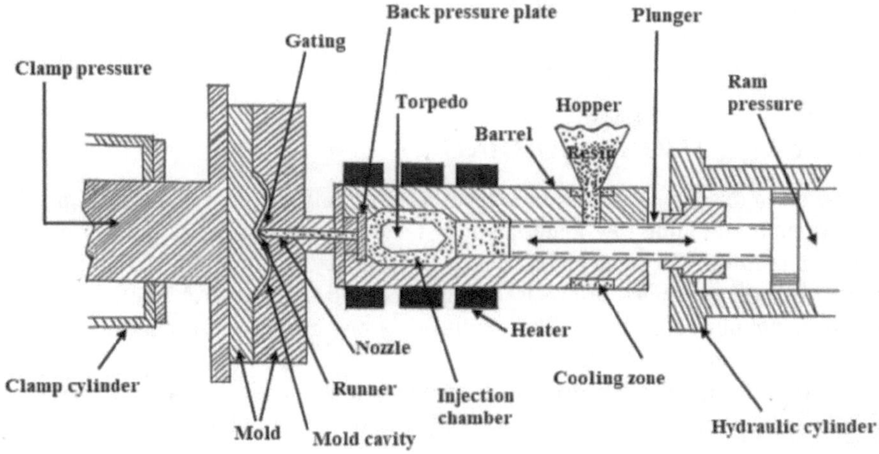

FIGURE 8.6 Plunger type injection moulding.

die's sprue opening, down the runner, through the lock, and into the mould cavity. The clamping action of the press platen keeps the mould firmly closed. The molten polymer is then pushed into all of the mould cavities, resulting in a perfect replica of the mould.

Advantages

- All types of polymers (thermosetting resin and thermoplastics) can be manufactured using the injection moulding process.
- Fibres appear to align as they move through the nozzle during injection into the mould cavity; this characteristic can be used in the design of composite parts for optimistic directional properties.
- Reaction injection moulding needs no heat energy and has a low mould cost.
- Structural reaction moulding is ideal for manufacturing structural parts in large quantities at low cost. This technique can generate composite parts with small to large complex geometries.

Disadvantages

- Mostly used for chopped fibre reinforcement.
- The initial capital investment is substantial.
- Significant fibre damage can occur as a result of the high shearing action into the barrel and nozzle.
- It is not suitable for limited production runs due to the high tooling and operating costs.
- Polymer burning can occur on the cylinder walls, followed by peeling into the melt, resulting in black spots on the surface of the composite component.

- If a small amount of moisture is present in the fibre or polymer, it will appear as a bubble on the finished component.
- This method does not allow for high fibre reinforcement.

8.4 COMPRESSION MOULDING

Compression moulding is used in the thermosetting plastics moulding process. Compression moulding raw materials may take any shape, such as powder, pellets, liquid or preform. To achieve repeatable consistency in the moulded product, the amount of polymer must be precisely regulated. Normally, the charge is preheated before being placed in the mould. Preheating raw materials softens plastics and reduces the manufacturing cycle time.

Infrared heaters and ovens are used for preheating. Reheating is often accomplished with the use of a hot spinning screw in the barrel of the moulding machine.

(a) Construction

A compression moulding machine is a type of press that is vertically oriented and has two platens. Figure 8.7 shows details of the compression moulding process. Moulding halves are fastened to these platens. There are two types of actuations that can be used to apply pressure: bottom platen upstroke, or using the top platen's down stroke.

Pressing is usually operated by hydraulics. Hydraulic power can generate pressures of up to a few hundred tons. In addition to pressing pressure, a clamping pressure of 100 tons should be used for proper clamping to prevent flushing. Compression moulding moulds are easier to use than the injection moulding process. Since the material is not powered to the cavity, a getting device is not needed. There is almost no possibility of defects such as powered shorts occurring. Mould cavities should be built with the ability of the charge to flow in the mould while pressing.

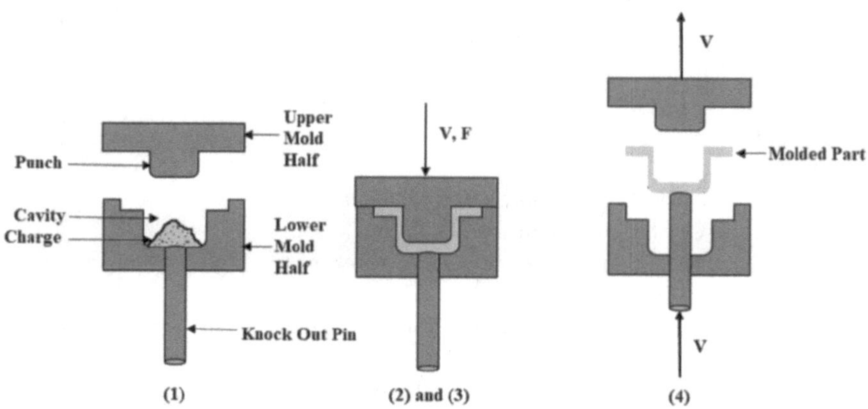

FIGURE 8.7 Plunger type injection moulding.

(b) Working

Normally, the procedure is carried out in three stages. The charge (raw material or plastic) is placed into the mould's lower half. As is well known, the compression moulding process's mould is divided into two sections. This is significant since the mould must be opened after the moulded pieces have solidified. The quantity of the charge is measured precisely so that it is utilized fully during the moulding process. If the charge is not held in place, the moulded component would be of poor quality. As a result, the precise quantity of charge is determined at the time of designing and establishing the procedure. The lower half of the mould is first heated to keep the charge temperature constant throughout the moulding process. By placing both halves of the mould close together, the imposed charge to the lower half of the mould is compressed. The charge is pressed under pressures that allow it to flow and take on the form of the mould cavity with extreme precision and dimensional accuracy. Precision and dimensional accuracy are determined by the consistency of the mould design and the precise measurement of the amount of charge to be moulded. The movement is carried out concurrently with the previous step. The charge is heated by the hot mould in order to polymerize and cure it into the desired formed moulded plastic part. The final compression moulding step is to open the moulding halves and remove the moulded plastic portion by pushing the knockout pin towards the inside. This is achieved after the moulded component has been moulded, solidified and cooled.

Advantages

- The mould cycle time is just a few minutes, so the production rate is high.
- It is possible to achieve a good surface finish with varying textures and styling.
- The compression moulding process achieves high component uniformity.
- Component design versatility is possible.
- During the manufacturing process, extra features such as inserts, bosses and attachments may be moulded.
- Raw material waste is kept to a minimum.
- The cost of maintenance is minimal.
- Residual stresses in the moulded part are absent or insignificant.
- Twisting and shrinkage in the product are minimized, resulting in good dimensional accuracy.

Disadvantages

- The initial capital investment is high due to the high cost of equipment and components.
- The process is ideal for high volume processing; however, it is not cost-effective for producing small quantities of parts or for prototyping applications.
- A high level of manpower is required.

- After compression moulding, the product can need secondary processing (trimming, machining).
- Uneven parting lines occur from time to time.
- The depth of the mould is limited.

8.5 TRANSFER MOULDING – TYPICAL INDUSTRIAL APPLICATIONS

The transfer moulding process incorporates the compression and transfer of the polymer charge principles. Figure 8.8 shows details of the transfer moulding process.

The polymer charge is moved from the transfer pot to the mould during transfer moulding. After the mould has cooled, the moulded portion is expelled. The method is divided into two variants: (a) pot transfer moulding, in which the charge is injected from a "pot" into the cavity through a vertical sprue tube and (b) plunger transfer moulding, in which the charge is injected via a plunger from a heated well via lateral channels into the mould cavity. As the polymer charge is inserted, the mould cavity

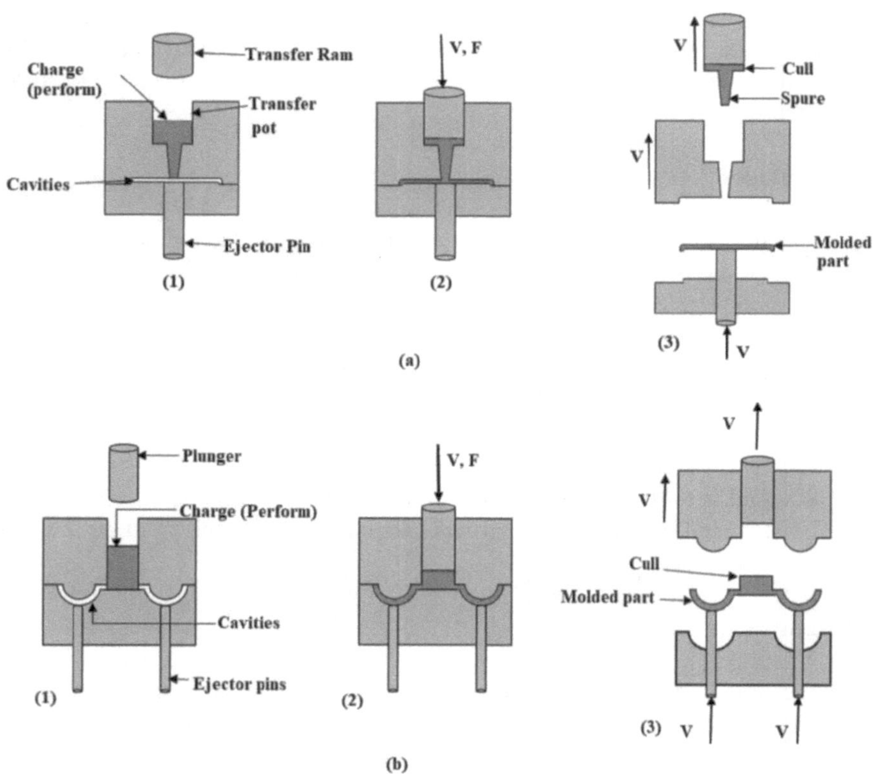

FIGURE 8.8 Transfer moulding: (a) pot transfer moulding and (b) plunger transfer moulding. The cycle in both processes is: (1) charge is loaded into pot, (2) softened polymer is pressed into mould cavity and cured and (3) part is ejected.

stays closed. The mould cavity is held closed until the resin dries. The mould cavity is opened, and the moulded portion can be removed with the aid of an ejector pin once it has hardened. After the process is completed, the sprue and gate attached to the moulded component must be trimmed.

Benefits of transfer moulding

- Reduced setup costs and faster setup time
- Low upkeep costs
- Plastic parts with metal inserts are possible
- Design adaptability
- Dimensional stability
- Pieces of uniform thickness
- High production rate.

Disadvantages

- Material wastage
- Lower production rate than injection moulding
- Air may become trapped inside the mould.

Applications

- This method is commonly used to encapsulate things like integrated circuits, plugs, connectors, pins, coils and studs.
- It is appropriate for moulding with ceramic or metallic inserts inserted in the mould cavity.
- Transfer moulding is also used to make radio and television cabinets, as well as car body shells.

8.6 BLOW MOULDING

Blow moulding is a moulding technique that uses air pressure to inflate soft plastic within a mould cavity. It is a critical industrial method for producing one-piece hollow plastic parts with thin walls, such as bottles and other containers.

1. Extrusion blow moulding

In extrusion blow moulding, a tube or perform (usually vertically oriented) is first extruded. It is then clamped into a mould with a cavity much greater than the diameter of the tube and blown outward to fill the mould cavity. Blowing is typically accomplished with a hot-air blast at a pressure ranging from 350 to 700 kPa. This method can produce plastic drums with volumes as high as 2000 litres. Steel, aluminium and beryllium copper are popular die materials. Figure 8.9 shows details of extrusion blow moulding.

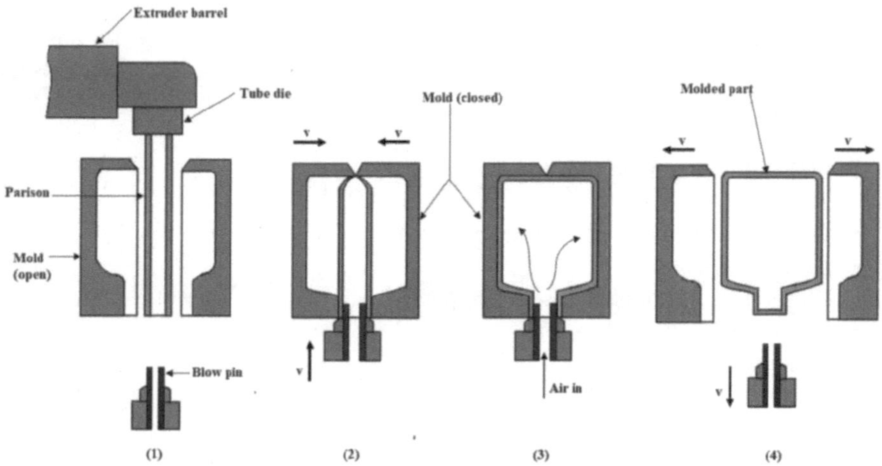

(1) extrusion of parison; (2) parison is pinched at the top and sealed at the bottom around a metal blow pin as the two halves of the mold come together; (3) the tube is inflated so that it takes the shape of the mold cavity; and (4) mold is opened to remove the solidified part.

FIGURE 8.9 Extrusion blow moulding: (1) extrusion of parison; (2) parison is pinched at the top and sealed at the bottom around a metal blow pin as the two halves of the mould come together; (3) the tube is inflated so that it takes the shape of the mould cavity; and (4) the mould is opened to remove the solidified part.

2. Injection blow moulding

A short tubular piece (comparison) is injection moulded in cool dies in injection blow moulding. The dies are then opened, and the parison is indexed and transferred to a blow-moulding die. Hot air is pumped into the parison, causing it to expand and make contact with the mould cavity's walls. Commonly manufactured items include plastic water bottles (typically made of polyethylene or polyetheretherketone, PEEK) and small, hollow containers. Figure 8.10 shows details of the injection blow moulding process.

3. Stretch blow moulding

Stretch blow moulding, in which the parison is simultaneously extended and elongated, subjects the polymer to biaxial stretching and thereby improves its properties.

4. Multilayer blow moulding

This makes use of coextruded tubes or parisons, allowing for the development of a multilayer structure.

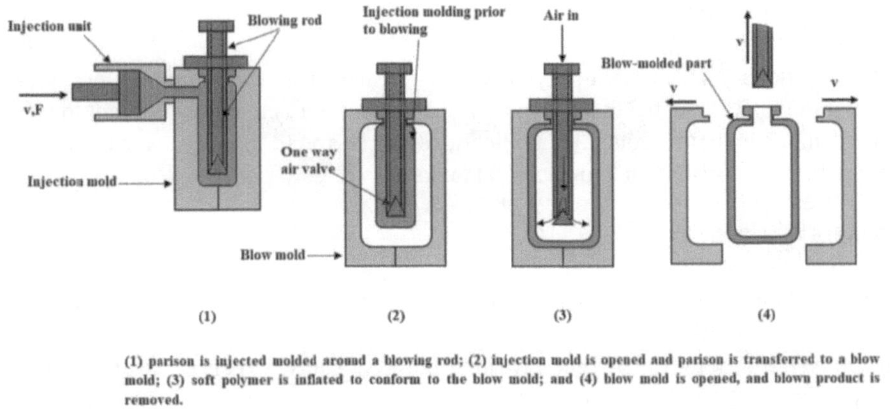

FIGURE 8.10 Injection blow moulding: (1) parison is injected moulded around a blowing rod; (2) injection mould is opened and parison is transferred to a blow mould; (3) soft polymer is inflated to conform to the blow mould; and (4) blow mould is opened, and blown product is removed.

8.7 ADVANCED PLASTICS USED IN THE AUTOMOBILE SECTOR

Advanced plastic processing is associated with the following factors:

- Longer vehicle life
- Minimal corrosion
- Freedom to design
- Comfort and safety
- Recyclability
- Substantial design freedom, allowing advanced creativity and innovation
- Flexibility in integrating components.

These high-performance plastics are used in automotive hardware. While all of them may easily be used in a single vehicle, just three types of plastic make up approximately 66% of the total high-performance plastics used in a car: polypropylene (32%), polyurethane (17%) and PVC (16%).

(1) Poly-vinyl chloride

PVC is known for its flexibility, has great thermal stability, is flame retardant and has an extremely low lead content. It can be injection moulded, compression moulded or blow moulded to create a multitude of rigid or flexible plastic products depending on the type and amount of plasticizers used. It is used to make automobile doors, sheathing electrical cables and instrument panels.

(2) Polypropylene

Polypropylene is a saturated polymer that is created from the monomer, propylene. One of the major benefits that polypropylene has to offer is that it is resistant to acids, bases and chemical solvents. Its applications can be found in gas cans, carpet fibres, bumpers, cable insulation and chemical tanks.

(3) ABS

ABS is popular due to its low production cost and the ease with which the material is machined by plastic manufacturers. It is made by polymerizing styrene and acrylonitrile. What's more, the styrene gives the plastic a shiny, impervious surface. It has outstanding high- and low-temperature performance, great insulation properties, and is easy to paint and glue. Its applications can be found in dashboards, wheel covers and automotive body parts.

(4) Polyamide

Polyamide is highly water absorbent, has high mechanical properties and is rigid in nature. It is a general-purpose polymer that can be extruded and moulded. The advantages that polyamides offer are high strength, abrasion resistance and resilience. Its applications in an automobile can be found in gears, cams, bearings and waterproof coatings.

(5) Polystyrene

Polystyrene exhibits incredible electrical and chemical resistance. It is easy to manufacture, is highly elastic and softens when heated beyond its glass transition temperature. Some of its mechanical properties include its strength, elongation, impact strength, toughness and modulus. It is used in car fittings, display bases and buttons.

(6) POM (polyoxymethylene)

POM is a type of thermoplastic material that is known for its excellent dimensional stability, creep resistance, high heat resistance and good electrical and dielectric properties. It is an engineering material used in parts that require precision. Applications for POM include high-performance components such as gears, interior and exterior trims and fuel systems.

(7) Polycarbonate

Polycarbonate offers excellent thermal, electrical, impact, optical, weathering properties. It offers a distinct combination of hardness, stiffness, and toughness. Due to its incredible impact strength, it is one of the top choices for headlamp lenses, bumpers, helmets, and bullet-proof glass.

(8) Polyethylene

Polyethylene exhibits high impact resistance, low density and good toughness. It is extremely cost-effective, moisture-resistant and useful in a multitude of thermoplastics processing methods. Its applications can be found in electrical insulation and reinforced glass.

High-performance plastics will continue to play a vital role in the automotive industry owing to the incredible benefits they offer. If you are in the automotive industry, then Plastivision is the platform for you as it will make things immensely easier for your business.

8.8 USAGE AND ADVANTAGES OF PLASTIC COMPONENTS IN THE AUTOMOTIVE INDUSTRY

1. **Fuel efficiency**: Due to their reduced weight, the inclusion of plastics in vehicle design boosts fuel efficiency and lowers emissions overall. Today, plastics make up nearly half of a vehicle's volume but only 10% of its weight.
2. **Innovation & design**: External plastic vehicle cladding allows car designers the ability to create innovative concepts that may not be possible when shaping metal. The use of plastics also reduces production and manufacturing costs, passing those savings on to the buyer. Additionally, when it comes to scrapes and dents it's much easier to repair or replace a plastic bumper or plastic car door cladding than metal, which rusts when scraped.
3. **Sustainability**: Fuel efficiency on its own lends to sustainability in the form of reduced emissions, but beyond that, as auto manufacturers create new renewable plastics to use in vehicle design, energy savings and car performance will increase. A large number of automobile manufacturers are already making use of recycled plastic for many of their applications such as seat cushions, replacement bumpers, splash guards and wheel liners.
4. **Safety**: The safety features in your car such as airbags and seatbelts are made from durable polyester. In an accident these polymer-based safety features and exterior items such as car door cladding and bumpers can take an impact far better than metal, often leading to a better outcome for you and your vehicle.
5. **Weather resistance**: Weather constantly takes a toll on your vehicle, which is why synthetic coatings on metal surfaces are used to reduce the chance of corrosion due to salt damage, extreme heat and water exposure. Even glass windshields are coated with plastic film to improve weather resistance, wear and tear, and reduce the chance of shattering. Under the hood, car parts are either coated or made entirely from plastics such as polyurethane which is heat-resistant, abrasion-resistant, yet flexible.

- **The future of plastic in the automobile industry**

Reducing fuel consumption and harmful emissions are on the minds of automobile manufacturers and consumers alike. As new green technologies become mainstream, so too will innovative bioplastics and polyamides. Electric vehicle makers will undoubtedly be the first to make the most of bioplastics since their lighter weight lowers energy consumption. EV manufacturers are already beginning to incorporate more recyclable plastics into vehicles. The Nissan Leaf, for example, is made from 25% recycled material, such as used pop bottles and recycled steel. Resins from recycled auto plastics are also being repurposed to make vehicle cladding for car doors and dash consoles in the Nissan Leaf. Even though many regular plastics are non-recyclable, the move toward bioplastics is a big step toward improving the sustainability of automotive manufacturing.

8.9 SUMMARY

Plastics have an appealing set of properties for many engineering applications where strength is not an issue.

They have wider application in various fields such as moulded components, extruded segments, films and sheets, electrical wire insulation coatings, automotive body parts and textile fibres. Plastics are often the main ingredient of other materials such as paints and varnishes, adhesives and various polymer matrix composites. There are various methods used for the preparation of plastic components like extrusion process, compression moulding, blow moulding, transfer moulding, etc. The application and use of plastics in automobiles is continuously increasing due to its various advantages like reduction in weight, fuel efficiency, safety and sustainability, etc.

8.10 REVIEW QUESTIONS

(1) State the advantages of plastics over metal.
(2) List the applications of compression moulding.
(3) List the applications of injection moulding.
(4) State the types of blow moulding.
(5) Define the extrusion process.
(6) List the various plastic manufacturing processes.
(7) State the advantages and disadvantages of compression moulding.
(8) State the advantages and disadvantages of transfer moulding.
(9) State the applications of compression moulding.
(10) Describe the extrusion process with the help of a neat sketch.
(11) Describe the screw type injection moulding process with the help of a neat sketch.
(12) Describe the plunger type injection moulding process with the help of a neat sketch.
(13) Describe the extrusion blow moulding process with the help of a neat sketch.
(14) Describe the injection blow moulding process with the help of a neat sketch
(15) Describe the compression moulding process with the help of a neat sketch
(16) Describe the transfer moulding process.

9 Powder Metallurgy

LEARNING OBJECTIVES

- To acquire the knowledge of powder metallurgy, history, applications and its importance.
- To understand the various power production techniques, tool equipments and their procedures.
- To understand the mechanism of sintering and secondary processes used in powder metallurgy for development of mechanical properties.
- To get details the applications powder metallurgy in automotive and allied industries.

9.1 INTRODUCTION

Powder metallurgy is a technology used to make products or articles out of powdered metals. In this process, powder is placed inside the mould and compacted with the use of a heavy compressive force, and then the parts are heated to temperatures well below their melting points to bind the particles together and improve their strength and other properties. Examples of such products include grinding wheels, filament wire, magnets, welding rods, tungsten carbide cutting equipment, self-lubricating bearings, electrical contacts and turbine blades with high-temperature strength.

The powder metallurgy process includes the development of powders, blending, compacting, sintering and a variety of secondary operations such as measuring, coining, impregnation, machining, plating, infiltration and heat treatment. The use of powder metallurgy is only economically viable for high-volume manufacturing.

The power metallurgy process consists of the following operations (Figure 9.1).

9.2 PRODUCTION OF METALLIC POWDER

In general, any metal may be ground into a powder. Metallic powder can be produced in three ways: atomisation, chemical and electrolytic methods. The microstructure, bulk and surface properties, chemical purity, porosity, form and size distribution of the particles are all affected by the process. These properties are important because they have a significant impact on flow and permeability during compaction and

DOI: 10.1201/9781003367321-9

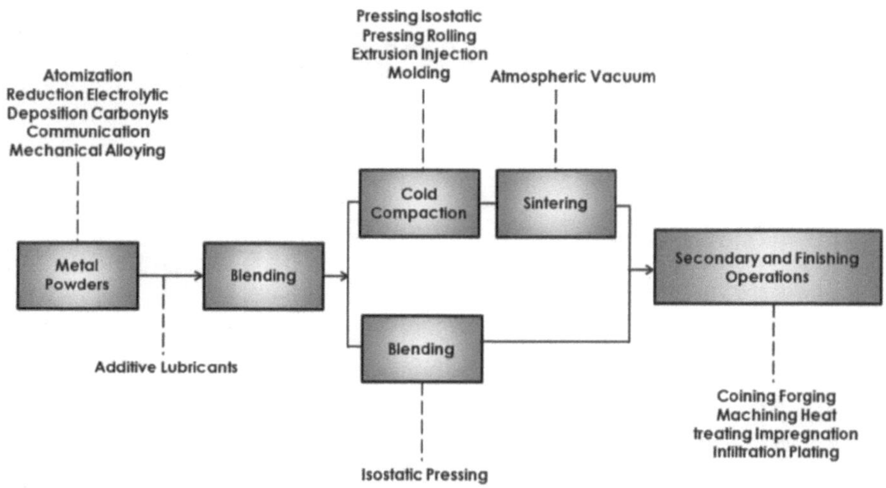

FIGURE 9.1 Processes involved in powder metallurgy.

subsequent sintering operations. The particle sizes produced range from 0.1 to 0.04 in. Each process can be further subdivided, and powders can be created by combining elements from each type. As a result, there is a diverse range of production routes available, each producing powders with distinct properties.

(A) Atomisation

Atomisation is the process of converting molten metal into smaller pieces by spraying or crushing it.

Atomisation is the dominant method for producing metal and pre-alloying powders from iron, steel, stainless steel, superalloy, titanium alloy, aluminium alloy and brass. It is the best method because it yields high production rates, favours economies of scale and because pre-alloying powders can only be produced by spraying or smashing molten metal into smaller particles. There are several industrial atomisation methods, some of which are as follows:

- Water atomisation
- Gas/air atomisation
- Centrifugal atomisation.

(a) Water atomisation

In water atomisation, the raw material is melted, and then the liquid metal is broken into individual particles. To do this, the melt stock, which is in the form of elemental, multi-element metallic alloys, is melted in an induction, arc or other type of furnace. Once the metal is molten and homogeneous, it is transferred to a tundish, which is a reservoir used to provide a constant, controllable supply of water. As the metal stream leaves the tundish, it is collided with a high-velocity water stream.

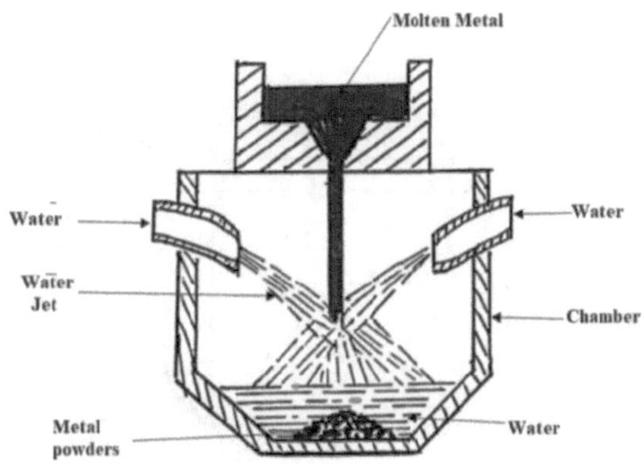

FIGURE 9.2 Water atomisation.

Water is sprayed at pressures ranging from 1 to 10 MPa (145 to 1450 psi). The molten metal stream is dispersed into fine droplets, which solidify and cool as they pass through the atomising tank. Particles accumulate at the tank's bottom. Water atomisation quickly cools the metal, causing the formation of irregular particles and it can also corrode certain metals. Figure 9.2 shows the water atomisation process.

(b) Gas atomisation

Atomisation is achieved in this step by forcing a molten metal stream into an orifice at moderate pressures.

An inert gas at pressures from 0.2 to 2.0 MPa (29 to 290 psi) is added into the metal stream shortly before it leaves the nozzle, causing turbulence as the entrained gas expands (due to heating) and exits into a large collection volume exterior to the orifice. The collection volume is filled with gas to facilitate more turbulence of the molten metal jet. As compared to water atomisation, this process produces a similarly wide distribution of very fine powders; however, the process must be performed more slowly than water atomisation, and the energy required to compress the gas is much greater than that required to pump water. Because it cools the metal at a slower rate, it produces a more spherical particle. The inert gas process has the advantage of providing greater control of oxygen levels in oxygen-sensitive materials. Figure 9.3 shows gas atomization process.

(c) Centrifugal atomisation

Centrifugal atomisation of molten metals is a low-cost method for producing powder and spray deposition. The properties of the powder and deposit formed by this method are mainly determined by the properties of the atomised droplets, which are largely dependent on the flow development of the melt on the atomiser. The rotating disk process and the rotating electrode process are the two main types of centrifugal atomisation processes.

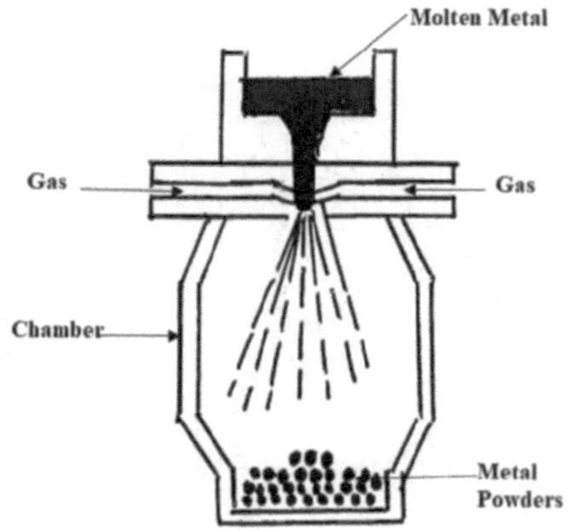

FIGURE 9.3 Gas atomisation.

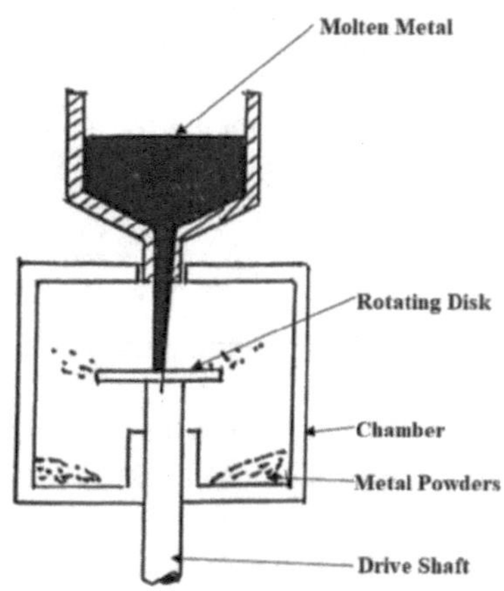

FIGURE 9.4 Rotating disc method.

(1) Rotating disk process

In this process, as shown in Figure 9.4, a molten metal stream falls onto a rapidly spinning disk or cup on a vertical axis, where centrifugal forces break up the molten metal stream to form a molten metal film and then particles. If very high speeds (over 10,000 rpm) are used, particle sizes are not very fine. An advantage of this method is that it can achieve much narrower distributions than gas or water atomisation. As a result, overly fine or oversized particles may be reduced.

(2) Rotating electrode process

For pulverisation, an arc is formed between a horizontal tungsten cathode and a spinning electrode of the desired metal in the rotating electrode process (REP).

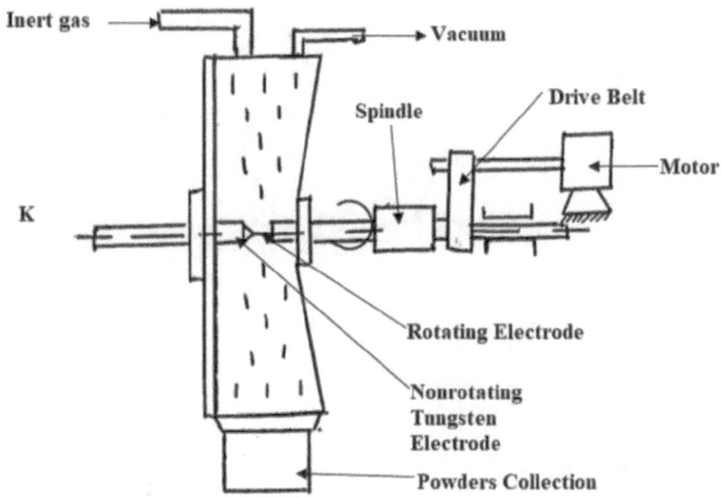

FIGURE 9.5 Rotating electrode process.

Figure 9.5 shows the rotating electrode process. The electric arc (or plasma in later versions) melts the spinning workplace electrode tip, causing centrifugal forces to produce a shower of molten metal droplets. The droplets solidify when they move through a cooling gas or vacuum and are deposited in a concentric chamber. One distinguishing feature of REP powder particles is their perfectly spherical form and lack of satellite particles adhering to them.

(B) Mechanical processes

There are basically two types of mechanical processes for the production of metal powders: the mechanical comminution process and the mechanical alloying process.

(a) Mechanical comminution process

Brittle materials, such as intermetallic compounds and ferro-alloys (ferro-chromium, ferro-silicon, and so on), are mechanically pulverised in ball mills. The cold stream process is increasingly being used to produce very fine powders, such as those required for injection moulding. Milling is the primary method for reducing the size of large particles and particle agglomerates. Commercially available comminuting devices include ball, hammer, vibratory, attrition and tumbler mills. Forces act on the feed metal during milling to change the resulting particles. Impact, attrition, shear and compression all have an effect on the size and shape of powder particles.

Lathe turning is a method used for materials such as magnesium to produce coarse particles from billets, which are then reduced in size by milling or grinding. Figure 9.6 shows the mechanical comminution process.

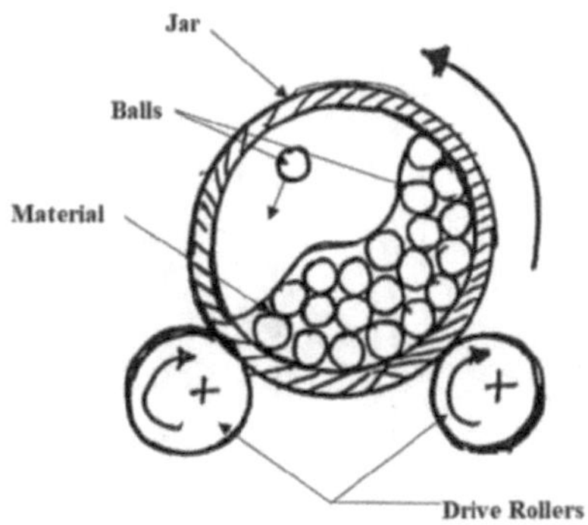

FIGURE 9.6 Mechanical comminution process.

(b) Mechanical alloying

Mechanical alloying is a method of creating new materials. It takes place in a high-energy mill where two separate powders are combined. Nanocrystalline structures that cannot be created by traditional melting methods arise as a result of the conversion of mechanical into chemical energy. Mechanical alloying was initially designed to overcome the drawbacks associated with using powder metallurgy to alloy difficult-to-combine components. Certain oxides are unable to dissolve in molten metals. Mechanical alloying allows these oxides to be dispersed through the metals. Examples are nickel-based superalloys strengthened with scattered thorium oxide or yttrium oxide (Y_2O_3). Because of their high strength and corrosion resistance at high temperatures, these superalloys are desirable candidate materials for use in jet engine turbine blades, vanes and combustors.

(C) Chemical processes

Chemical processes, in general, produce very fine powder particle sizes. Oxide reduction, precipitation from solutions, and thermal decomposition are the most popular chemical powder treatments. Chemical reduction entails a series of chemical reactions that convert the metal into elemental powders. A popular method is to separate metals from their oxides by using reducing agents, which bind to the oxygen in the oxide and make metal powders. The powders formed in this manner have a wide range of properties while maintaining tightly regulated particle sizes and shapes. Because of the pores present within individual particles, oxide-reduced powders are frequently described as "spongy." Solution-precipitated powders may have very small particle size distributions and are very pure. The most common method for processing carbonyls is thermal decomposition. After milling and annealing, the purity of these powders exceeds 99.5%.

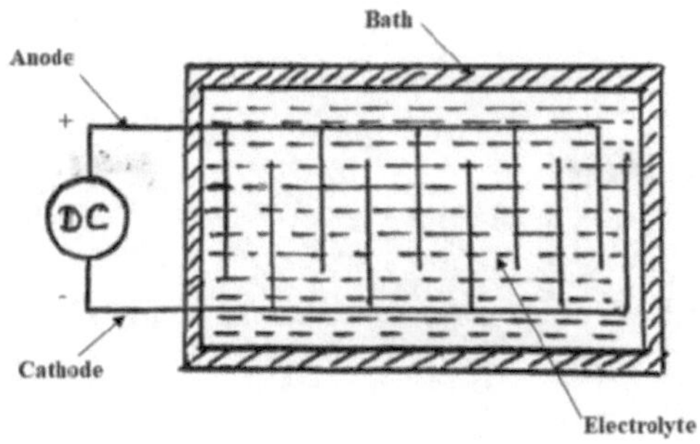

FIGURE 9.7 Electrolytic deposition.

(D) Electrolytic deposition

Electrolytic deposition is often classified as a fourth mode of powder fabrication, as shown in Figure 9.7. In this process, an electrolytic cell is set up with the anode as the source of the desired metal, and the process includes the precipitation of a metallic material at the cathode of an electrolytic cell. Many metals may be deposited in a spongy or powdery state by selecting appropriate conditions such as electrolyte composition and strength, temperature, current density and so on. Powder may be produced by additional processing such as washing, drying, reducing, annealing and crushing. Copper is the most common metal manufactured in this way, but chromium and manganese powders are also produced.

(E) Hybrid atomisation

There is a growing demand for fine spherical powder with uniform particle size for advanced powder metallurgy applications such as metal injection moulding, solder for electronics components, joining and conductive inks. However, traditional powder production technologies cannot easily manufacture such powders, necessitating the development of a new powder production technology. The National Institute of Materials Science (NIMS) created the world's first powder production process, hybrid atomisation, which can easily manufacture powder with 10 microns or smaller spherical particles, uniform size and low oxygen content.

Such powders cannot be manufactured using traditional powder manufacturing technology. Hybrid atomisation effectively incorporates gas and centrifugal atomization. Using a gas jet, gas atomisation splits molten metal into fragments ranging in size from tens to hundreds of micrometres.

A spinning disk located underneath the spray is moved at high speed (5000–66,000 rpm), and the molten metal is dispersed evenly over the rotating disk using a gas spray flow, creating a thin liquid film of 10 μm thickness or less, followed by fine droplets scattered from the rotating disk to produce a fine spherical powder. Figure 9.8 shows the hybrid atomization process.

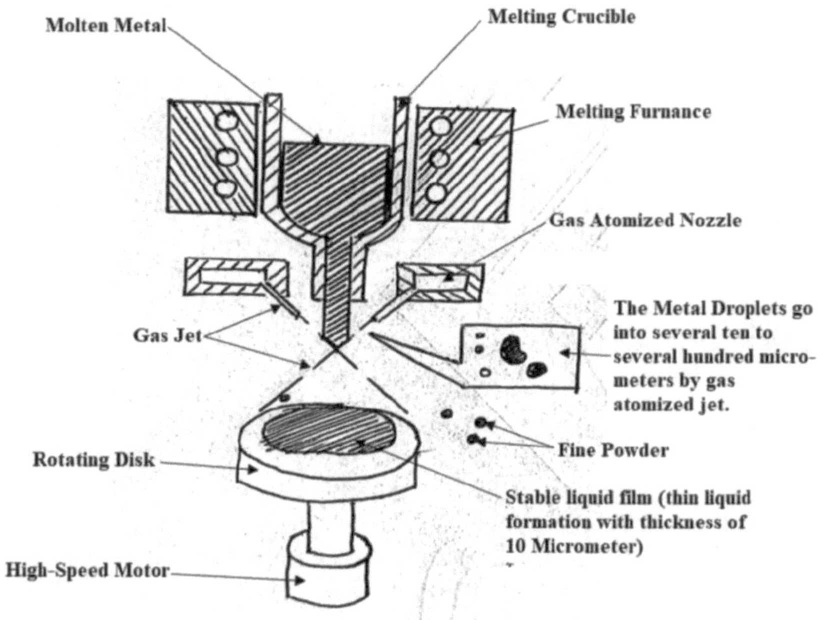

FIGURE 9.8 Hybrid atomisation.

9.3 PROCESSING METHODS

Following the production of metallic powders, the traditional PM sequence consists of three steps: powder blending and mixing; compaction, in which the powders are pressed into the desired component shape; and sintering, which requires heating to a temperature below the melting point to cause solid-state bonding of the particles and part strengthening.

The three stages, which are often referred to as primary operations in PM, are as follows: secondary operations are often conducted on occasion to boost dimensional accuracy, increase density and for other purposes.

(A) Blending and mixing of the powders

Blending is the mixing of powders of the same chemical composition but probably different particle sizes. To minimise porosity, different particle sizes are often blended. Mixing is the process of combining powders of different chemistries. The ability to combine different metals into alloys that would be difficult or impossible to manufacture using other methods is one of the benefits of PM technology. In industrial practice, the distinction between blending and mixing is not always precise. Blending and mixing can be carried out mechanically.

Figure 9.9 shows four alternatives: (a) rotation in a drum; (b) rotation in a double-cone container; (c) agitation in a screw mixer; and (d) stirring in a blade mixer. These

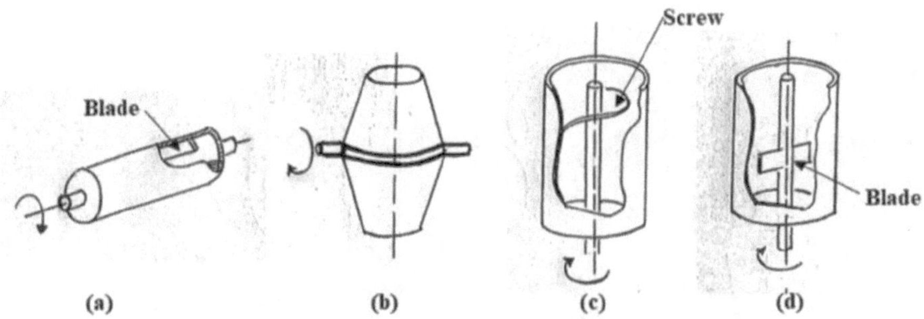

FIGURE 9.9 Blending and mixing of the powders: (a) rotation in a drum; (b) rotation in a double-cone container; (c) agitation in a screw mixer; and (d) stirring in a blade mixer.

instruments are more scientific than one would think. The best results seem to occur when the container is between 20% and 40% filled. The containers are typically constructed with internal baffles or other means of preventing free-fall during the blending of powders of varying sizes, so differences in settling rates between sizes result in segregation, which is exactly the opposite of what is desired in blending.

Powder vibration is undesirable since it induces segregation. During the blending and/or mixing phase, other ingredients are usually added to the metallic powders. These additives include lubricants, such as zinc and aluminium stearates, used in small amounts to reduce friction between particles and at the die wall during compaction; binders, which are used in certain cases to achieve adequate strength in pressed but unsintered parts; and deflocculants, which inhibit powder agglomeration for better flow characteristics during subsequent processes.

(B) Compaction

In compaction, high pressure is applied to the powders in order to shape them into the desired shape. The traditional compaction process is pressing, which involves opposing punches squeezing the powders stored in a die. Figure 9.10 depicts the phases in the pressing period. The work portion after pressing is referred to as a green compact, with the word green indicating that it has not yet been completely processed. As a result of pressing, the part's density, known as the green density, is much higher than the starting bulk density. When pressed, the green strength of the component is sufficient for handling but much less than that obtained after sintering.

As pressure is applied during compaction, the powders are initially repacked into a more efficient arrangement, eliminating "bridges" formed during filling, reducing pore space and increasing the number of contacting points between particles. When pressure increases, the particles are plastically deformed, increasing interparticle contact area and causing additional particles to congregate.

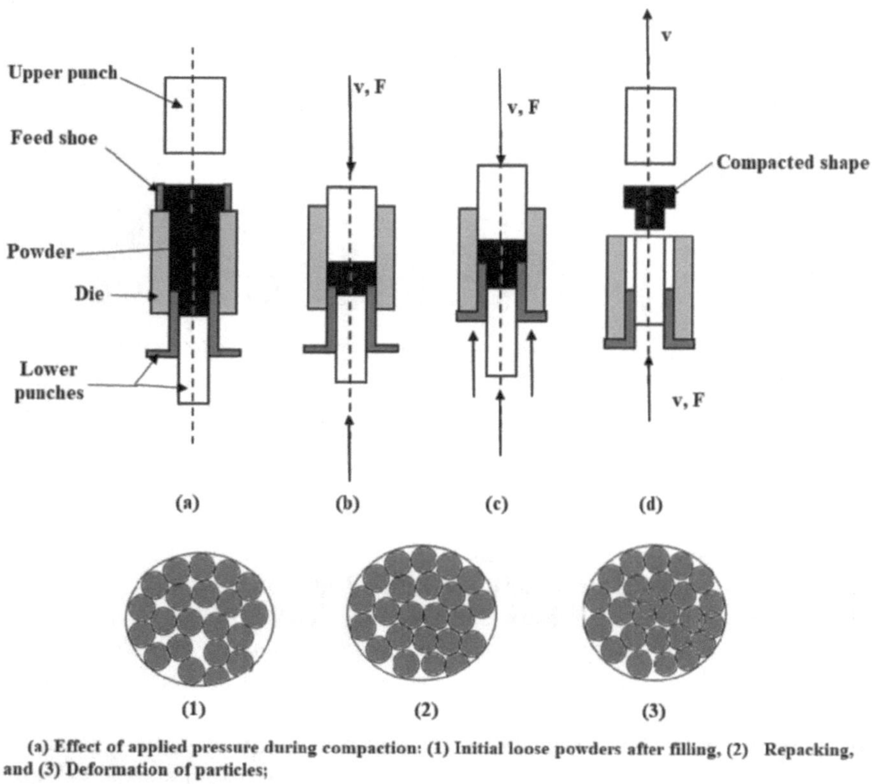

FIGURE 9.10 Compaction: (a) effect of applied pressure during compaction: (1) initial loose powders after filling, (2) repacking, and (3) deformation of particles.

(C) Sintering

The thermal treatment of a powder or compact is at a temperature below the melting point of the main constituent in order to increase its strength by bonding the particles together. The procedure is usually performed at temperatures ranging between 0.7 and 0.9 of the metal's melting point (absolute scale). Since the metal remains unmelted at these treatment temperatures, the terms solid-state sintering or solid-phase sintering are often used to describe this traditional sintering. After compaction, the components are sintered in a furnace. This usually has two heating zones, the first of which eliminates the lubricant and the second of which allows for diffusion and bonding between powder particles.

9.4 ADVANTAGES AND DISADVANTAGES

Advantages

- Powder metallurgy processes are quiet and clean, and any complex or complicated form can be manufactured.
- Since the dimensional precision and surface finish are much stronger for many applications, machining can be avoided.
- Unlike casting, no material is discarded as scrap in the press forming machining process, and the process utilises the entire raw material.
- Through this method, difficult-to-process materials such as diamond can be transformed into functional components and equipment.
- High output rates are simple to achieve.
- The phase diagram constraints that prevent the formation of alloys between mutually insoluble constituents in liquid state, such as copper and lead, are eliminated in this process, and mixtures of such metal powders can be easily processed and formed.
- This process allows for the manufacture of several parts that would otherwise be impossible to make, such as sintered carbides and self-lubricating bearings.
- The process allows for effective control of many properties in the parts provided by this process, such as purity, density, porosity, particle size, and so on.
- The components generated by this process are highly pure and have a longer shelf life.
- It allows the manufacture of parts from alloys with low castability.
- Since exact proportions of constituent metal powders can be used, composition uniformity can be ensured.
- The preparation and processing of powdered iron and nonferrous parts manufactured in this manner exhibit excellent properties that cannot be obtained in any other way.
- Simple shaped parts can be rendered to size with 100 micron accuracy and waste-free.
- Porous parts that could not be made any other way can be created.
- Parts with a wide range of compositions and materials can be manufactured.
- Structure and properties can be more precisely regulated than in other fabricating methods.
- There is no need for highly trained or professional labour during the powder metallurgy process.
- This method can quickly produce super-hard cutting tool pieces, which are difficult to produce using other manufacturing techniques.
- The obtained component shapes are highly reproducible.
- Grain size control, relatively uniform structure, and defects such as voids and blowholes in structure can be removed.

Limitations/disadvantages

- The powder metallurgy process is not economical for small-scale manufacturing because the cost of the tool and die of powder metallurgical set-up is relatively high.
- The size of products is limited as compared to casting because large presses and costly equipment are needed for compacting.
- Metal powders are costly and can be difficult to store without oxidation in some situations.
- Since metallic powders lack the ability to flow to the degree that molten metals do, intricate or complex shapes created by casting cannot be produced by powder metallurgy.
- Powder metallurgy articles, in most instances, do not have the same physical properties as wrought or cast pieces.
- Obtaining specific alloy powders can be difficult at times. Parts pressed from the top have a lower density at the bottom.
- This method cannot yield a fully deep structure.
- The procedure is not cost-effective for small-scale production.
- Powdering brass, bronze and a variety of steels is a difficult process.

9.5 SECONDARY OPERATIONS

While PM parts may be used as finished parts, they can require some additional operations after sintering to improve their properties or impart special characteristics to the parts.

1. Repressing

A second pressing process is one in which the component is compressed in a closed die to reduce porosity (i.e., increase density) for applications where density is critical to achieving the requisite mechanical or other physical properties. The pieces are sintered a second time after being repressed.

2. Sizing and coining

Sizing is the process of pressing a part through a finish die to ensure dimensional accuracy. Coining is a press working procedure performed on a sintering part to enhance the surface finish by further densification or to press information into its surface.

3. Finishing procedures

Some powder metal parts need additional finishing operations, which include: machining to produce holes or threaded holes, various other geometric or other processes, grinding to improve dimensional accordance and surface finish; plating to improve resistance to wear and corrosion and, in some cases, appearance; and heat treating to improve hardness and other mechanical properties.

4. Joining

This allows for the development of larger parts and more complex shapes. Many joining techniques exist, including diffusion bonding, sinter brazing and laser welding. The interconnected porosity is filled with an alloy whose melting point is lower than the sintering temperature of the metal from which the component is produced, for example, copper-based alloys penetrate ferrous components, typically during the sintering process.

5. Infiltration

Infiltration renders the components impermeable, and it improves mechanical properties at the expense of dimensional precision. Infiltration makes some heat treatments easier. For example, without interconnected porosity, it is easier to obtain a given case depth.

6. Impregnation

By impregnating sintered parts with oil or another non-metallic material, they are more resistant to corrosion. Self-lubricating bearings are created by impregnating porous sintered bearings with lubricants; these bearings are only possible by powder metallurgy.

9.6 AUTOMOTIVE APPLICATIONS OF POWDER METALLURGY

Powder metallurgy has offered a practical solution to the issue of manufacturing refractory metals, which have now become the foundation for producing heat-resistant materials and cutting tools of extreme hardness. Porous self-lubricating bearings are another common and useful commodity made from powdered metals. In short, modern technology would be impossible to imagine without powder metallurgy products, the numerous fields of application for which are expanding year after year.

In terms of increased efficiency, reduced fuel consumption and environmental effects, and improved safety, the automotive industry has undergone significant changes. PM has played an important role in these changes, both in terms of enhancing existing components and allowing for the integration of novel components into new systems. The advent of eco-cars may be seen as a challenge, but it is actually opening up opportunities for the use of PM in the development of new products in the areas of practical electromagnetic controls and drives. Some of the powder metal products are mentioned below.

- Porous components, such as bearings and filters.
- Tungsten carbide gauges, wire drawing dies, wire-guides, stamping and blanking equipment, blocks, hammers, rock drilling parts, and so on.
- Wrought iron powder is used to make a variety of machine pieces. Cutting tools made of tungsten carbide powders and titanium carbide powders that are highly heat and wear resistant are used in die manufacturing.

- Tungsten, tantalum and molybdenum components are used in electric lamps, radio valves, oscillator valves, X-ray tubes in the form of filament, cathodes, anodes, control grids, electric contact points, and so on.
- Products with complex shapes that require extensive machining when manufactured using other techniques, such as toothed components such as gears.
- Electrical connections, crankshaft drive or camshaft sprocket, piston rings and rocker shaft braces, door mechanisms, connecting rods and brake linings, clutch facings, welding rods, and so on.
- Products that include the combined properties of two metals or metals and non-metals, such as non-porous bearings, electric motor brushes, and so on.
- The development of porous metal bearings that are later impregnated with lubricants. Copper and graphite powders are used in the production of automotive parts and brushes.
- Cermets are metal–ceramic composites that are bonded in the same way that metal powders are. They combine the useful properties of ceramics' high refractoriness and metals' toughness.
- They come in two varieties: oxide-based and carbide-based.

9.7 SUMMARY

Powder metallurgy is a technology used to make products or articles out of powdered metals. Powder metallurgy processes are quiet and clean, and any complex or complicated form can be manufactured. Metallic powder can be produced in three ways: atomisation, chemical and electrolytic methods. The powder metallurgy process includes the development of powders, blending, compacting, sintering and a variety of secondary operations such as measuring, coining, impregnation, machining, plating, infiltration and heat treatment. The powder metallurgy process is not economical for small-scale manufacturing because the cost of the tool and die of a powder metallurgical set-up is relatively high. Various critical components having complex shapes used in different systems can be easily manufactured with the help of powder metallurgy technique.

9.8 REVIEW QUESTIONS

(1) List the different processes used to manufacture metal powder.
(2) List the various steps used to manufacture components using powder metallurgy.
(3) List the various primary operations carried out in the powder manufacturing process.
(4) List the various processes used to produce metallic powder.
(5) List the advantages of powder metallurgy.
(6) List the disadvantages of powder metallurgy.
(7) List the various applications of powder metallurgy.
(8) State the various limitations of powder metallurgy.

(9) Describe the mechanical processes used for the production of metal powder with necessary processes.

(10) Describe the following processes with a neat sketch:
   1. Mixing and blending
   2. Compaction
   3. Sintering.

(11) Describe the processing methods used for powder metallurgy.

(12) Describe secondary operations used in powder metallurgy.

(13) Describe the chemical processes used for the production of metal powder.

(14) Describe the atomisation process used for the production of metal powder.

(15) "Structure and properties can be controlled more closely in powder metallurgy than in other fabricating processes." Justify this statement with an example.

(16) "Powder metallurgy is important in the manufacturing process." Justify this statement.

(17) "Selection of suitable powder production method is important in powder metallurgy." Justify this statement.

10 Surface Treatment

LEARNING OBJECTIVES

- To Recognize the importance of surface processing operations in the automotive industry.
- To get details of classification of surface processing operations, their details, tools, equipments and procedures.
- To understand automotive painting process, paints, their constituents and application procedures.
- To understand the performance driven surface treatment technologies used in automotive applications.

10.1 INTRODUCTION

Surface processing operations are classified as

(1) Cleaning
(2) Surface treatments
(3) Thin-film deposition and coating.

Cleaning processes involve the removal of soils and contaminants that have accumulated as a result of previous processing or the factory environment. Surface treatments are mechanical and physical operations that modify the surface of a part in some way, such as improving the finish or impregnating it with atoms of a foreign material to change its chemistry and physical properties. Coating and thin-film deposition refers to a variety of processes that involve the application of a layer of material to a surface. They include chemical as well as mechanical methods.

DOI: 10.1201/9781003367321-10

10.2 SELECTION AND USE OF SURFACE TREATMENT AND CLEANING PROCESSES

Factors in selection of the surface treatment and cleaning process include:

Purpose

- The contaminant to be removed
- Substrate material to be cleaned
- Environmental and safety factors
- Degree of cleanliness required
- Size and geometry of the part
- Production and cost requirements.

Uses

- To prepare the surface for subsequent industrial processing, such as coating application or adhesive bonding
- To improve worker and customer hygiene conditions
- To remove contaminants that might chemically react with the surface
- To enhance the appearance and performance of the product.

10.3 SURFACE CLEANING PROCESSES: SAND BLASTING, TUMBLING, ALKALINE, ACID AND ELECTROLYTIC CLEANING

Metal surfaces must often be cleaned before subsequent operations to remove unwanted substances such as pigmented drawing compounds, unpigmented oil and grease, chips and cutting fluids, polishing and buffing compounds, rust and scale, and miscellaneous contaminants. This section describes common cleaning processes. Surface contaminants can be removed effectively by a suitable surface cleaning method, so surface cleaning is of great significance to solve the various problems in manufacturing.

(A) Mechanical cleaning process

Mechanical cleaning entails physically removing soils, scales or films from the work surface of a workpart using abrasives or other mechanical action. Mechanical cleaning processes frequently perform additional functions in addition to cleaning, such as deburring and improving the surface finish. Shot peening involves the application of a high-velocity stream of small cast steel pellets (called shot) to a metallic surface, which has the effect of cold working and inducing compressive stresses into the surface layers. Shot peening is primarily used to increase the fatigue strength of metal parts. Its purpose is thus distinct from that of blast finishing, though surface cleaning is achieved as a by-product of the operation.

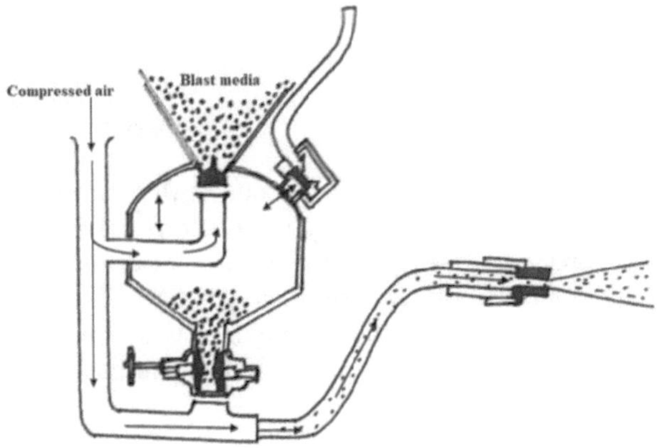

FIGURE 10.1 Abrasive cleaning.

(1) Abrasive blast cleaning/sand blasting

Depending on the finish requirements, blasting may be used to remove all classes of scale and rust from forgings, castings, weldments and heat-treated parts. Depending on the finish requirements, blasting may be used to remove the majority of scale, with pickling used to remove the remainder. Figure 10.1 shows abrasive blast cleaning.

Process

In general, abrasive particles such as sand, steel grit or shot are propelled against the surfaces to be cleaned.

If abrasive particles are used, it is referred to as abrasive blast cleaning; if sand particles are used, it is referred to as sand blasting. Some cleaning is done using a high-velocity air blast, which is guided by hand. The abrasive is fed from an overhead storage hopper to the centre of a radially rotating wheel, where the metallic shot or grit is hurled in a regulated stream onto the work to be cleaned. All traces of sand, scale, oxides and other impurities are removed all the way down to the virgin metal, creating an ideal surface for bonding finishing coatings. Applications include engine blocks, crankshafts, train cars, car wheels, oil and gas pipes, steel strips and a variety of other items.

Power brushing

For eliminating weld flux, heat-treatment scale, burns and other contaminants, the method employs both fibre and wire wheels (both powered). These brushes spin at a high velocity. They can provide a fine finish on metal surfaces when combined with fine abrasive compounds. Power brushing is often used to smooth sharp outlines left by grinding, allowing for homogeneous plating. Natural fibre brushes, horse hair, cord materials and other materials are also utilised. Synthetic fibres are also employed. Manual wire brushing is often used to clean weldments, castings, forgings and other metal surfaces.

Belt sanding and buffing

Belt sanding is a basic way of achieving a smooth finish on metal parts by utilising a belt sander in which an infinite abrasive coated belt rotates on two pulleys while the components are pressed against it. Buffing creates a gleaming and shiny surface. Buffing wheels are composed of very soft materials like felt, muslin or linen and are loaded with very fine abrasive particles combined with wax. A glossy surface is obtained when components are held against the rotating buffing wheel.

Cleaning by vibration

The components are filled with abrasive material and vibrated at 900–3600 cycles per minute in containers. The amplitude and frequency of vibration are chosen based on the shape and size of the components to be cleaned, the abrasive material used, the noise level, and so on.

(2) Tumbling

Tumbling is frequently the least expensive method of cleaning rust and scale from metal items. It is a type of mass finishing method. Barrel finishing, vibratory finishing, centrifugal disc finishing, centrifugal barrel finishing, spindle finishing, drag finishing and other fundamental mass finishing methods are listed below.

This procedure is used to clean, descale, deburr, radius edges and corners, modify surface conditions, and so on, and also to get rid of surface roughness, brighten the components to prevent corrosion, dry, alleviate tension, impart compressive stress, and so on. This method is also known as barrel finishing or tumbling barrel finishing, as shown in Figure 10.2.

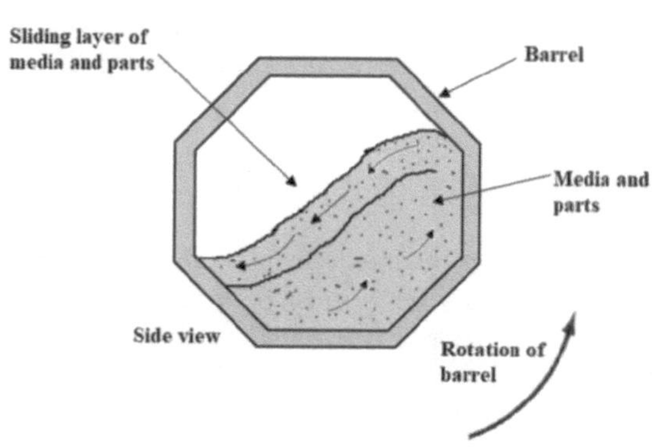

FIGURE 10.2 Tumbling.

- **Process**

This approach uses a horizontally oriented barrel with a hexagonal or octagonal cross-section, in which parts are mixed by spinning the barrel at speeds ranging from 10 to 50 revolutions per minute. Normally, the barrel is loaded to around 60% capacity with a mixture of components, media, compound and water. The load climbs upward to a turnover point as the barrel spins; then the force of gravity overcomes the inclination of the mass to stay together, and the top layer slides toward the lower section of the barrel. Work pieces are placed in a drum or barrel, together with stars, jacks, slugs or abrasive materials, to complete the operation. Sand, granite chips, slag or aluminium oxide pellets can be used as abrasive materials. During operation, the barrel rotates, and the movement of the workpieces and accompanying slugs or abrasive material against each other provides a fine cutting action through friction, which removes the fins, flashes and scale from the goods.

Drawbacks

- When compared to other mass finishing procedures, it is a relatively sluggish process.
- It is common for the processing to take many hours of tumbling.
- There is a greater need for floor space due to increased noise generation.

(B) Chemical cleaning process

Chemical cleaning is used to remove contaminants such as oils and grease from the surfaces of components using one or more of the procedures listed below.

(i) A solution in which the contaminant is dissolved in the cleaning solution.
(ii) Saponification is a chemical reaction that turns animal or vegetable oils into water-soluble soap.
(iii) Emulsification, in which the cleaning solution combines with the contaminant to produce an emulsion, with the contaminant and emulsifier suspended in the emulsion.
(iv) Dispersion, in which the surface-active ingredients in the cleaning solution reduce the concentration of the pollutant.
(v) Aggregation, in which lubricating impurities are removed from the surface by various cleaning solvent agents and gathered as big dirt particles.

Chemical cleaning procedures are classified into three broad categories as follows:

(i) Acidic cleaning
(ii) Alkaline cleaning
(iii) Solvent cleaning.

Alkaline solutions, emulsions, solvents, heated vapours, and different acids, salts and organic compound mixes are among the cleaning fluids used in chemical cleaning.

Chemical cleaning uses various types of chemicals to effect contaminant removal from the surface. The major chemical cleaning methods are

(1) Solvent cleaning
(2) Emulsion cleaning
(3) Alkaline cleaning
(4) Acid cleaning
(5) Ultrasonic cleaning.

In certain circumstances, chemical action is supplemented by other sources of energy; for example, ultrasonic cleaning employs high-frequency mechanical vibrations in conjunction with chemical cleaning. Here are a few examples:

1. Alkaline cleaning

The most common type of cleaning is with alkali. It is efficient and economical in removing oil and grease by saponification, emulsification or both.

Process

A bath is made with cleaning agents such as caustic soda or sodium metasilicate, which are mixed with some sort of soap to aid in emulsification. The mixture creates alkali, which is used as a cleaning agent.

Except for zinc, lead, tin, brass and aluminium, this procedure is utilised on all metals.

- **Emulsion cleaning**

Emulsifiable solvents are compounds formed by combining an organic solvent with a water-soluble emulsifying agent. An excellent emulsion is made by combining soap, an organic solvent and water. They are less harmful to the environment than chlorinated solvents. It is the least expensive technique used on the majority of metals. Aside from the procedures mentioned above, other chemical cleaning processes utilised by industry include vapour degreasing, electrolytic cleaning and electrolytic polishing.

Vapour degreasing is a type of cold solvent cleaning in which the surfaces to be cleaned are subjected to chemical vapours. Chlorinated solvents are used in this application. A hot alkaline cleaning solution is utilised in electrolytic cleaning, and a DC electric current is passed through the solution bath. It emits oxygen at the positive pole and hydrogen at the negative pole. This movement shatters the oil coating that holds dirt to the metal surface. The material of the component will influence whether it is an anode or cathode. For example, steel components clean better as an anode, whereas soft metals like zinc and tin clean better as a cathode. As the current runs through the metal, hundreds of small oxygen bubbles form on the surface, and as they burst, they produce a scrubbing, scouring or agitating motion, resulting in the cleanest surface.

(C) Solvent cleaning

Solvent cleaning is the most effective method of cleaning heavy oils, grease, dirt and fats from metal surfaces, using either straight solvents or emulsifiable solvents. Straight solvents may be either

(a) Petroleum based
(b) Chlorinated type.

Petroleum solvents are high flash point distillates (38–93^{0}C) that allow them to be used at room temperature. They are less expensive than chlorinated hydrocarbons but must be handled with caution due to their fire hazard. The soil (contaminants) dissolves in the solution after the pieces are immersed in the solvent bath and brushed clean. As a result, these solvents become polluted quickly. Chlorinated solvents (trichloro-ethylene and perchloroethylene) do not have the fire hazard of petroleum solvents and are commonly employed in vapour degreasers where the cold components are suspended in the vapour of the boiling solvent.

1. Acidic cleaning

After oil, grease and other non-acid-soluble contaminants have been removed from the metal surface, acid cleaners are used for pickling, deoxidising and bright dipping (usually by alkaline cleaning).

Acid pickling is a harsher technique used to remove heavier oxides, rusts and scales; it usually results in some etching of the metallic surface, which helps enhance organic paint adhesion.

Process:

Diluted sulphuric, hydrochloric or phosphoric acid is sprayed on the part, or it is dipped into a tank, agitated, then thoroughly cleaned and rinsed. Acid pickling is also used to remove oil and grease, and it is utilised to remove minor rust in some situations. When steel parts are cleaned with acid, hydrogen is produced, which is absorbed by the steel and causes "hydrogen embrittlement". Of course, the hydrogen in steel can be lowered by heating the pieces after pickling.

2. Electrolytic cleaning

When extreme cleanliness is required, this is an efficient final cleaning method for eliminating oil and grease from machined surfaces. It is nearly always used for the final cleaning of steel items before electroplating. It's also known as electro-cleaning.

Process

An alkaline cleaning solution is utilised in electrolytic cleaning. An alkaline cleaning solution is subjected to a 3–12-V direct current. As a result, oxygen is emitted at the positive pole while hydrogen is emitted at the negative pole. The electrolytic action generates gas bubbles at the component surface, resulting in a scouring action that aids in the removal of sticky dirt coatings. The materials used to make the part, as well as the required cleaning action, dictate whether the part should be anode or cathode. Parts made of soft metals, such as lead, zinc and tin, must be cleaned cathodically because they would be etched anodically if cleaned anodically.
Stainless steel can be cleaned anodically or cathodically.

3. Ultrasonic cleaning

In rare cases, it may be required to remove insoluble particles from difficult-to-reach cavities such as small holes, slots and indentations.
Ultrasonic scrubbing action is one of the most effective ways to clean them.

Ultrasonic cleaning is increasingly used to clean aircraft, automobiles, electrical equipment, computers, precision machine parts, optical and magnetic parts, and jewellery and watch parts. Small components are best suited to the procedure. A generator generates high-frequency electric energy, which is transformed into high-frequency sound waves by a transducer in the cleaning system. These high-frequency sound waves vibrate the cleaning solution and the component (to be cleaned) that is dipped in it. Acids, solvents and detergents are examples of cleaning agents. When sound waves move through a liquid, it cavitates or ruptures (due to its inelastic nature), creating microscopic vacuum pockets that almost instantly collapse, resulting in a powerful scrubbing motion. High-frequency systems (above 25 kHz) are employed, despite the fact that lower frequency cavitation occurs, although those plants are difficult to manage.

10.4 SURFACE COATING PROCESSES: ELECTROPLATING, GALVANISING, METAL SPRAYING, PAINTING

Coating is a process for bonding items, providing new functionality and improving surfaces that involves soaking the target's surface (substrate or target) with various adhesives or compounds. The approach has been implemented for a variety of applications in a wide range of industrial industries. A variety of coatings are applied to metal surfaces to improve their appearance, appeal and acceptability by the consumer, as well as to protect them against contamination caused by human touch or the environment. Coatings also improve functional properties such as insulation, corrosion resistance, fire resistance, and so on.

Coatings on the market vary in terms of how long they last, how easy they are to apply, how successful they are at protecting and how much they cost. Typically, there is a direct relationship between the cost of a coating and the length of time it will last. Coatings are classified as (a) interim coatings, (b) durable coatings and (c) permanent coatings based on this assessment. (a) Interim coatings: These prevent rust on components that are in the production process. Surface lubrication is occasionally used to aid subsequent forming operations, notably deep forming and wire drawing. Chemical conversion is the most well-known and commonly utilised intermediate coating method (or conversion coatings). Phosphate coatings, chromate coatings and anodic coatings are subdivided further. (b) Durable coatings: Durable coatings last for a longer amount of time. These are further divided into two types: (i) organic coatings formed from resins such as alkyds, vinyls and epoxies and (ii) inorganic coatings derived from porcelain enamels and different plasma-sprayed materials. (c) Permanent coatings or metallic coatings: Permanent coatings are metal or inorganic coatings. Zinc, tin, lead, copper, cadmium, chromium, brass, aluminium and stainless steel are the most common coating materials. These are applied to work surfaces using a variety of techniques including hot dipping, plating, metal spraying, vacuum metallising and sputtering. Various types of important coatings used in industry are detailed in the following subsections: (a) conversion coating and (b) organic coating.

Use of metal coating process

- Provide corrosion protection
- Enhance product appearance (e.g., providing a specified colour or texture)
- Increase wear resistance and/or reduce friction of the surface
- Increase electrical resistance
- Increase electrical conductivity
- Prepare a metallic surface for subsequent processing
- Rebuild surfaces worn or eroded during service.

Selection of the metal coating process

Selection of metal coating process usually involves various factors, hence it is essential to understand the process of selection of metal coating.

Process selection includes:

- General factors such as process availability and quantity required.
- Job factors such as size, weight and machinability.
- Surface preparation factors such as undercutting and tolerances.
- Finishing factors such as required finish and post-surfacing actions.

Materials selection includes:

General considerations such as familiarity with similar applications and pricing.
The operating environment, such as wear kinds and lubrication.
Required properties include wear resistance, hardness and machinability.
Substrate characteristics such as past surface treatments and size.

Permanent coatings are made of metal or inorganic materials. The melting point of the base metal (on which the coating is applied) should be higher than that of the coating or plating metal. Lead, copper, tin, nickel, zinc, cadmium, silver, gold, chromium, brass or bronze can be used to create a wide range of metal coatings. The procedures listed below are routinely employed to provide metallic coatings:

- Hot dipping
- Electroplating
- Galvanising
- Tin plating
- Metallising
- Sputtering.
- Hot dipping

This procedure begins with a thorough cleaning of the base metal using acid or other means, followed by suitable fluxing (typically with zinc ammonium chloride), and finally immersing the base metal in a bath of molten coating metal. It is the most widely utilised approach because it is the least expensive. The most popular coating materials are aluminium, lead, tin and zinc. Aluminium, zinc and tin coatings on steel and lead and zinc alloys on nonferrous metals are examples of hot-dip processes.

1. Electroplating process

Electroplating, also known as electrochemical plating, is an electrolytic technique that deposits metal ions in an electrolyte solution onto a cathode. In the electrolytic process, the anode is typically formed of plated metal and acts as a source of coating metal. The workpiece on which the coating will be applied is made as a cathode. Figure 10.3 shows the electroplating process.

Faraday's two basic laws underpin electrochemical plating. For our purposes, the laws are as follows:

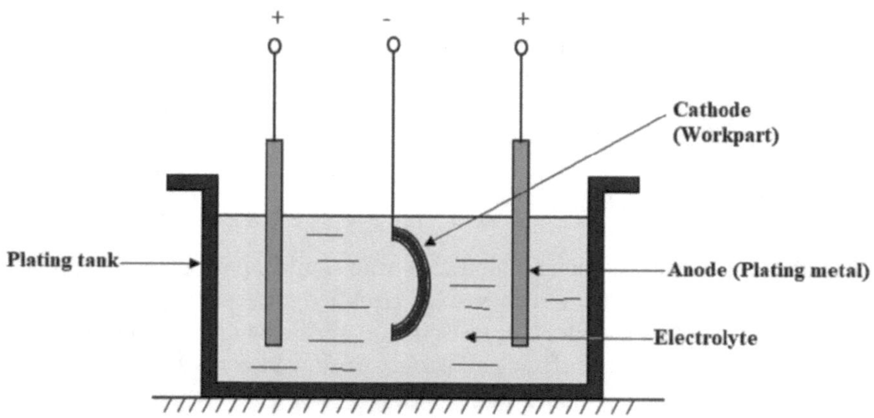

FIGURE 10.3 Electroplating process.

(1) The mass of a substance liberated in electrolysis is proportional to the amount of electricity passed through the cell; and
(2) The mass of the material liberated is proportionate to its electrochemical equivalent (ratio of atomic weight to valence). The following equation summarises the impacts:

$$V = K\,I\,t$$

where V is the volume of metal plated in m^3
I is the flowing current in ampere,
T is the time for which current passes through, and
K is a constant depending on electrochemical equivalent
and density of electrolyte is $m^2/A - S$.

- An external power source sends direct current through an electrolyte solution, which is an aqueous solution of acids, bases or salts.
- An electrolyte transmits electric current in solution by the mobility of plate metal ions.
- To achieve the best results, the items to be plated should be chemically cleansed.

Methods and applications

For electroplating, a range of equipment is available, with the choice based on item size and shape, throughput requirements and plating metal, including:

(1) Barrel plating
(2) Rack plating
(3) Strip plating.

Plating on barrels

This is carried out in rotating barrels that are either horizontally or at an oblique inclination.

The process is well-suited for batch plating of many small parts. Electrical contact is maintained by the tumbling action of the pieces as well as an externally attached conductor that projects into the barrel.

Rack plating

This is utilised for parts that are too big, heavy or complicated for barrel plating. The racks are made of heavy-gauge copper wire that has been shaped into appropriate forms to hold the parts and conduct current to them. The racks are designed so that workpieces can be hung on hooks, held in place by clips, or put into baskets.

Strip plating

This is a high-production process in which the work consists of a continuous strip drawn through the plating solution by a take-up reel. One example of a possible use is plated wire. This process can also be used to plate small sheet-metal parts held in a long strip. Zinc, nickel, tin, copper and chromium are common electroplating coating metals. Steel is the most widely used substrate metal. Jewellery is plated with precious metals (gold, silver and platinum). Electrical connections are also made of gold. Fasteners, wire goods, electric switch boxes and various sheet-metal parts are examples of zinc-plated steel products. The zinc coating acts as a sacrificial barrier against corrosion of the steel underneath. Nickel plating is applied to steel, brass, zinc die castings, and other metals for corrosion protection and ornamental purposes. Automotive trim and other consumer goods are examples of applications. In addition, nickel is employed as a base coat behind a considerably thinner chrome plate.

Tinplate is still commonly used to protect against corrosion in "tin cans" and other food containers. Tin plate is also used to increase electrical component solderability. As a plating metal, copper has a variety of significant applications. It is extensively used as a decorative coating on steel and zinc, either alone or as a brass plate alloyed with zinc. It is also widely used in plating printed circuit boards. Chromium plate (often referred to as chrome plate) is prized for its beautiful appearance and is frequently used in automobile products, office furnishings and kitchen appliances. It also provides one of the toughest electroplated coatings, making it popular for parts requiring wear resistance (e.g., hydraulic pistons and cylinders, piston rings, aircraft engine components, and thread guides in textile machinery).

2. Galvanising

This is a hot-dipping method that involves immersing a metal substrate in a molten bath of a second metal and then coating the second metal onto the first. The first metal, of course, must have a higher melting temperature than the second. Steel and iron are the most common substrate metals. The most popular coating metals are zinc, aluminium, tin and lead. Depending on the coating metal, hot dipping is known by several names: galvanising is the process of coating steel or iron with zinc (Zn); aluminising is the coating of aluminium (Al) onto a substrate; tinning is the coating of tin (Sn); and terneplating is the plating of a lead–tin alloy onto steel. The following are the steps in the galvanising process. An alkali wash is used to remove oil and grease from the metal sheet. The metal is then pickled in dilute sulphuric acid for 15–20 minutes at 60–90°C to eliminate rust and scale.

After that, the surface is prepared by treating it with 5% hydrochloric acid to dissolve any sand grains, etc., and then storing it under water to prevent oxidation. The metal is then put through a 5–20% zinc ammonium chloride solution to remove any surface oxide. It is then immersed in a bath of molten zinc at around 450°C for zinc metal coating, before being put through a pair of heated rollers to ensure that the coating is uniform. It is then annealed and slowly cooled.

Application

Galvanising is extensively used to protect iron that is exposed to the elements, such as roofs, wire fences and pipes.

3. Spraying of metals

The coating metal is sprayed on the cool surface of the base metal while it is still molten. The procedure is carried out with the assistance of a specialised spraying gun. An oxy-hydrogen flame or an electric arc is used to melt the sprayed metal. The molten metal is then sprayed with a compressed air jet. Because some of the metal is oxidised, the final spray contains a mixture of metal and its oxide. Most guns employ motorised rollers to feed metal in the form of wire to the flame, although some use powder or granulated metal. Compressed air is used to totally atomise the molten metal or oxides and project them against a prepared surface where they are implanted, ensuring good mechanical adherence. This is depicted in Figure 10.4.

The surface must first be roughened and free of dirt, oil and grease. Compressed air aids in cooling the work parts, allowing asbestos and some polymers to be used as coatings. A metal spray gun can be used manually or installed on a machine. Spraying is used to coat metals such as aluminium and copper. Spraying protects and coats wood, plastic and worn-out machine and building parts.

Advantages

- The coating can be applied to finished goods.
- Because coating can be applied to pieces after they have been assembled, there is no risk of failure, or being harmed during the assembling of pieces.

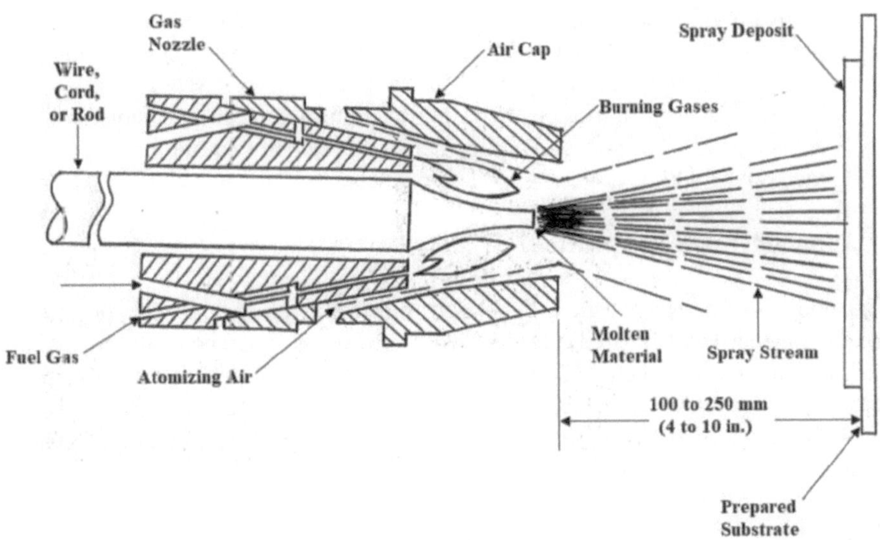

FIGURE 10.4 Metal spraying.

- Coating can also be applied to large and oddly shaped items.
- Increased productivity.

Disadvantages

- The coating is inconsistent and slightly porous.
- The coating's adhesive strength is relatively poor (as compared with hot-dipping or electroplated).

Applications

- Sprayed coatings can be applied to non-metallic surfaces such as wood, plastic, and so on.
- Metal spraying is a technique for reclaiming worn-out machine parts.

Sputtering

Sputtering is yet another method of depositing thin coatings on a substrate. When a discharge is initiated in an argon atmosphere, atoms are sputtered or ejected from a cathode source (material to be utilised in plating) and deposited on the workpiece surfaces (anode). Other minerals deposited include quartz, alumina, carbides, nitrides and glasses, in addition to elemental metals.

10.5 AUTOMOTIVE PAINTING PROCESS, INGREDIENTS, PAINTING PROCEDURE

Paint performs the following important functions:

- Paint provides a "skin" to protect the body substrate (steel, aluminium and plastics) from the elements. Most motor vehicles are constructed primarily of steel sheet metal.
- If this steel were left uncovered, the reaction of oxygen and moisture in the air would cause it to rust. Painting serves to prevent rust, therefore protecting the body.

Automotive painting process

The body is basically washed, degreased, electrodipped in zinc phosphate, oven baked, sanded, sealed, dried, cleaned (with chemicals), primered, baked, painted, baked, clear coated, baked, and finally examined for defects (which are repaired). All high-volume manufacturers have completely automated lines that finish a single body in 14–18 hours. Even though the paint process varies by manufacturer and sector, the basic procedure is as follows (Figure 10.5).

As a result, layers of meticulously set out coatings result, each of which is only slightly thicker than typical human hair.

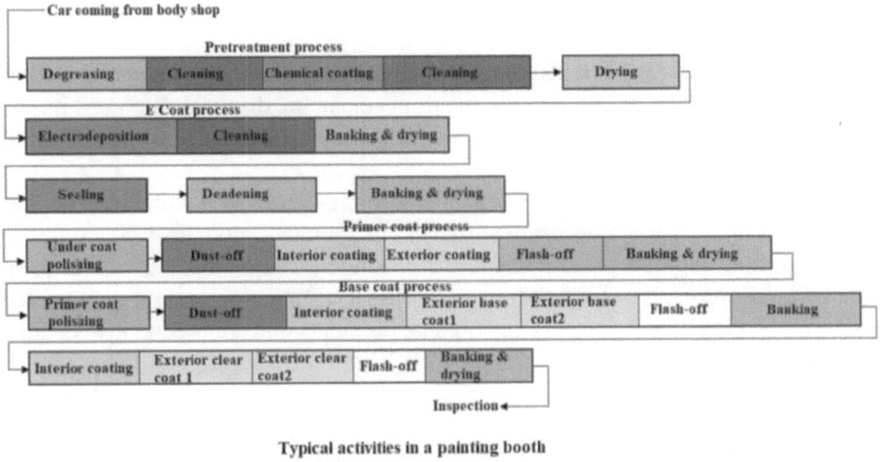

FIGURE 10.5 Typical activities in a painting process.

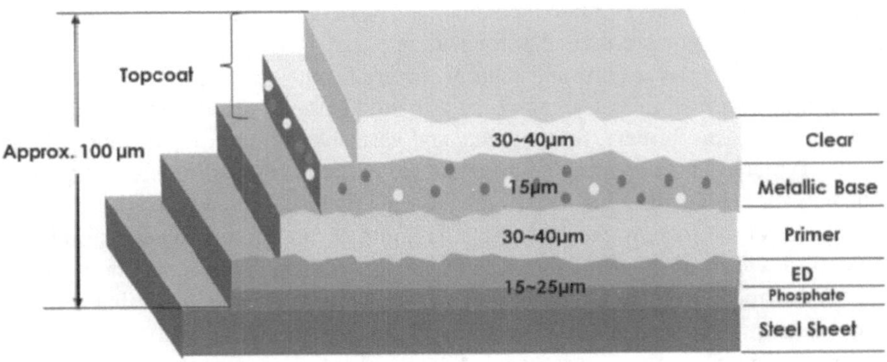

FIGURE 10.6A Automotive coating process.

Pre-treatment is the process of removing and cleaning superfluous metal and forming an acceptable surface structure allowing for the bonding of a corrosion prevention layer. The electrodeposition (ED) of the anticorrosion or rust prevention layer is the next stage. For anticorrosion and the elimination of water leaks, a sealant such as poly vinyl chloride (PVC) is used as it also reduces the chipping and vibrating noise.

A primer is then used to increase adhesion between the surface and the basecoat; it also serves as a barrier between the surface and the base coat. It also provides a smoother surface for successive layers and is anti-chipping. Finally, topcoats, which include a basecoat and a clearcoat, are applied; they give colour, look, gloss, smoothness and other desirable surface attributes. Figure 10.6a shows various layers of the coating.

Ingredients of the paint

The selection of components used to manufacture paint will affect its stability (shelf life), application characteristics, handling, clean up, disposal, and most importantly, the performance of the product on which it is applied. Paint formulas usually include the following.

A pigment is a powdered component of paint that determines its tone and colour. Pigments' primary role is decorative, that is, to provide colour and covering capacity to the film. Pigments also improve the coating's strength and performance, and some of them have corrosion inhibitor qualities. Pigments are natural or manmade oxides, metal salts, metal powders and organic pigments that are used to give coatings vivid hues.

Special pigments enable the creation of effects such as "chameleon", "metallic", "nacre" and "luminous surface". A binder is a pure or dispersion solution whose primary role is to hold the painting material together; that is, it is the element that gives the paint a key quality – adhesion, or the capacity to stay on the painted surface. After painting, the binder firmly retains the colouring pigment and creates a smooth, glossy surface. Binders are present in all paint and varnish materials (coatings). Powder paintwork coatings may or may not contain solvents, and clear automobile varnishes may or may not contain coloured pigments.

A filming agent is a pure and natural material, but it can also be synthetic. In the paint and varnish industry, many natural and synthetic resins and vegetable oils are utilised as film formers. Some paints and varnishes contain two or more filming agents. Typically, film-forming compounds are either excessively viscous liquids or delicate solids. Solvents are added to the painting material to lower its viscosity.

Because of the solvent, the paint remains liquid during production and storage. As solvents, a wide variety of organic liquids (hydrocarbons, ketones, alcohols, ethers) and combinations of them are utilised. Thus, the solvent is an element that gives the paint liquid qualities that are required for uniform application. The solvent evaporates throughout the painting process, leaving a two-component mixture of pigment and binder.

The final volume and density of a painting are created by a filler. Fillers are white or faintly tinted powders of cheap natural minerals in their composition (talc, gypsum, mica, kaolin, chalk, etc.). Some fillers, such as mica or asbestos, are used to improve heat resistance.

Modern painting materials are multicomponent mixtures that include a variety of target additives designed to improve specific product qualities. In particular, additives can improve coating elasticity (plasticisers), resistance to aging (antioxidants), fire resistance (flame retardants), absorb UV radiation, prevent pigment from precipitating, prevent the creation of a surface layer during storage, and improve bottling.

Some additives, such as drying accelerators, plasticisers, matting and structuring additives, can be added to the painting material right before usage.

There are four main methods of applying paint:

- By spreading, e.g. by brush, roller, paint pad or doctor blade

- By spraying, e.g. air-fed spray, airless spray, hot spray and electrostatic spray
- By flow coating, e.g. dipping, curtain coating, roller coating and reverse roller coating
- By electro-deposition.

10.6 SURFACE FINISHING PROCESS: POLISHING, BUFFING, BURNISHING, SUPER FINISHING

Surface finish is a term that refers to the process used to alter a metal's surface by adding, removing or reshaping. The goal is to protect the metal and improve the aesthetic side. The result depends on the metal finishing method. Often, there are a few different ways to achieve the same or similar results.

1. Polishing

Polishing is a surface finishing technique that uses a polishing wheel to remove appreciable metal from rough surfaces in order to remove scratches, tool marks, pits and other imperfections. Generally, the size and form of the completed surface are unimportant in polishing, but tolerances of 0.025 mm or less can be reached in machine polishing. Leather, papers, canvas, felt or wool are used to make polishing wheels. Except for honing, lapping and super finishing, polishing can be performed after any of the machining procedures. Typically, many processes are required to remove the faults and then apply the desired shine to the surface. Polishing is a technique that is very similar to grinding.

The work can be manually pressed to wheels installed on floor stand grinders. They are broadly divided into two types: endless-belt machines and coated abrasive wheels. Polishing is a technique for removing scratches and burrs and smoothing rough surfaces that uses abrasive grains attached to a polishing wheel that rotates at a high speed of around 2300 m/min (7500 ft/min). Because the wheels are composed of canvas, leather, felt and even paper, they are relatively flexible. The abrasive granules are attached to the wheel's outer rim.

The wheel is supplied with new grits when the abrasives have been worn down and used up. Rough polishing is done with grit sizes ranging from 20 to 80, finish polishing with grit sizes ranging from 90 to 120, and fine finishing with grit sizes greater than 120. Polishing activities are frequently performed by hand.

2. Buffing

Buffing looks similar to polishing but has a different purpose. Buffing is used to create aesthetically pleasing surfaces with a high sheen. Buffing wheels are constructed of materials comparable to polishing wheels (leather, felt, cotton, etc.), but they are often softer. The abrasives are very fine and are housed in a buffing compound that is forced into the wheel's exterior surface while it revolves, as opposed to polishing, where the abrasive grains are bonded to the wheel surface. The abrasive particles, like those used in polishing, must be renewed on a regular basis. Buffing is typically done

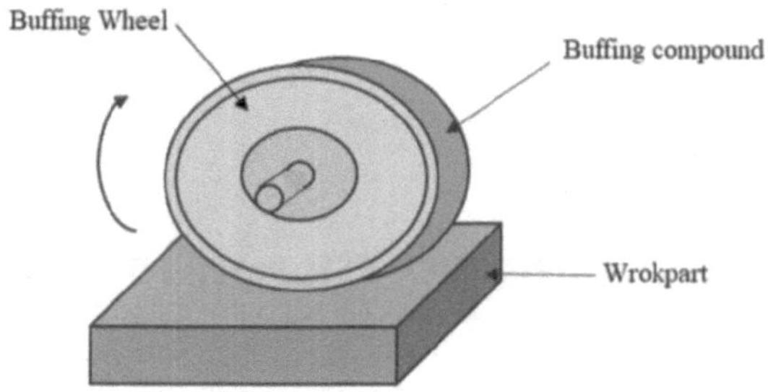

FIGURE 10.6B Buffing process.

by hand, though machines have been devised to automate the process. Speeds range from 2400 to 5200 m/s (8000 to 17,000 ft/s). Figure 10.6b shows a schematic diagram of the buffing operation.

- Applications
- These include boats, automobiles, motor-cycles, sporting items, store fixtures, bicycles, tools, commercial and residential hardware and household utensils and appliances.

3. Burnishing

Burnishing is a cold working procedure that causes plastic deformation on metallic surfaces by exerting pressure with a ball or roller. It is a process of completion and strengthening. The use of this procedure can improve surface finish, surface hardness, wear resistance, fatigue resistance, yield and tensile strength, and corrosion resistance. Burnishing is one of the most essential finishing techniques used to improve the fatigue resistance of components. The following are examples of burnishing tools applied to give a gloss or fine surface finish, typically in operations involving cold treatment of metal surfaces.

A burnishing tool creates a polished surface on a turned or bored metal surface by rotating hardened rolls in a continuous planetary rotation. The rotation of the rolls raises the yield point of the metal surface's soft section at the point of contact. The point of contact causes the metal surface to deform, resulting in a completed metal surface. The point of contact causes the metal surface to deform, resulting in a completed metal surface. Burnishing tools come in a variety of shapes and sizes. Burnishing is not a metal cutting technique. Chips are not formed during the burnishing process. It is essentially a cold forming process in which metal close to a machined surface is displaced from protrusions to fill depressions. Because of the work hardening of the surface during burnishing, there will be a hardened layer on the surface, which is believed to boost the component's fatigue resistance. In addition to improving surface polish and fatigue strength, the burnishing process improves wear and corrosion resistance.

Dimensional accuracy and surface polish are given significant consideration in today's manufacturing business. Thus, measuring and describing surface roughness can be used to forecast machining performance.

Burnishing is a procedure that uses a small degree of plastic deformation to produce an exact change in the surface roughness of a workpiece. The metal on the surface of the work item is redistributed without material loss during the burnishing process. Aside from generating a fine surface quality, the burnishing process has additional benefits such as enhanced hardness, corrosion resistance and fatigue life due to the produced compressive residual stress.

Other methods of burnishing are discussed below.

(a) Barrel burnishing

This is the process of polishing the inside of a barrel. It is similar to barrel rolling in that medium balls, shots or round pins are put to work in the barrel instead of an abrasive medium. Burnishing has no cutting action. Burnishing does not normally eliminate obvious scratches or pits, but it does generate a smooth, uniform surface and minimise surface porosity.

(b) Roller/ball burnishing

Internal and external flat, cylindrical or conical surfaces are polished using hardened steel or cemented carbide rollers or steel balls set in a holder. Figure 10.7 shows the ball burnishing process.

Applications

These include hydraulic system components, seals, valves, spindles and fillets on shafts.

4. Super finishing

Super finishing is a process that uses bonded abrasive stones in a specific way to achieve an extremely high-quality surface finish with nearly no faults in the surface layer. In super finishing, a very thin layer of metal (0.005–0.02 mm) is removed. This operation can be used on the external and internal surfaces of steel, cast iron and non-ferrous alloy parts. In super finishing, an abrasive stick with a very fine grit (grain size 400–6000 is held in an appropriate holder and applied to the surface of the workpiece with light spring pressure. The stick is fed and oscillated, and the workpiece is rotated or reciprocated according to the needs of the superfinished shape. The work rotational speed in this process is low (2–20 m/min), and the longitudinal feed ranges from 0.1 to 0.15 mm per workpiece revolution. The abrasive stick oscillates fast in short strokes [2–5 mm] with a frequency ranging from 500 to 1,800 strokes per minute, and springs hold the stick against the work with a force ranging from 2 to 10 kg.

For super finishing, special general-purpose machine tools are available. Other types of regular machines, particularly lathes, are occasionally used for this purpose.

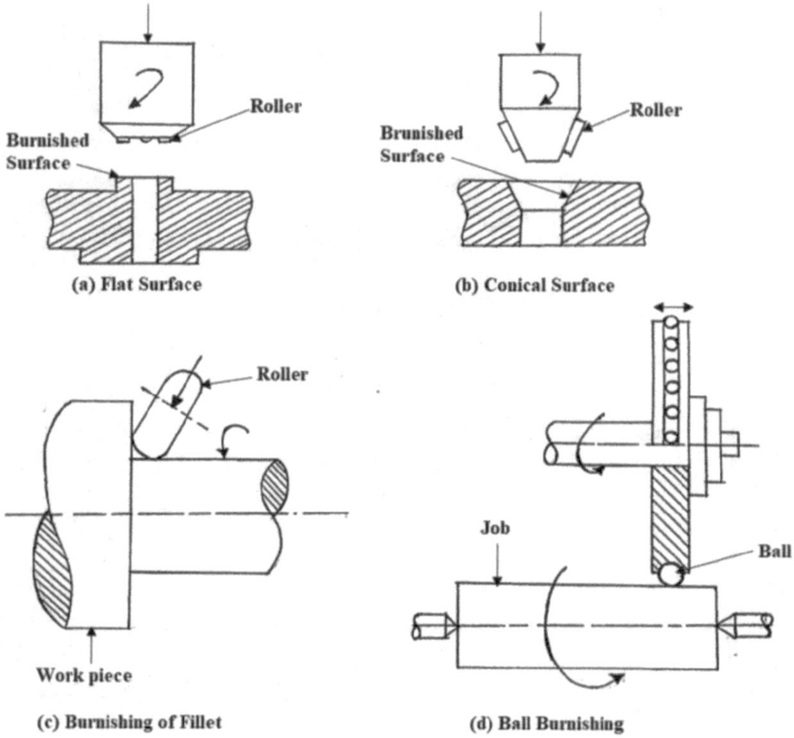

FIGURE 10.7 Ball burnishing.

Single-purpose machine tools, such as those used to finish crankshaft journals, camshafts, and so on, are also used.

10.7 SURFACE TREATMENTS FOR AUTOMOTIVE MANUFACTURING

Surface treatments used in the daily manufacturing of automotive parts are chosen to meet the functional and decorative requirements of mass production. Modern automotive systems require increased loads (mechanical, thermal, etc.), longer life time, weight reduction, friction reduction and corrosion resistance. Improved and innovative deposition processes in PVD, PECVD, thermochemical heat treatment and thermal spraying have been developed over the past decade. These novel treatments are increasingly being used in powertrain and engine applications. Creating optimum surfaces for various substrate materials (e.g., Al-alloys, case hardened steels, etc.) and geometries (e.g., bores) has an impact on operating costs.

Surface solutions based on new or enhanced deposition processes, such as PVD (physical vapour deposition) and PACVD (plasma-assisted chemical vapour deposition), thermochemical heat treatment and thermal spraying, have been developed and industrially used in recent decades.

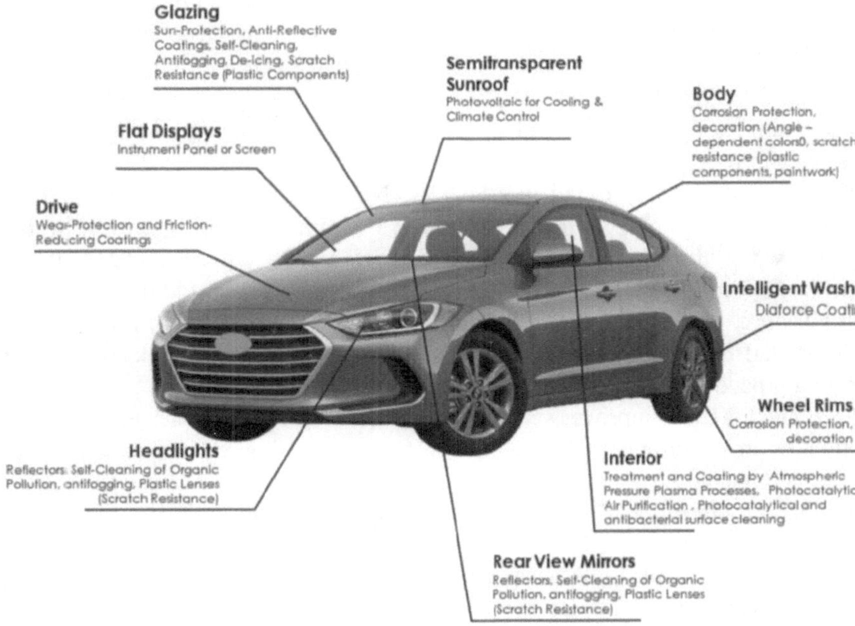

FIGURE 10.8 Application of surface treatments in the car industry.

- **Plasma-assisted surface treatments**

Plasma-assisted surface treatments encompass various plasma generation techniques in terms of the plasma employed, the coating material generated, and the interaction between the material flow and the substrate surface (coating) or the substrate surface near region (diffusion processes). There are two kinds of plasma used: local thermodynamic equilibrium (LTE) plasma and thermodynamic non-equilibrium plasma (cold plasma). At atmospheric pressures, the basis for plasma spraying is the local thermodynamic equilibrium plasma, also known as short thermal plasma. The substance used in the coating process is melted by this heated plasma. During the coating growth process, the plasma does not play a significant role. Cold plasmas are used in plasma-assisted PVD and CVD processes, as well as plasma-assisted thermochemical heat treatment (TCHT, such as plasma nitriding operations). Cold plasma is defined by the presence of both hot electrons and cold ions. During the coating development or diffusion treatment, energised molecules, atoms and ions are employed.

- **Thermochemical heat treatment (TCHT)**

Nitriding, nitrocarburising, oxidation thenitriding and nitrocarburising processes are based on the diffusion of nitrogen and nitrogen plus carbon at temperatures ranging from 350–600°C, primarily into ferrous materials such as steel grade and cast iron, but also into Ti alloys, Al alloys and other metallic surfaces. To get the best nitriding outcomes, special nitriding steel grades were designed. Nitrogen and carbon absorption form a functioning surface zone on top of the treated material. That zone could be a pure diffusion zone with specific nitrogen content. Depending on the steel grade, nitrides of alloying elements, such as CrN, may occur.

A high enough nitrogen concentration at the surface, on the other hand (depending on the treatment method, time and temperature, as well as steel grade), results in the development of iron nitride phases, γ′-Fe4N or ε-Fe2-3N.

Carbon is commonly found in ε-Fe2-3N. Nitrides forming the compound zone on top of the diffusion zone (thickness up to 0.8 mm) are visible as a white layer (also known as the white zone, compound layer) with a typical thickness ranging from 2 to 25 m. The compound zone can be oxidised as an additional step. Salt bath nitriding, gas nitriding and plasma nitriding are the most often used industrial nitriding and nitrocarburising processes. Sulphur is occasionally added to the gas nitriding procedures, resulting in sulphurnitriding.

In general, TCHT improves surface hardness, increases wear resistance, decreases friction, improves fatigue behaviour and increases corrosion resistance.

- **Plasma nitriding (PN) and plasma nitrocarburising (PNC)**

Plasma nitriding and nitrocarburising are vacuum processes (mbar range in conventional nitriding furnaces) used to treat steels, cast iron and Ti alloys for diffusion treatment in automotive applications. To energise, atomise and ionise the gas containing the diffusion components, such as nitrogen for plasma nitriding and nitrogen and carbon for nitrocarburising, a glow discharge (cold plasma) is utilised. The gas mixes also have a high concentration of hydrogen and, in certain cases, Ar. At a typical temperature of 350–600°C, active nitrogen (and carbon) infiltrates into the steel as a result of impinging on the heated surface. The layer is a diffusion layer with a specific nitrogen concentration in the hydrogen/nitrogen gas mixture. However, compound layers of the γ′ or ε′ types can also be created. In comparison to salt bath nitriding, GN and GNC, the easy management of compound zone growth is a benefit. The majority of nitriding procedures for automotive parts are performed in hot-wall furnace plasma nitriding systems.

- **Gas nitriding (GN) and gas nitrocarburising (GNC)**

The gas nitriding process, also known as ferritic gas nitriding, is based on nitrogen diffusion, often using ammonia with or without atmospheric dilution. The majority of steel components are heated to 500–520°C. Because of the temperature and the catalytic effect of the steel surface, the ammonia is dissociated. To treat spring steels, a unique process operates at temperatures lower than 500°C. As demonstrated in the following equation, ammonia dissociates into atomic nitrogen and hydrogen:

$$2NH_3 > + 2N + 6H$$

The instable atomic nitrogen is immediately combining to form molecular nitrogen according to the following equation:

$$2N + 6H > N_2 + 3H_2$$

10.8 APPLICATIONS OF SURFACE TREATMENTS IN THE CAR INDUSTRY

- **Fasteners**

Nuts, bolts, screws, washers, and so on are the things that hold things together. A fastener's finish is crucial to its function. Most fasteners are coated to prevent corrosion and to achieve special properties like controlling the amount of torque required to tighten a threaded fastener. The automotive industry is the single largest fastener consumer of any industry, using 26 billion parts per year out of the 200 billion produced, or roughly 42% of the industry. Because of this consumption, the automotive industry is a significant player and trendsetter in the fastener industry. Because fasteners are used on all parts of a vehicle, each part necessitates a different set of tolerances. The requirement of the auto industry to meet government regulations has an impact on the product mix available to fastener manufacturers and finishers. However, automakers continue to demand the same or higher quality standards in part performance, challenging finishers to come up with better alternatives. The use of ZnNi on fasteners is steadily increasing as the demand for greater corrosion resistance grows.

- **Brake callipers**

Cast iron brake callipers are an essential part of modern braking systems. They are electrolytically zinc plated for improved corrosion resistance and aesthetic appeal. Sealers can be used to improve corrosion resistance even further. It is necessary to use inorganic-based sealers that do not contain any organic lacquers for brake fluid compatibility. Steering knuckles are another cast iron application that can be electroplated for increased corrosion resistance.

- **Tubes for fluid delivery**

Fluid supply tubing for automobiles, such as power steering, air conditioning and brake and gasoline lines, is typically made of low-carbon steel, necessitating a corrosion-resistant coating. Many fluid-delivery tubes are processed in straight conditions and then subjected to a post-forming process for productivity reasons. As a result, a highly ductile coating is required that retains its corrosion-protective qualities even after bending.

- **Tie rod ends**

Tie rod ends are often plated after the rubber coating has been applied. As a result, the plating process must be kept to a maximum temperature of 80°C.

- **Anti-vibration components**

A car contains a huge number of anti-vibration components, such as suspension bushes and engine mounts. Depending on the kind and needs, several surface finishes are applied. Anti-vibration components are coated by: (1) phosphating; (2) plating before/after vulcanisation Traditionally, those parts receive a zinc phosphate coating for corrosion protection and effective rubber-to-steel bonding. As the need for corrosion resistance grows, an increasing proportion of anti-vibration components are zinc-nickel plated. The main challenge of this application is to plate on rubber and steel parts. There are two methods: (1) x plating before vulcanisation – companies choose plating before vulcanisation to avoid contamination of the plating bath and complete coverage of the crevice area around the rubber; no sealer can be used because the heat treatment for vulcanisation would destroy it; proper adhesion of the rubber on the plated layer is critical. (2) x plating after vulcanisation – plating of already vulcanised parts allows the employment of a sealer for greater corrosion protection; plating temperatures are kept below 80°C to avoid rubber attack.

- **Shock rods**

Piston rods used in shock absorbers, strut rods and gas springs require a strong chromium-plated surface to provide exceptional wear and corrosion resistance as well as a low coefficient of friction. Without this coating, the service life of these components would be extremely short. The components are installed on cars in regions where they are subject to significant environmental impact, such as caustic saltwater solutions from de-icing road surfaces and stones smashing against them from the road.

- **Piston rings**

Piston rings form a seal between the engine piston and the cylinder wall. The service life of the rings has been significantly increased since the introduction of hard chromium plating on these components. Hard chromium has great wear resistance and a low coefficient of friction, which is very significant in engine applications.

- **Engine valves**

Engine valve systems are hard chromium plated to give superior wear and a low coefficient of friction. The requirement to replace these components has been substantially decreased with the introduction of this type of surface coating.

- **Car door handles**

The automobile industry wants ever-lower weight and, as a result, fuel savings, which is a key factor in car manufacture. Placing on plastics (POP) is the optimum solution since it combines relatively low weight and production costs of plastic parts with exquisite metallic appearance. Today, the majority of modern car door handles are constructed of plated plastic.

- **Emblems**

Today, practically all OEM insignia are manufactured of plated ABS or ABS/PC blends. Several procedures are required to apply a corrosion-resistant and ornamental metallic coating to a piece of plastic.

- **Front grills**

The majority of current car front grills are composed of plated plastic, usually ABS polymers. Several procedures are required to turn this piece of plastic into a functional and ornamental grill. Plated cast aluminium wheels are fitted to about half of all new cars worldwide. While the usual finish is likely to be paint or powder coating, there is an increasing demand for the bright nickel/chromium-plated finish.

- **Fuel injection housing**

Injection pumps used in diesel engines are typically built of cast aluminium covered with an electroless nickel coating. The pumps work at extremely high pressures and are powered indirectly by gears or chains from the crankshaft. Electroless nickel coatings are utilised when excellent corrosion and wear protection is required. Furthermore, electroless nickel deposits a consistent thickness across the whole item without the requirement for costly anode fixturing. Diesel cylinder liners are strong chromium plated to give high wear resistance and a low coefficient of friction. The requirement to replace these components has been considerably decreased with the introduction of this surface coating.

10.9 SUMMARY

The surface processing operation involves cleaning, surface treatments and thin-film deposition and coating processes. Mechanical cleaning entails physically removing soils, scales or films from the work surface of a workpart using abrasives or other mechanical action. The chemical cleaning process involves solvent cleaning, emulsion cleaning, alkaline cleaning, acid cleaning and ultrasonic cleaning. Acid pickling is a harsher technique used to remove heavier oxides, rusts and scales; it usually results in some etching of the metallic surface, which helps enhance organic paint adhesion. Coating is a process for bonding items, providing new functionality and improving surfaces that involves soaking the target's surface (substrate or target) with various adhesives or compounds. Modern painting materials are multicomponent mixtures that include a variety of target additives designed to improve specific product qualities. Surface treatments used in the daily manufacturing of automotive parts are chosen to meet the functional and decorative requirements of mass production.

10.10 REVIEW QUESTIONS

(1) State the purpose of surface cleaning processes.
(2) Describe the use of surface cleaning.
(3) List the factors responsible for the selection of the surface cleaning process.
(4) State the purpose of the surface coating processes.
(5) Describe the use of surface coating.
(6) List the factors responsible for the selection of the surface coating process.
(7) List the various surface cleaning methods.
(8) State the applications of buffing, burnishing, polishing, galvanizing and electroplating.
(9) List the various mechanical cleaning methods. Draw neat sketches of the following
 (i) Tumbling
 (ii) Metal spraying
 (iii) Galvanising
 (iv) Electroplating
 (v) Burnishing
 (vi) Buffing.
(10) Describe the tumbling process with a neat sketch.
(11) Describe the galvanising process with a neat sketch.
(12) Describe the buffing process with a neat sketch.
(13) Describe the acid pickling process with a neat sketch.
(14) Describe the electroplating process with a neat sketch.
(15) Describe the electro-cleaning process with a neat sketch.
(16) Describe the metal spraying process with a neat sketch.
(17) Describe the alkaline process with a neat sketch.
(18) Describe the burnishing process with a neat sketch.
(19) Describe the painting procedure used in a paint shop.
(20) "Metal coating improves the life of components". Justify this statement.
(21) "Industrial cleaning and coating processes are important operations in manufacturing". Justify this statement.

11 Press Shop Process

LEARNING OBJECTIVES

- To understand the details of the principles and fundamentals of sheet metal operations.
- To get details of types of press, their constructional details and press tools- dies and punches.
- To get acquainted with mechanics of forming process.
- To understand the automotive stamping process as well as various automotive body components and their materials.

11.1 INTRODUCTION

Press shop processes are also known as sheet-metal forming processes. These are among the most versatile of all metalworking operations. They generally are used on workpieces having high ratios of surface area to thickness.

Unlike bulk deformation processes, such as forging and extrusion, sheet-metal forming operations often prevent the sheet thickness from being reduced. Sheet metal working operations are used for both low- and high-run production.

Advantages of sheet metal working include high productivity rate, highly efficient use of material, the easy servicing of the machines used and the ability to employ less skilled workers. Parts made from sheet metal have many attractive qualities, as well, including: good dimensional accuracy, adequate strength, light weight and a broad range of possible dimensions, from the miniature parts used in electronics to the large parts used in airplane structures.

In sheet forming, sheet blanks are plastically deformed into a complex three-dimensional geometry, usually without any significant change in sheet thickness or surface characteristics. The surface area-to-volume of the initial metal is high; therefore, this ratio is a useful means to distinguish bulk deformation from sheet metal processes. Sheet metal operations are nearly always performed as cold working processes and are accomplished using a tool set consisting of a punch and a die, which are the positive (male) and negative (female) portions of the tool set, respectively. The characteristics of sheet metal forming processes are as outlined below.

DOI: 10.1201/9781003367321-11

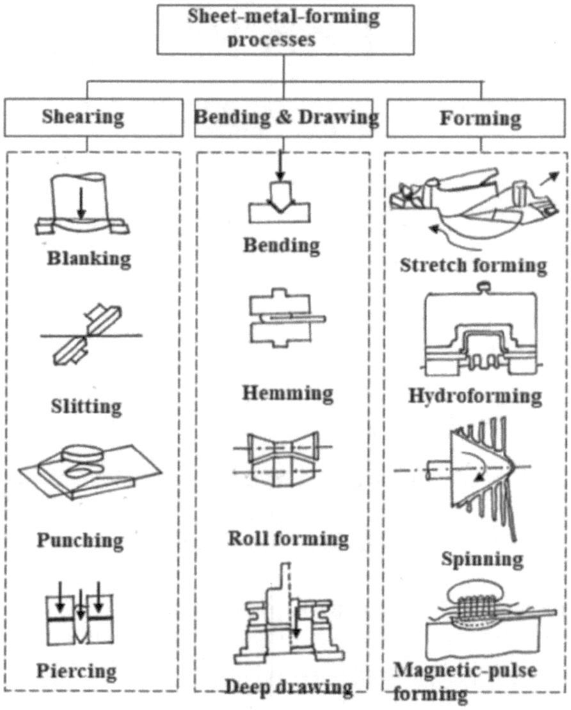

FIGURE 11.1 Sheet metal working process.

The workpiece is a sheet or a part fabricated from a sheet. The deformation usually has the objective to cause significant changes in shape, but not in cross-section, of the sheet. Reduction in sheet thickness is usually not desirable, but it is an unavoidable consequence of the process. In some cases, the magnitudes of permanent (plastic) and recoverable (elastic) deformations are comparable; thus, elastic recovery or springback may be significant.

Most sheet-metal operations are performed on machine tools called presses. The term stamping press is used to distinguish these presses from forging and extrusion presses.

The tooling that performs sheet metalwork is called a punch-and-die; the term stamping die is also used. The sheet-metal products are called stampings. Figure 11.1 shows the various sheet metal forming processes.

11.2 SHEET METAL WORKING: BENDING, FORMING AND DEEP DRAWING

Sheet metalworking covers cutting and forming operations conducted on relatively thin sheets of metal. The typical thickness of sheet metal is between 0.4 mm (1/64 in) and 6 mm (1/4 in). When the thickness of the stock surpasses roughly 6 mm, it is commonly referred to as plate rather than sheet. Flat rolling is used to create sheet or plate stock for sheet metalworking. Low-carbon steel (0.06–0.15% C typically) is the most often utilised sheet metal.

11.2.1 Bending

Bending is the process of deforming a flat sheet along a straight line to form the needed angle. The metal on the inside of the neutral plane is compressed, while the metal on the outside of the neutral plane is stretched (Figure 11.2).

Bending is a popular forming activity, as illustrated by automotive bodies, exhaust pipes, appliances, paper clips or file cabinets.

Bending is a procedure that deforms metal in such a way that the length and thickness before and after bending remain the same. It just alters the shape of the workpiece. Actually, this is an ideal condition, but the length and thickness fluctuate to some extent in the bend area. Every metal bends when subjected to a moment or couple of forces, every metal bends. Sheet metals have a great bending capacity, making them adaptable for a variety of shaping and forming processes.

- **Neutral axis**
  This is a fictitious axis that is not stressed during bending.
- **Outer fibres**
  Outer fibres are those that are under tension during the bending process. These can be found on one side of the neutral axis. It is shown in Figure 11.3.

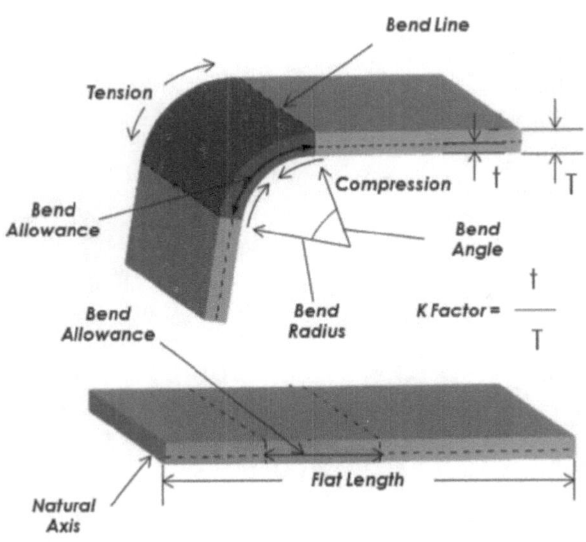

FIGURE 11.2 Bending process.

- **Inner fibres**
  Inner fibres are the fibres that are compressed during the bending process. The blue colour in the image represents these fibres.
- **Bend allowance**
  Bend allowance is the length of the neutral axis in the bend zone.
- **Bend angle**
  The bend angle is the angle formed by the bend area at the centre of the curve.
- **Bend radius**
  Bend radius is defined as the distance between the bend centre and the neutral axis. It is represented by the symbol r.
- **Minimum bend radius**
  The minimum bend radius is the bend radius at which a crack emerges on the bend's outer surface. It is typically represented in terms of sheet thickness, such as 2T, 3T, 4T, and so on. It varies depending on the material.
- **Spring back**
  When the load is released, the sheet metal exhibits some elastic recovery and returns to its original position. This is referred to as spring back. After springing back, it will increase the final bend radius and decrease the bend angle.

Bending operations are carried out with the use of punch and die tooling. V-bending, conducted with a V-die, and edge bending, performed with a wiping die, are the two most frequent bending processes and accompanying tooling, as shown in Figure 11.3. Sheet metal is bent between a V-shaped punch and die in V-bending. V-dies can be used to create included angles ranging from very obtuse to very acute. V-bending is typically utilised for low-volume applications. Sheet metal cantilever loading is used in edge bending. A pressure pad is used to impart force to the part's base against the die, while the punch forces the part to yield and bend over the die's edge.

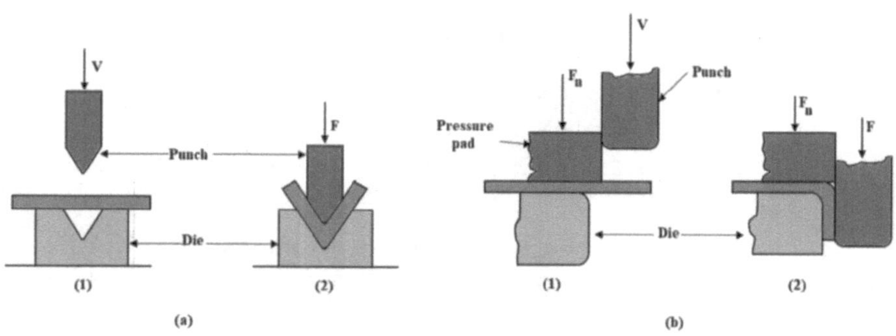

FIGURE 11.3 Two common bending methods: (a) V-bending and (b) edge bending; (1) before and (2) after bending.

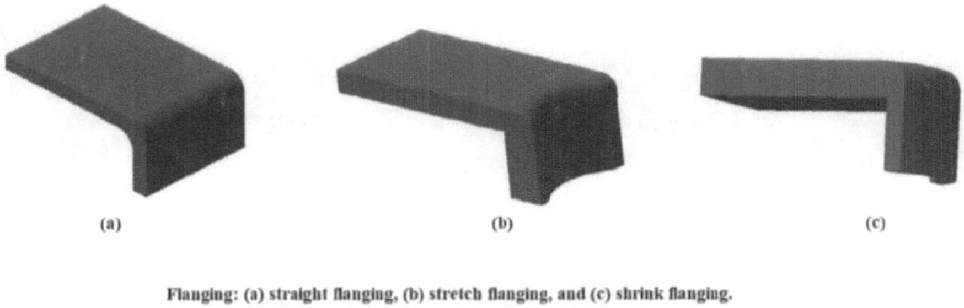

FIGURE 11.4 Flanging: (a) straight flanging, (b) stretch flanging and (c) shrink flanging.

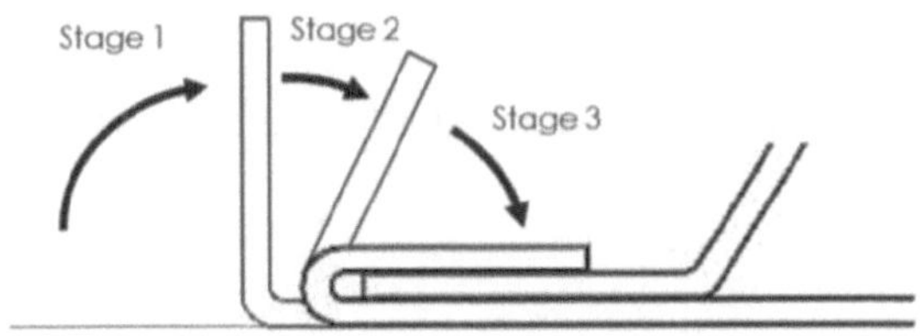

FIGURE 11.5 Hemming process.

Flanging, hemming, seaming and curling

- **Flanging**

This is a bending operation in which the edge of a sheet-metal part is bent at a 90^0 angle (usually) to form a rim or flange.

It is frequently used to reinforce or stiffen sheet metal. The flange can be created along a straight bend axis, as shown in Figure 11.4(a), or it can entail some stretching or shrinking of the metal, as shown in Figures 11.4(b) and (c)

- **Hemming**

Figure 11.5 shows details of hemming process. Hemming, in addition to increasing parts' rigidity, removes acute edges, enhances aesthetics and joins parts. Bending and flanging are used to create a variety of car interior and exterior panels, which are subsequently joined by hemming. Hems come in a variety of styles (Figure 11.4). Flattened, open and teardrop hems are used for edge finish and appearance, whereas radius flat, modified flat, rope and modified rope hems are used to seam two sheet metal sections together. Modified flat hems have superior fitting behaviour and shape fixation than radius hems. For brittle materials, rope hems are employed. Hemming operations are typically performed in three stages: bending or flanging to 90°, prehemming to around 135° and hemming to 180° or more, known as flattening.

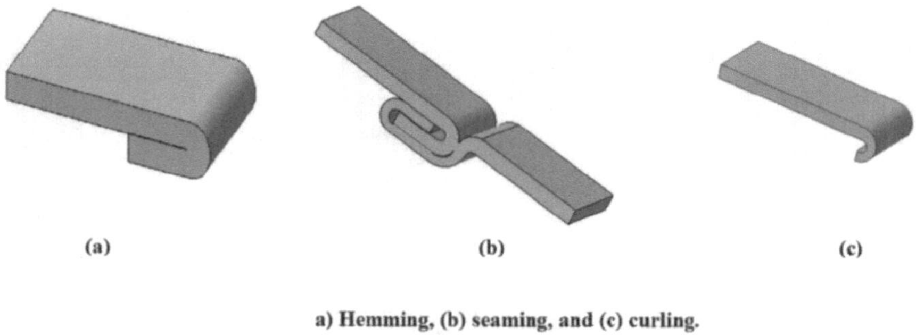

FIGURE 11.6 (a) Hemming, (b) seaming and (c) curling.

This is frequently done to remove the sharp edge from the component, increase stiffness and improve aesthetics.

- **Seaming**

In this operation, two sheet-metal edges are joined together (see Figure 11.6b). Seaming is used to create stovepipes, food cans, drums, barrels and other flat sheet metal products. In this situation, the lockseam is used to attach the ends of the same sheet metal piece.

- **Curling**

Bending, which also forms the edges of the portion into a roll or curl, is another term for curling (Figure 11.6c).

- **Miscellaneous Bending Operations**

Figure 11.7 depicts a variety of different bending operations to demonstrate the wide range of shapes that can be bent. The majority of these operations are accomplished in relatively simple dies comparable to V-dies.

11.2.2 Forming

During the forming process, sheet metal is stressed beyond its yield point, causing it to take a permanent set and keep the new shape (Figure 11.8). The shape of the punch and die surface is directly duplicated in this method, with no metal flow. The operation is employed in the production of door panels, steel furniture, aircraft bodies, and so on.

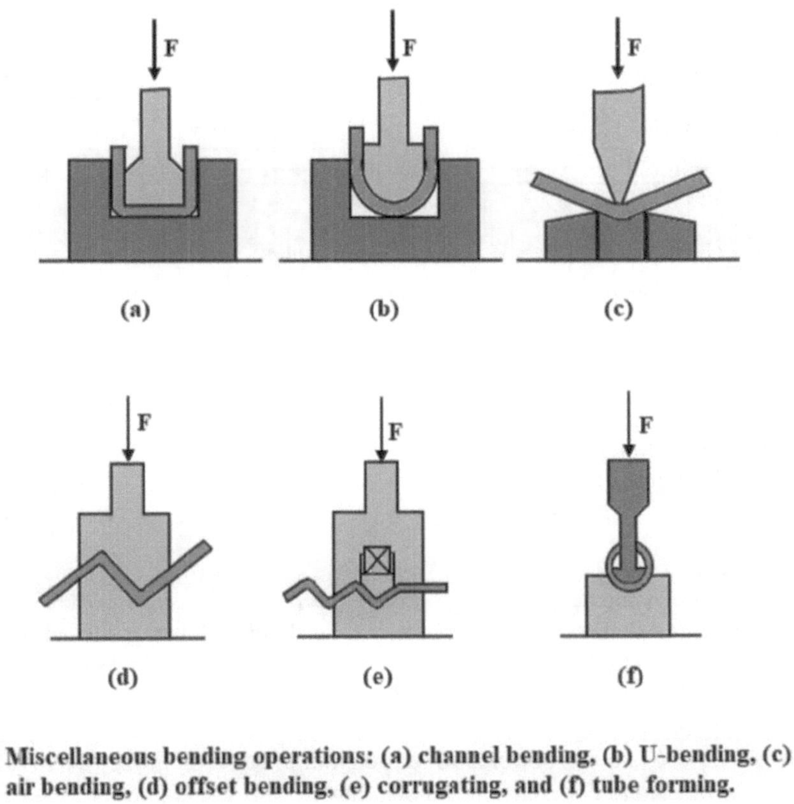

FIGURE 11.7 Miscellaneous bending operations: (a) channel bending, (b) U-bending, (c) air bending, (d) offset bending, (e) corrugating and (f) tube forming.

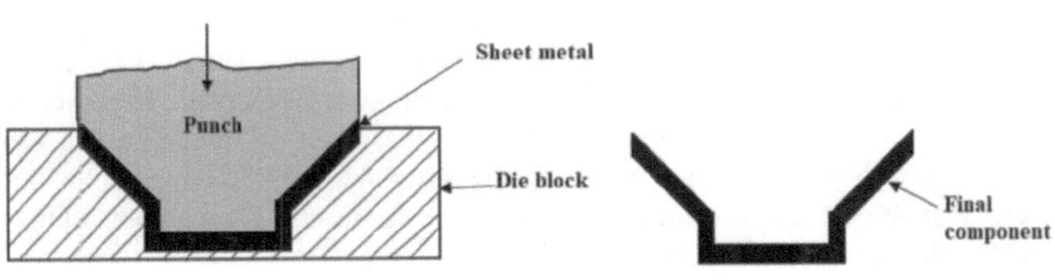

FIGURE 11.8 Forming process.

11.2.3 Deep Drawing

In the deep drawing process, we begin with a flat metal plate or sheet and shape it into a cup by pressing the sheet in the centre with a circular punch that fits into a cup-shaped die. In the home kitchen, we utilise a variety of containers, including deep saucepans (or bhagona), which are manufactured by a deep drawing method.

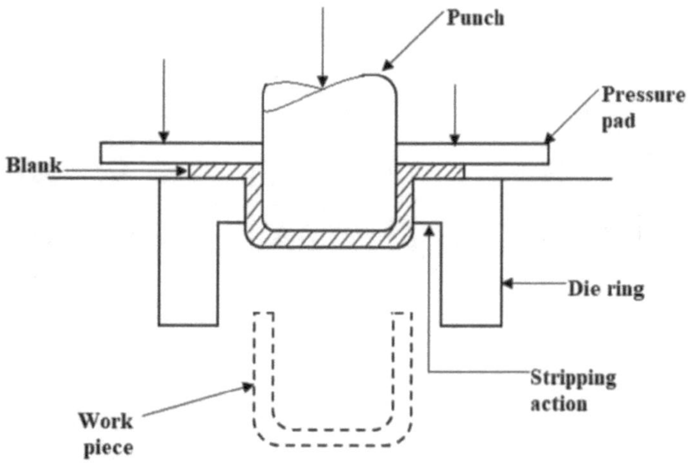

FIGURE 11.9 Deep drawing.

Figure 11.9 shows the deep drawing process. Deep drawing is used when the depth of the cup is greater than half its diameter, while shallow drawing is used when the depth to diameter ratio is less than half. The drawing process produces parts of varied geometries and shapes. The sheet metal portion is subjected to a complex pattern of stress during the drawing process. The portion of the blank between the die wall and the punch surface is only subjected to tension, but the portion further down near the bottom is subjected to both tension and bending.

The part of the metal blank that forms the flange at the top of the cup is subjected to circumferential compressive stress and buckling, causing it to thicken. As a result, the flange must be held down by a pressure pad, or its surface will buckle and become uneven, much like an orange peel. Deep drawing is a difficult operation and the material used should be specially malleable and ductile, otherwise it will crack under the induced stresses. The wall thickness of a deep-drawn component does not remain consistent, and the vertical walls get thinner as tensile stresses increase. However, the narrowest piece is all around the bottom corner of the cup. This thinning of the sheet at certain spots is referred to as "necking." And after that deep drawing, the component may be subjected to finishing procedures such as "ironing," the goal of which is to achieve more uniform wall thickness.

11.3 DIFFERENCE BETWEEN PUNCHING, PIERCING AND BLANKING

Piercing, punching and blanking are three common machining processes that are used to manipulate raw metal, such as sheet metal. All three processes require the use of a machine, which in some way deforms or alters the physical properties of the raw metal. While similar, piercing, punching and blanking have different functions. An overview of each of these machining processes and how they differ is presented here.

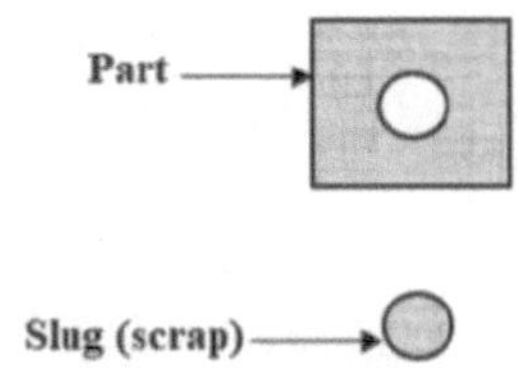

FIGURE 11.10 Punching process.

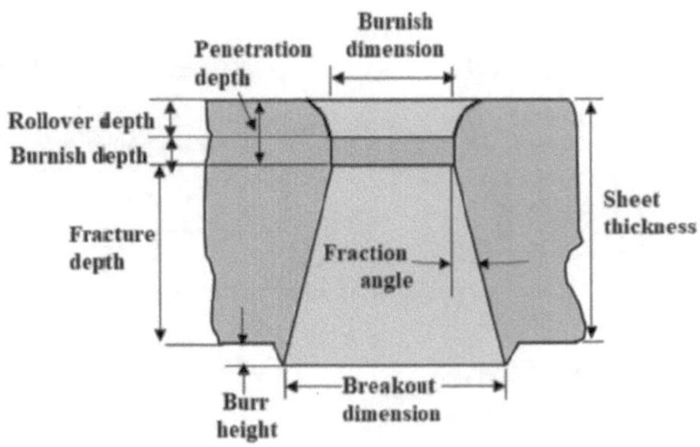

FIGURE 11.11 Piercing.

11.3.1 Punching

The punching process is identical to the piercing procedure. The development of a hole during punching is the desired result (Figure 11.10). The difference between punching and piercing is that with punching, a cylindrical hole is made, whereas with piercing, the hole generated can be of any shape.

11.3.2 Piercing

Piercing is the operation of making a hole in a sheet metal with a punch and a die. The materials punched out to make the hole are waste. Piercing is typically the quickest method of producing holes in a steel sheet or strip, as well as the most cost-effective approach for medium to high output. Piercing is employed primarily when precise holes are required and the production lot is large enough to justify the tooling costs. As shown in Figure 11.11, the piercing operation normally begins with a cut that produces a burnished surface on the hole wall and some rollover (curved surface generated by deformation of the workpiece prior to cutting).

The punch completes its stroke by breaking and tearing away the metal that was not cut during the piercing operation's initial stage.

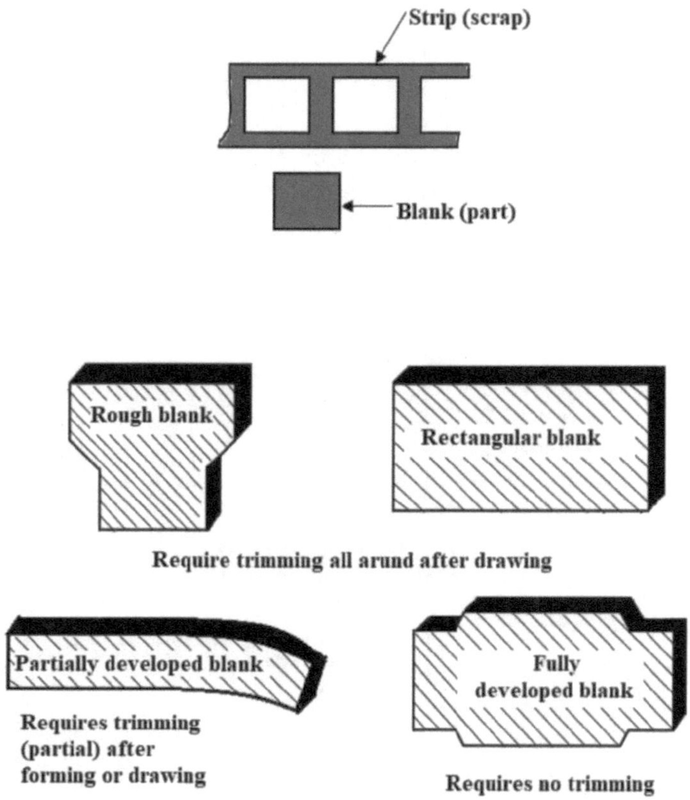

FIGURE 11.12 Blanking.

11.3.3 Blanking

Blanking is the process of cutting or shearing a piece of metal with a predetermined contour from sheet metal stock in order to prepare it for following operations. The piece of metal formed as a result of this operation is known as a blank, and the remainder can be waste, as illustrated in Figure 11.12. Blanking is the process of cutting out a workpiece of suitable design and sufficient metal to provide the desired part.

The sorts of blanks used in sheet-forming operations are divided into four categories for ease of use. Rectangular blanks, rough blanks, partially developed blanks and fully developed blanks are the four classes. A rectangular blank is the simplest to make and has the least complexity. A rectangular blank, as the name implies, is rectangular in shape and has four straight sides. A fuel-tank half for a vehicle is an example of a part manufactured from a rectangle blank.

A rough blank can be carved into a variety of shapes. It could have straight sides and be shaped like a triangle, diamond, trapezoid, hexagon, octagon or even a "T" (Figure 11.13). Rough blanks are used to produce drawn pieces. A developed blank is a blank of any shape, symmetrical or asymmetrical, that has been precision sized so that no trimming is required after the part has been shaped. This category includes

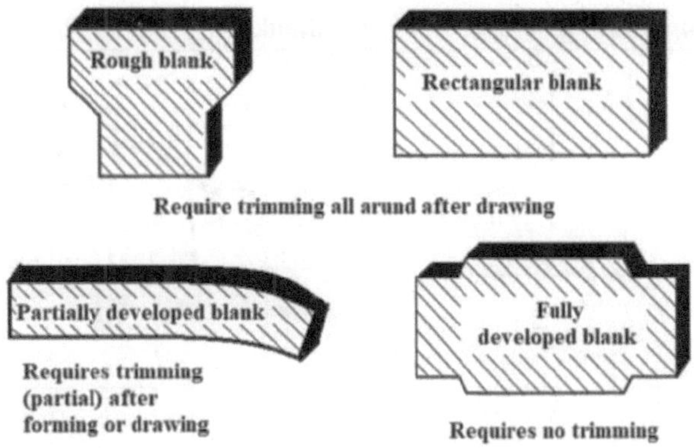

FIGURE 11.13A Forming operation. Blanks of various shapes have been generated involving rough blank, rectangular blank, partially developed blank and fully developed blank.

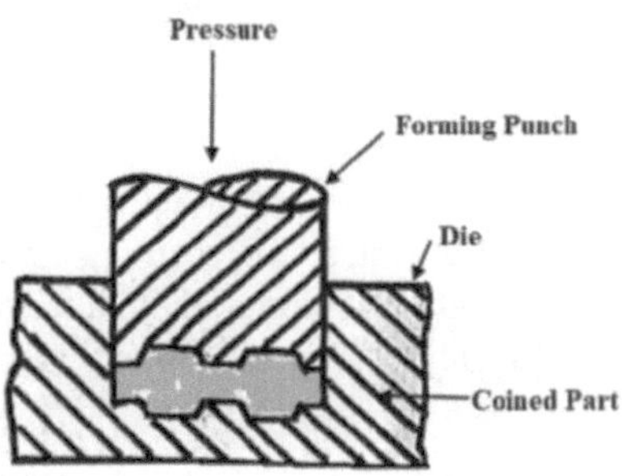

FIGURE 11.13B Forming operation.

almost all stampings used in chassis frame construction, as well as many internal parts of automotive bodywork and front-end assemblies. A partially developed blank is one that has been precisely sized on two opposite sides of a part, with the remaining two sides supplying surplus metal for holding. This sort of blank is used for sketching parts that only require holding on two sides. Roof rails are a common example of a part for which partially formed blanks might be used. Because there are so few automobile stampings of this sort, partially formed blanks are rarely employed.

- **Differences between punching/piercing and blanking**

Sr. No	Punching/Piercing	Blanking
1	In this operation clearance is given to the die.	In this operation clearance is given to the punch.
2	The component size is equal to the punch.	The component size is equal to the die.
3	Slug is scrap.	Slug is the desired part.

Sr. No	Punching/Piercing	Blanking
4	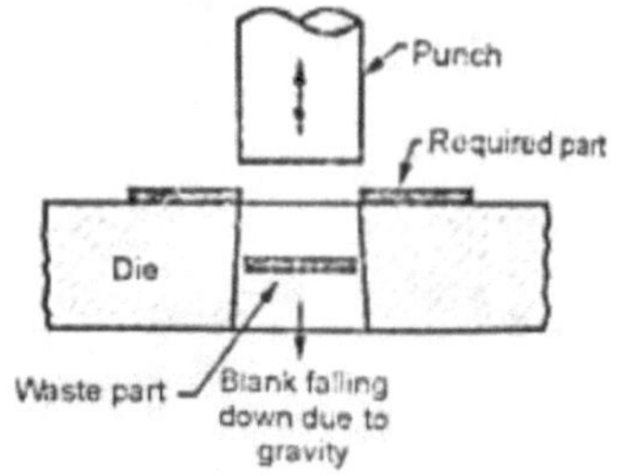	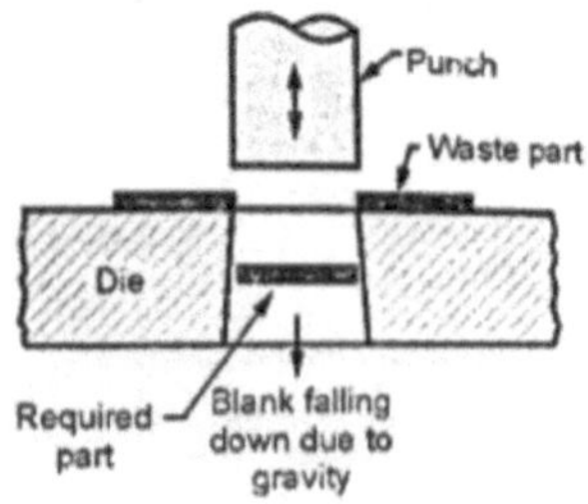

- **Other sheet metal operations**

Sheet metal embossing is a stamping process for producing raised or sunken designs or relief in sheet metal. This process can be made by means of matched male and female roller dies, or by passing a sheet or strip of metal between rolls of the desired pattern. Other sheet metal operations are explained in the following sections.

- **Slitting**

Slitting is a shearing operation, but instead of creating cuts at the end of a workpiece like shearing, it is used to cut a wide coil of metal into a series of narrower coils while the main coil moves through the slitter's circular blades (Figure 11.14a).

- **Notching**

This is a shearing technique that removes a metal scrap piece off the outside edge of a metal workpiece. Notching is primarily a manual, low-production process. During a notching process, an outside edge of a metal workpiece is removed by using numerous shear blades arranged at right angles to each other (Figure 11.14c).

- **Perforating**

This is the punching of a large number of holes, usually similar and placed in a regular pattern, in a sheet, workstation blank or previously produced part. The holes are normally round, but they can be any shape (Figure 11.14b).

- **Nibbing**

This is the process of cutting out a contour or other shape by punching a sequence of overlapping round or square holes along the edge of the part; blanking is commonly used to replace this operation. It is intended for cutting off flat parts from sheet metal, with basic and complex shapes. It is only utilised for a small number of components (Figure 11.14d).

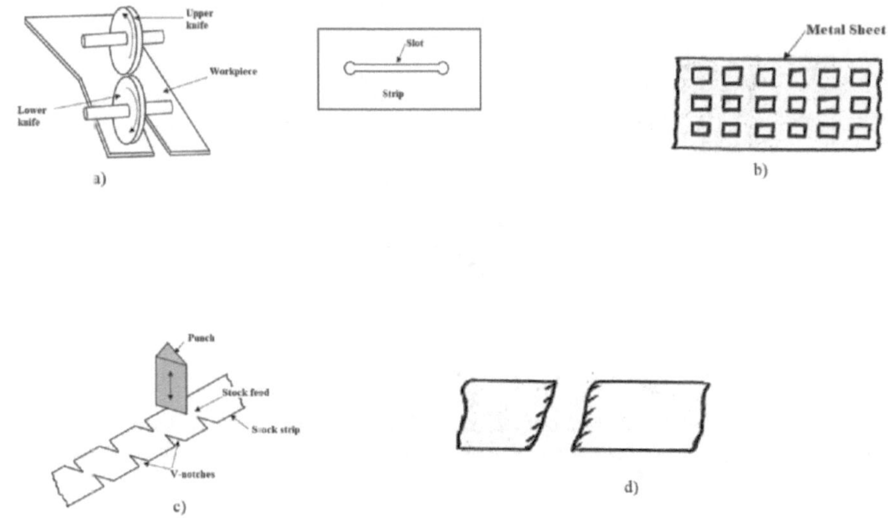

FIGURE 11.14 (a) Slitting operation, (b) perforating, (c) notching, (d) nibbing.

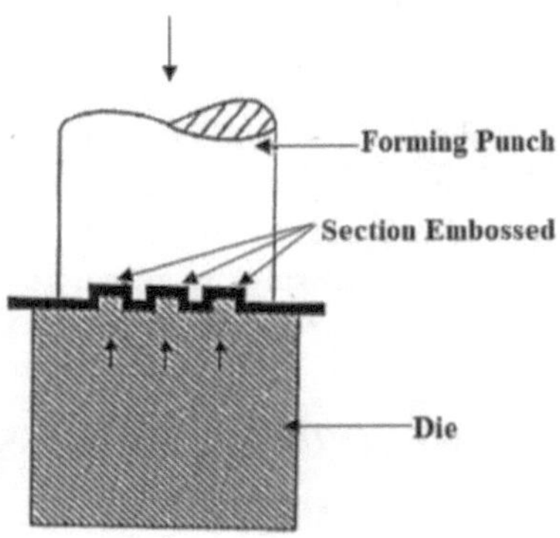

FIGURE 11.15 Embossing.

- **Lancing**

Lancing is a one-step operation that combines cutting and bending or cutting and shaping to partially remove the metal from the sheet. Lancing is used to form louvers in sheet metal air vents for heating and air conditioning systems in buildings, among other things.

- **Embossing**

This is a technique in which blanks of sheet metal are stretched to shape under pressure using a punch and a die (Figure 11.15). The punch moves at a slow pace to

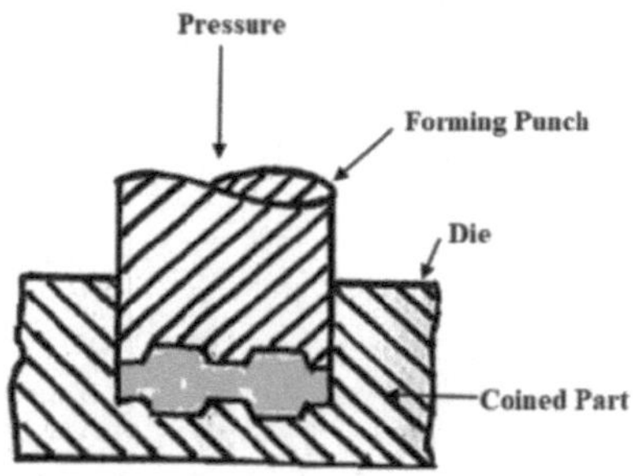

FIGURE 11.16 Coining.

allow for proper stretching. The operation stiffens the metal that is being embossed. A vast variety of ornamental products, such as plates in sheet metal, are made in order to alleviate stress in the material.

- **Coining**

This is essentially a cold working procedure that takes place in dies where the metal blank is confined and its lateral flow is controlled (Figure 11.16). It is mostly employed in the manufacture of essential objects such as medals, coins, stickers and other similar items with shallow configurations on their surfaces. The operation entails inserting a metal slug into a die and applying intense pressure with a punch. The metal flows plastically and is squeezed into the shape formed by the punch and die. Because of the extremely high pressures required, the method is limited to soft metals with high plasticity.

11.4 MECHANICS OF THE FORMING PROCESS

In a drawing operation, in addition to the work load and power required, the maximum possible reduction without any tearing failure of the workpiece is an important parameter. In the analysis that we give here we shall determine these quantities. Since the drawing operation is mostly performed with rods and wire we shall assume the workpiece to be cylindrical as shown in Figure 11.17 a typical drawing die consists of four regions:

(i) A bell-shaped entrance zone for proper guidance of the workpiece
(ii) A conical work zone
(iii) A straight and short cylindrical zone for adding stability to the operation
(iv) A bell-shaped exit zone.

The final size of the product is determined by the diameter of the stabilising zone (d_f), with the other important dimensions being the half cone angle (α) and sometimes the black tension (F_b), which is provided to keep the input workpiece straight. The workload, i.e. the drawing force F, is applied on the exit side as shown in Figure 11.17.

A die can handle a job having a different initial diameter d_i which in turn determines

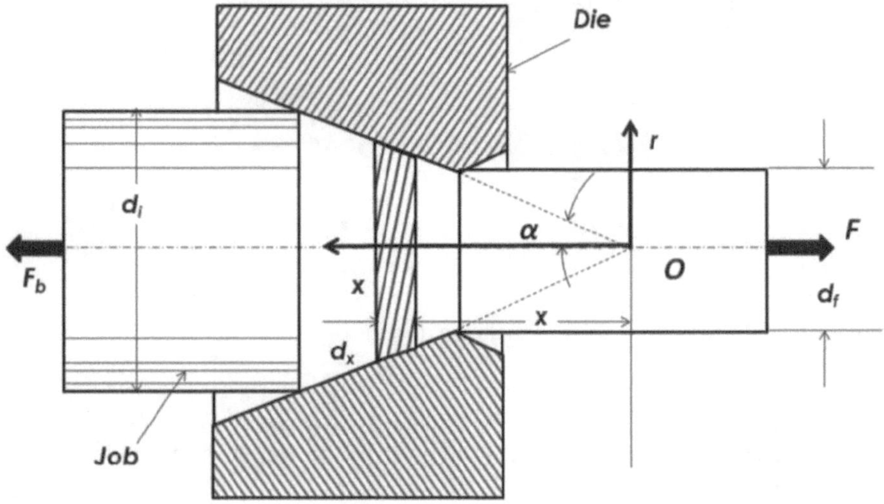

FIGURE 11.17 Drawing of a cylindrical rod.

the length of the job die interface. The degree of drawing operation (D) is normally expressed in terms of the expression reduction factor in the cross-sectional area thus

$$\mathrm{D} = \frac{\left(A_i - A_f\right)}{A_i} = \frac{\left(d_i^2 - d_f^{\,2}\right)}{d_i^2}$$

When the true strain is

$$\varepsilon = ln\frac{A_i}{A_f} = ln\left(\frac{1}{1-D}\right)$$

with A_i and A_f being the initial and final cross-sectional areas of the workpiece, respectively.

11.5 DETERMINATION OF THE DRAWING FORCE AND POWER

The case is an axi-symmetric one-cylindrical coordinate system with r, x being of no importance and is chosen with its origin O at the vertex of the die cone. This case is presented in Figure 11.18. An element of the length dx at the distance x along with the stresses is acting on it. For the sake of simplification we make the following assumptions:

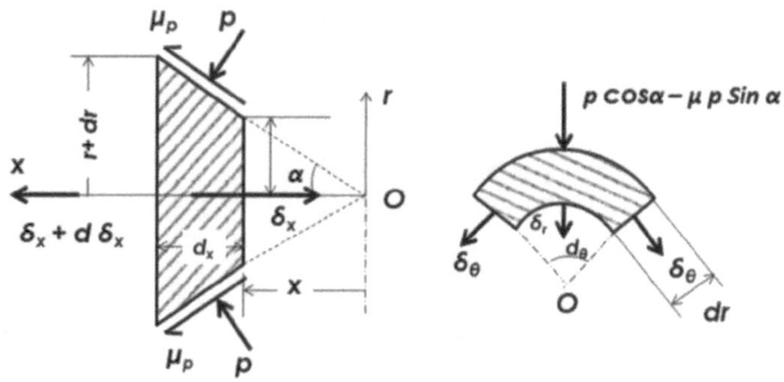

FIGURE 11.18 Stresses in an element during drawing.

(i) The coefficient of friction μ and the half cone angle α are small.
(ii) The yield stress σ_Y is constant and given by the average of the initial and final values.
(iii) $-p$ and σ_x are the principal stresses.
(iv) It does not vary in the radial directions and it should be noted that both p and μp act on the whole conical surface of the element.

$$tan\alpha = \frac{(r+dr)-r}{dx} = \frac{dr}{dx}$$

$$dr = tan\alpha dx$$

Considering the equilibrium of the element in the direction in the x direction we have

$$\left(\sigma_x + d\sigma_x\right)\pi(r+dr)^2 - \sigma_x \pi r^2 + \mu p 2\pi r \frac{dx}{cos\alpha} cos\alpha + p2\pi r \frac{dx}{cos\alpha} sin\alpha = 0$$

Neglecting the higher order terms and using

$$rd\sigma_x + 2\left[\sigma_x + p\left(1 + \frac{\mu}{tan\ \alpha}\right)\right]dr = 0 \tag{11.3}$$

where σ_x and p are related through the yield criteria.

A segment of an annular element of the thickness δ_r at the surface is acting and the resultant radial stress at the surface is composed of the radial components of both p and μp. Now considering the radial equilibrium of the segment we get

$$-(pcos\alpha - \mu psin\alpha)(r+\delta r)d\theta dx - \sigma_r r d\theta dx - 2\sigma_\theta sin\frac{d\theta}{2}\delta r dx = 0$$

Neglecting the higher order terms and remembering that α is small, this equation reduces to

$$[p(1-\mu sin\alpha+\sigma_r)rd\theta dx=0$$

or

$$\sigma_r=-p(1-\mu sin\alpha)\sim -p$$

Again, since the circumferential strain rate is the same as the radial strain rate (the circumference being proportional to the radius) according to the von mises stress strain rate law $\sigma_r=\sigma_\theta$ thus taking σ_x, σ_r, and σ_θ as the principal stresses we have

$$\sigma_1=\sigma_x, \sigma_2=\sigma_r=\sigma_\theta=\sigma_3=-p$$

Using the von Mises yield criterion given by the equations we get

$$\sigma_x+p=\sigma_Y, \text{or } p=\sigma_Y-\sigma_x$$

$$rd\sigma_x+2\left[\sigma_x+(\sigma_Y-\sigma_x)\phi\right]dr=0$$

where

$$\phi=\left(1+\frac{\mu}{tan\alpha}\right)$$

therefore

$$\int\frac{dr}{r}=-\int\frac{d\sigma_x}{2\left[\phi\sigma_Y+(1-\phi)\sigma_x\right]}$$

$$lnr=+\frac{1}{2(\phi-1)}lnl\left[\phi\sigma_Y+(1-\phi)\sigma_x\right]+C$$

When

$$r=\frac{d_i}{2}, \sigma_x=\frac{F_b}{A_i}$$

So

$$C = ln\frac{d_i}{2} - ln\frac{1}{2(\phi-1)}ln\left[\phi\sigma_Y + (\phi-1)\frac{F_b}{A_i}\right]$$

Substituting C in the equation relating r and σ_x we have

$$lnln\left(\frac{2r}{d_i}\right) = \frac{1}{2(\phi-1)}ln\left[\frac{\phi\sigma_Y + (1-\phi)\sigma_x}{\phi\sigma_Y + (\phi-1)\frac{F_b}{A_i}}\right]$$

$$\left(\frac{2r}{d_i}\right)^{2(\phi-1)} = \left[\frac{\phi\sigma_Y + (1-\phi)\sigma_x}{\phi\sigma_Y + (\phi-1)\frac{F_b}{A_i}}\right]$$

Rearranging we get

$$\frac{\sigma_x}{\sigma_Y} = \frac{F_b}{\sigma_Y A_i}\left(\frac{2r}{d_i}\right)^{2(\phi-1)} - \left(\frac{\phi}{\phi-1}\right)\left[\left(\frac{2r}{d_i}\right)^{2(\phi-1)} - 1\right]$$

Hence the drawing stress σ_{xf} at $r = \frac{d_f}{2}$ is given by

$$\frac{\sigma_{xf}}{\sigma_Y} = \frac{F_b}{\sigma_Y A_i}\left(\frac{d_f}{d_i}\right)^{2(\phi-1)} - \left(\frac{\phi}{\phi-1}\right)\left[1 - \left(\frac{d_f}{d_i}\right)^{2(\phi-1)}\right]$$

Finally the drawing force is

$$F = \sigma_{xf} A_f$$

Where σ_{xf} from the equation is the drawing speed, the exit velocity is V, and the power required for the drawing operation is

$$P = FV \qquad (11.6)$$

11.6 DETERMINATION OF MAXIMUM ALLOWABLE REDUCTION

The maximum allowable reduction of the maximum degree of a drawing operation (D) is determined from the constraint that the pulling stress σ_x cannot be more than the tensile yield stress of the work material so at this limiting condition

$$\frac{\sigma_{xf}}{\sigma_Y} = 1$$

Using this equation the relation we obtain is

$$D_{max} = 1 - \frac{1}{\left[\phi - \frac{F_{b(\phi-1)}}{\sigma_Y A_i}\right]^{1/(\phi-1)}}$$

- **Bending**

While determining the work load, an estimate of the amount of elastic recovery (spring back) is required in a bending operation. When the final shape is prescribed a suitable amount of overbending is required to take care of this spring block. In this section we shall work out these quantities and also illustrate how the stock size for a given job is completed

Figure 11.19 shows a bending operation with characteristic dimensions. A radius r_p is provided at the nose of the punch and accordingly the die centre has radius (r_p+t)

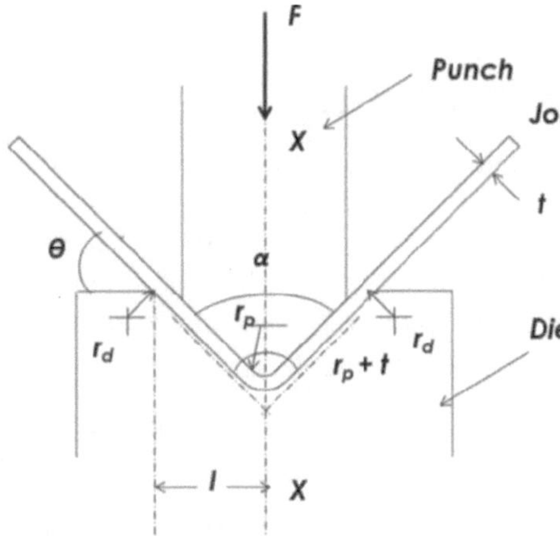

FIGURE 11.19 Details in bending.

where t is the job thickness. The portions of the die in contact with the job during the operation are also provided with some radius, r_d.

The angle between the two faces of the punch and the die is α at the instant shown, and the angle between the two bent surfaces of the job is (π – 2θ) As we shall subsequently show, the bending force F is maximum at some intermediate stage depending on the frictional characteristics. The degree of a bending operation is normally specified in terms of the strain in the outer fibre. The width of the job w (in the direction perpendicular to the plane of the paper) is much larger as compared with t, and hence a plane strain condition can be assumed. It is obvious that the stock length should be calculated on the basis of the length of the neutral plane of the job. Since the radius of curvature involved in a bending operation is normally small, the neutral plane shifts towards the centre of curvature. Usually a shift of 5–10% of the thickness is assumed for the calculation of strain and stock length. Thus the strain in the outer fibre of the bend is given by

$$\varepsilon_{max} = ln\left[1+\frac{\left(r_p+t\right)-\left(r_p+0.45t\right)}{\left(r_p+0.45t\right)}\right]$$

$$= \ln\left[1+\frac{1}{1.82\left(\frac{r_p}{t}\right)+0.82}\right]$$

assuming a 5% shift of the neutral plane. Depending upon the ductility of the job material ε_{max} has a limiting value beyond which a fracture takes place. So, from the above equation and the limiting value $\varepsilon_{max} = \varepsilon_{fracture}$ we can determine the smallest punch radius for a given job thickness.

11.7 DETERMINATION OF WORK LOAD

Since the job undergoes plastic bending, the stress distribution at the cross-section along the centre line a is shown in Figure 11.20. This distribution is obtained by neglecting all other effects of curvature except the shift of neutral line. It is obvious the in the zone on either side of the neutral plane the strain level is within the elastic range. When the strain (both in the tensile and compressive zones) reaches the yield limit, plastic deformation starts. Assuming the yield stress to be σ_{Y_0} (same in both tension and compression) and linear strain hardening, the stress distribution will be as shown in Figure. The magnitudes of σ_{Y_1} and σ_{Y_2} are different due to the shift of the neutral plane. For the sake of simplicity, the stress distribution for large plastic bending is idealised as shown in the figure when the strain hardening rate is n, then

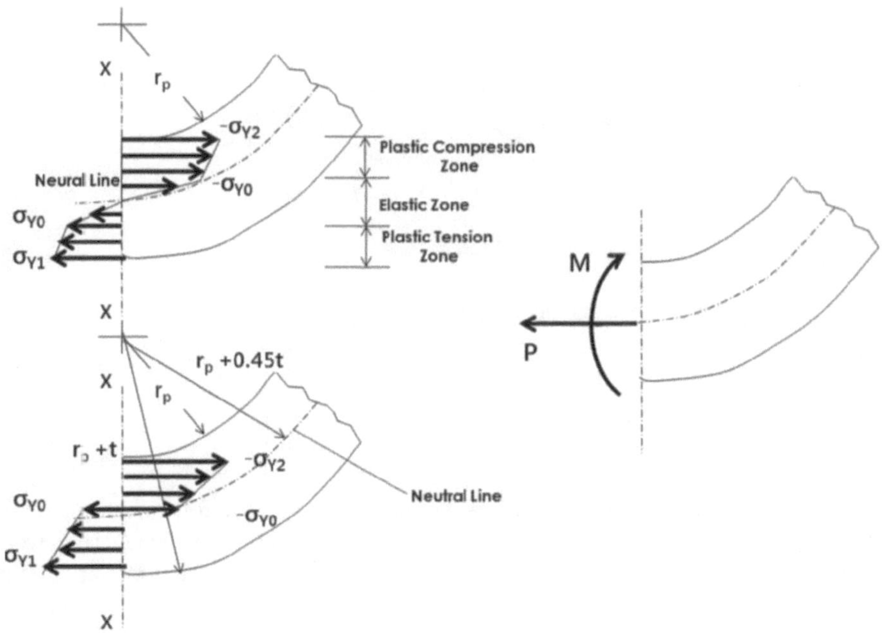

FIGURE 11.20 Mechanics in bending.

$$\sigma_{Y_0} = \sigma_{Y_0} + n\varepsilon_{max}$$

$$\sigma_{Y_2} = \sigma_{Y_0} + nln\left[1 + \frac{\left(r_p + 0.45t\right) - \left(r_p t\right)}{\left(r_p + 0.45t\right)}\right]$$

$$\sigma_{Y_0} = \sigma_{Y_0} + nln\left[1 + \frac{1}{2.22\left(\frac{r_p}{t}\right) + 1}\right]$$

- **Mechanics of bending**

The loading due to this stress distribution can be represented by a bending moment M and the force p (per unit width of the job)

$$M = (0.55t)^2\left(\frac{\sigma_{Y_0}}{6} + \frac{\sigma_{Y_1}}{3}\right) + (0.45t)^2\left(\frac{\sigma_{Y_0}}{6} + \frac{\sigma_{Y_2}}{3}\right)$$

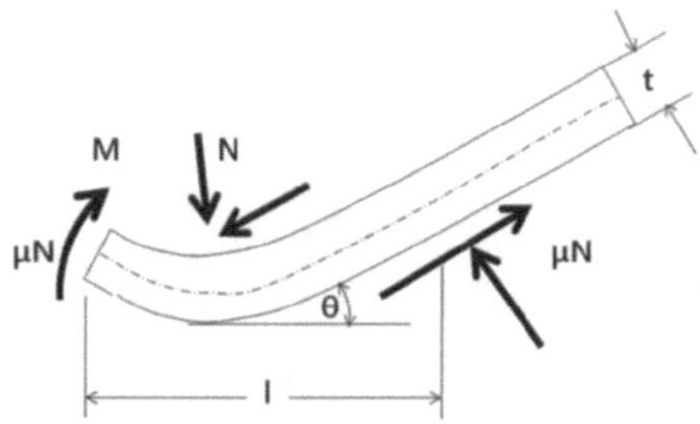

FIGURE 11.21 Free body diagram of the half job.

$$P = \frac{t}{2}\left[0.1\sigma_{Y_0} + .055\sigma_{Y_1} - 0.45\sigma_{Y_2}\right]$$

Now let us consider the right half of the unit width and the forces and moments acting on it. Figure 11.21 shows free body diagram of the half job. Since P arises from the shift of the neutral plane which is very small, it can be neglected in comparison with the other forces. The normal and frictional forces exerted by the die and the punch at their contact lines (since r_p is small as compared with the other dimensions the finite contact of the job and the punch can be idealised as a line) are N and μN, respectively

As t is small, the moment due to μN is negligible. Hence $M = \frac{Nl}{Cos\theta}$, one half of the bending force per unit width is given as

$$\frac{F}{2} = Ncos\ \theta + \mu N\, sin\theta$$

$$F = 2N(cos\,\theta + \mu\ sin\theta)$$

Substituting N in terms of M we obtain F

$$F = \frac{2M}{l}\left(cos^2\theta + \mu sin\,\theta cos\theta\right)$$

Now differentiating F with respect to θ we get

$$\frac{dF}{d\theta} = \frac{2M}{l}(-sin2\theta + \mu cos\ 2\theta)$$

Since M is independent of θ it is obvious that F reaches a maximum when

$$\theta = \theta_{cr} = \frac{1}{2} tan^{-1}\mu$$

Thus the maximum work load per unit width is given as

$$F_{max} = \frac{M}{l}\left[1 + cos\left(tan^{-1}\mu\right) + \mu sin\left(tan^{-1}\mu\right)\right]\backslash$$

- **Estimation of spring back**

Figure 11.22 shows the stress–strain characteristics of a linearly strain-hardened material. When the material is unloaded from point A the path of unloading is given by the line AB. The amount of recovered strain obviously is

$$\varepsilon_{rec} = \text{BC} = \frac{\sigma_A}{E}$$

So the amount of bending strain recovered is given by the elastic bending strain resulting from M which is removed when the operation is over. The elastic strain of a beam due to bending moment M results in an included angle, as shown in Figure 11.22

or $$\frac{M}{I} = \frac{E}{R} = \frac{E\phi}{L}$$

where I and R are the second moment of the beam cross-section and the radius of curvature, respectively, and L is the length of the neutral plane.

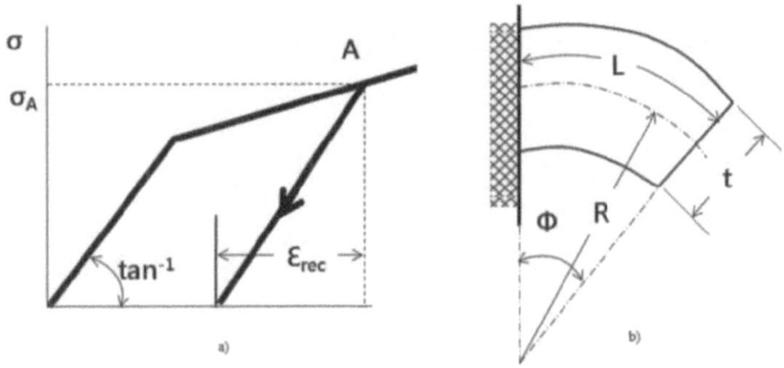

FIGURE 11.22 Analysis of spring back.

Thus $\phi = ML/EI$, which is nothing but the amount of springback, so, with the original included angle α/2 (for the half portion)

$$\frac{2\phi}{\alpha} = \frac{2ML}{EI\alpha}$$

Since the 5% shift in the neutral axis

$$\frac{\alpha}{2} = \frac{L}{\left(r_p + 0.45t\right)}$$

$$\frac{2\phi}{\alpha} = \frac{M\left(r_p + 0.45t\right)}{EI}$$

For rectangular cross-section with unit width

$$I = \frac{1}{12}t^3$$

Using equation the total springback finally obtained is

$$\frac{2\phi}{\alpha} = \frac{12M\left(r_p + 0.45t\right)}{Et^3}$$

11.8 SOLVED EXAMPLES

1. A steel wire is drawn from an initial diameter 12.5 mm to a final diameter of 10.5 mm at a speed of 80 m/min. Coefficient of friction at the job die interface is 0.1 and half cone angle of the die is 6^0. The specimen has a tensile yield strength $209\,\mathrm{N/mm^2}$. Tensile yield stress is of $414\,\mathrm{N/mm^2}$ at a strain of 0.5. Without considering the back tension and assuming linear stress–strain relationship, determine the drawing power and the maximum possible reduction with the same die.

We know that

$$\varepsilon = ln\frac{A_i}{A_f} = 2ln\frac{d_i}{d_f} = 2ln\frac{10.5}{12.5} = 0.84$$

$$\sigma_{Y_f} = \left(209 + \frac{414 - 209}{0.5} * 0.84\right) = 533.4\ N/mm^2$$

The average yield stress

$$\sigma_y = \frac{209+553.4}{2} = 381.2\ N/mm^2$$

From the given data

$$\phi = \left(1+\frac{\mu}{tan\alpha}\right) =$$

$$\phi = \left(1+\frac{0.1}{tan\alpha}\right) = 2$$

From the following equation, with $F_b = 0$

$$\frac{\sigma_{xf}}{\sigma_Y} = \left[\frac{F_b}{\sigma_Y A_i}\left(\frac{d_f}{d_i}\right)^{2(\phi-1)} - \left(\frac{\phi}{\phi-1}\right)\left[1-\left(\frac{d_f}{d_i}\right)^{2(\phi-1)}\right]\right]$$

$$\sigma_{xf} = 381.2*2\left[1-\left(\frac{10.5}{12.5}\right)^2\right] = 175.168\,N/mm^2$$

By using the equation

$$F = \sigma_{xf} A_f$$

The exit velocity is V, and the power required for the drawing operation is

$$P = FV$$

$$P = 175.168*\frac{\Pi}{4}*10.5^2*\frac{80}{60} = 20.223\text{ kW}$$

2. A mild steel plate having 100 mm length, thickness 6 mm and 25 mm width, is bent exactly at the centre at 90^0. The data given are $E = 207\,\text{kN}/\text{mm}^2$, = 0.1, $E = 207\,\text{kN}/\text{mm}^2$, μ = 0.1, $l = 0.25$ mm, $\sigma_{Y_0} = 345\frac{\text{N}}{\text{mm}^2}$, $\text{n} = 517\text{N/mm}^2$. What will be the minimum possible corner radius r if the fracture strain $\varepsilon_{strain} = 0.2$ for the given material. By considering r to be minimum, find out (1) the maximum bending force, (2) the required punch angle and (3) the stock length.

For the calculation of minimum possible radius of the corner by using the equation

$$\varepsilon_{max} = ln\left[1 + \frac{(r_p + t) - (r_p + 0.45t)}{(r_p + 0.45t)}\right]$$

$$0.2 = \ln\left[\frac{1}{1.82\left(\frac{r_p}{t}\right) + 0.82}\right]$$

$$r_p = r_{pmin} = 12.18\,\text{mm, say}\,12.2\,\text{mm}$$

By using

$$\sigma_{Y_1} = \sigma_{Y_0} + n\varepsilon_{max}$$

$$\sigma_{Y_1} = [345 + 517 * .02] = 448.4\ \text{N/mm}^{2}.$$

$$\sigma_{Y_2} = \sigma_{Y_0} + nln\left[1 + \frac{(r_p + t) - (r_p + 0.45t)}{(r_p + 0.45t)}\right]$$

$$\sigma_{Y_2} = \sigma_{Y_0} + nln\left[1 + \frac{1}{2.22\left(\frac{r_p}{t}\right) + 1}\right]$$

$$\sigma_{Y_2} = [345 + 517 ln l\left[1 + \frac{1}{2.22\left(\frac{12.2}{6}\right) + 1}\right] = 431.16 N/mm^2,$$

From these values the bending movement can be calculated as follows

$$M = (0.55t)^2\left(\frac{\sigma_{Y_0}}{6} + \frac{\sigma_{Y_1}}{3}\right) + (0.45t)^2\left(\frac{\sigma_{Y_0}}{6} + \frac{\sigma_{Y_2}}{3}\right)$$

$$P = \frac{t}{2}\left[0.1\sigma_{Y_0} + .055\sigma_{Y_1} - 0.45\sigma_{Y_2}\right]$$

$$M = (0.55*6)^2\left(\frac{345}{6} + \frac{448.4}{3}\right) + (0.45*6)^2\left(\frac{345}{6} + \frac{431.16}{3}\right) = 1684.75\text{N}$$

The maximum banding force per mm width of the job from the value

$$F_{max} = \frac{M}{l}\left[1 + cos\left(tan^{-1}\mu\right) + \mu sin\left(tan^{-1}\mu\right)\right]$$

$$F_{max} = \frac{1684.75}{25}\left[1 + cos\left(tan^{-1}0.1\right) + 0.1sin\left(tan^{-1}0.1\right)\right] = 135.11N/mm$$

Actual bending force will be

$$F_{max} = \frac{1684.75}{25} = 135.11\ N/mm$$

$$F_{max} * 20N = 135.11 * 20 = 2702.2\ N$$

To find out the punch angle to obtain a 90^0 bend to determine the spring back angle 2

$$\frac{2\phi}{\alpha} = \frac{M\left(r_p + 0.45t\right)}{EI}$$

$$\frac{2\phi}{\alpha} = \frac{12(12.2 + 0.45*6)}{207*10^3*6^3} * 1684.75 = 0.00637$$

$$2\phi = 0.00637\alpha$$

Since α is the punch angle

$$\alpha = 90^0 - 2\phi = 90^0 - 0.00637\alpha$$

$$1.00637\alpha = 90$$

$$\alpha = 89.43^0$$

Total length of the neutral plane which is required for the stock length

$$L_s = 2\left[\left(r_p + 0.45t\right)\frac{\Pi}{4} + 50 - \left(r_p + t\right)\right] = 87.00\ mm$$

3. As per the job described in Example 2, where the maximum available force is 4000 N, what should be the minimum value of the die length, $2l$. Also find the capacity of the machine to produce the same job.

$$F_{max} = \frac{M}{l}\left[1 + cos\left(tan^{-1}\mu\right) + \mu sin\left(tan^{-1}\mu\right)\right]$$

From the above equation it is clear that the parameters

$$F_{max} * l = 425 * 25 = 103600\ N - mm$$

Since here F_{max} is limited to 4000 N the minimum possible value of l is given by

$$l_{min} = \frac{103600}{4000} = 25.9\ mm$$

$$2l_{min} = 51.8\ mm$$

For the calculation of the minimum capacity of the machine, the value of l has to be found. When $2l$ exceeds the stock length L, from the above example, the total length of the neutral plane which is required for the stock length is

$$L_s = 2\left[\left(r_p + 0.45t\right)\frac{\Pi}{4} + 50 - \left(r_p + t\right)\right] = 87.00\ mm$$

or $$2l_{max} = L = 87.00\ mm \text{ or}$$

$$l_{max} = 43.5\ mm$$

The minimum capacity of the machine required is

$$\left(F_{max}\right)_{min} = \frac{103,600}{43.5} = 2381.60\,\text{N}$$

Unsolved example

1. An aluminium strip of 60 mm width and 4 mm thickness is bent into a 90^0 angle. The data given are fracture strain = 0.25, $= 70\,kN/mm^2$, $E = 75\,kN/mm^2$, $\mu = 0.1$, $l = 0.25\,mm$, $\tilde{A}_{Y_0} = 14\,N/mm^2$, $n = 30\,N/mm^2$, die opening = 30 mm.

   Determine (1) minimum possible bending radius, (2) the angle of the bending punch with spring back and (3) the peak bending force.
2. A sheet-metal part 3.0 mm thick and 20.0 mm long is bent to an included angle 0^0 and a bend radius of 8 mm in a V-die. The metal has a yield strength of 220 MPa and tensile strength = 340MPa. Compute the required force to bend the part, given that the die opening dimension = 15 mm.

11.9 PRESS – TYPES, CONSTRUCTION AND WORKING

A hydraulic press is a mechanical device which is based on "*Pascal's law*," which states that equal intensity of pressure exerts in all directions in a closed system. It is applicable here in such a way that if there is any pressure change at one point in a closed system then the same intensity of pressure will change at another point in the same system. A hydraulic press is able to develop high forces with the application of less effort. There are two main parts in the hydraulic press assembly, one is called the ram and the second is known as the plunger. The ram works as an output medium, whereas the plunger gives input. In between the ram and plunger hydraulic fluid is filled in a closed container which is responsible for the whole operation, i.e. force and pressure transmission. In a practical situation both the plunger and ram have different areas.

11.9.1 Types of Presses

Presses are classified according to the following characteristics:

(1) Source of power
- Hand press or fly press
- Power press.

(2) Type and design of frames
- Inclined press
- Arch press
- Gap press
- Horn press
- Straight side press.

(3) According to action
- Single action
- Double action
- Triple action.

(4) Mechanism used for applying power to ram
- Crank
- Eccentric
- Knuckle
- Cam
- Toggle
- Rack and pinion
- Hydraulic.

(5) Number of drive gears
- Single drive
- Twin drive
- Quadruple drive.

(6) Method of power transmission from motor to crankshaft
- Direct
- Non-geared.

(7) Based on tonnage
- Mechanical press
- Hydraulic press.

11.9.2 Construction

In a practical hydraulic press system, generally multiple rams are assembled together. The number of rams used depends upon the working load. In a hydraulic press multiple rams of small sizes are preferred instead of a single large ram to control the thrust forces, because it is easy to control the thrust forces on a small size as compared to a large size. In a press assembly, one side/table is always fixed, while the other moves due to the application of ram force, and pressing operations take place between the fixed and free sides. Ram is operated by the hydraulic pressure of fluid. The high-pressure liquid is supplied using a pump and hydraulic accumulator. The hydraulic accumulator works as the junction between the pump and the rams. The hydraulic accumulator stores the high-pressure liquid when the press is in a stationary position. The hydraulic press is used where high thrust is required for operation.

- **Fly press**

This is the most basic sort of press, and is also known as a hand press, ball press or one-side fly press, and is made of a sturdy cast iron frame. Figure 11.23 shows details of fly press. The nut is formed by the top section of the frame. A vertical screw can pass through the nut, which carries an arm. At each end of the arm, two cast iron weights (balls) are supported. The handle is used to rotate the arm.

- Extend the frame below the nut to form guides. Ram is hooked to the screw's bottom.
- Ram carries punch at the bottom of its body. The die is secured to the press base.

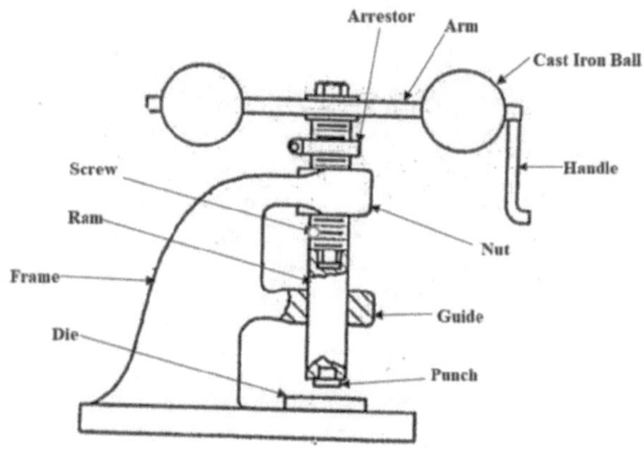

FIGURE 11.23 Fly press.

- **Working**

I. Sheet metal is draped over the die.
II. With the help of the handle, the arm rotates quickly.
III. Heavy balls store kinetic energy for long-term screw movement. Movement of the screw causes the ram and punch to move lower.
IV. The collar's stroke can be adjusted with the use of a stop collar/arrestor.
V. The most advanced sort of fly press is the double-sided press.

Power press

- **Main parts of the typical power press**

This is the all-purpose machine tool, and the base is one of the components of a press. Figure 11.24 shows details of power press. It is the main supporting part for workpiece-holding dies and various press-controlling mechanisms. The size of the workpiece that can be processed on a press is limited by the size of the table. In the case of some special presses, the base also contains a mechanism for tilting the frame in any desired inclination position.

- **Frame**

The frame is the primary body of the press and is placed on one edge of its base. It houses the ram, the driving mechanism and the control mechanisms. Some presses have a column-shaped frame.

- **Ram**

This is the major functioning part of the press that is directly involved in the processing of a workpiece. The ram moves back and forth within its guideways with a

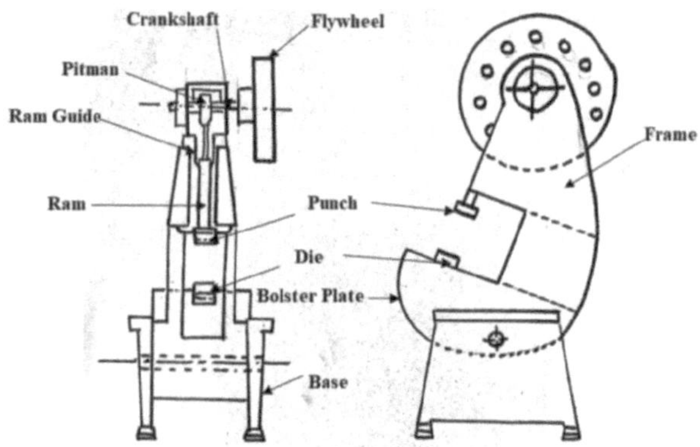

FIGURE 11.24 Power press.

predetermined stroke length and power. The stroke length and power imparted can be modified to meet the needs of the user. The lower end of the ram carries punch to process the workpiece.

- **Pitman**

This is the component that connects the ram to the crankshaft, also known as the ram eccentric.

- **Driving mechanism**

Driving mechanisms are utilised in various types of presses, such as cylinder and piston arrangements in hydraulic presses, crankshaft and eccentric mechanisms in mechanical presses, and so on. These mechanisms transfer power from the motor to the ram to drive it.

- **Control mechanism**

Control mechanisms are used to operate a press under controlled settings. Normally, two parameters are changed via controlling mechanisms: the length of the ram's stroke and the power of the stroke. Power transfer can be withdrawn using a clutch equipped with drive mechanisms as needed. In most presses, the controlling mechanisms are integrated with the driving mechanisms. Nowadays, computationally controlled presses are employed, with control directed by a microprocessor. With automation, these presses provide dependable and accurate control.

- **Flywheel**

In most presses, the driven gear or driven pulley is formed in the shape of a flywheel, which is used for storing the energy (reservoir of energy) to maintain the constant speed of the ram while the punch is pressed against the workpiece. The flywheel is installed in the driving mechanism directly before the clutch in the power transmission sequence.

- **Brakes**

In any mobile system, brakes are critical. In general, two types of brakes are used: regular brakes, which can swiftly bring the driven shaft to rest after disengaging it from the flywheel, and anti-lock brakes. Another example is emergency brakes, which are available as a foot brake on any machine. These brakes contain a power-off switch as well as standard enhanced braking to quickly bring all motions to a halt.

- **Balster plate**

This is a thick plate attached to the press's bed or base, and it is used to tightly fasten the die assembly to hold the workpiece. Because the die used in press operations may contain more than one part, the phrase die assembly is used instead of die.

Working of power press

The ram of a mechanical power press is powered, and the rotating motion produced by the electrical motor is translated into a reciprocating action of the ram by means of various mechanical mechanisms. The flywheel is attached to the driving shaft through a clutch and is mounted at the end of it. The energy is stored in the flywheel during idle periods and is consumed to keep the ram at a constant speed when the punch is forced into the job. The electric motor is directly linked to the flywheel.

Advantages of a hydraulic press

- More adaptable and simple to use.
- Tonnage capacity changeable from zero to maximum.
- Constant pressure may be maintained during the stroke.
- Throughout the stroke, force and speed can be modified.
- Stronger than a mechanical press.
- It is safe since it will stop at a preset pressure.
- Stroke length can be adjusted to any length within the confines of hydraulic cylinder travel.
- At any point throughout the ram stroke, the press can exert its entire tonnage.

11.10 PRESS TOOLS – DIES, PUNCHES

Punches are simple tools that are forced by a punch press through a workpiece, commonly sheet metal, to create quick, precise holes by shearing. Punches are typically made of carbides or tool steel. Most presses are operated mechanically, but simple hand punches also are used. A punch frequently passes through the material and into a die. A die holds the workpiece, and determines the shape produced on it by the punch. Dies are usually customised to the particular item being produced.

The press tool (commonly known as die/dies) is an assembly of die, punch, punch plate, punch backplate, stripper plate, etc. to produce sheet metal components/ stamped parts from the flat metal sheet.

11.10.1 Dies

The die set is the unit assembly which incorporates a lower and upper shoe, two or more guide posts and guide post bushings. It consists of the following parts (Figure 11.25).

- **Die**: The die is the feminine component of a full tool used to produce work in a press. It is sometimes referred to as a full tool that consists of two mating elements for creating work in press.
- **Die block**: The die cavity is contained within the block or plate.
- **Lower shoe**: In most cases, the lower shoe of a die set is fixed on the upper plate of a press. The die block is secured to the bottom shoe. It also has the guiding posts fixed in it.
- **Punch**: The punch is the male component of the die assembly that is moved or fastened to the press rams or slides directly or indirectly.
- **Upper shoe**: Die post bushings are located in the upper portion of the die set.
- **Back up plate**: This is also known as a pressure plate. It is positioned such that the intensity of pressure on the punch holder does not become excessive. To avoid crushing, the plate spreads pressure over a wide region and reduces the degree of pressure on the punch holder. **Stripper:** A stripper is a plate that is used to remove the metal strip from a punch or die, whether it is cutting or non-cutting. It may also serve as a guide for the strip. Dies used for drawing sheet metal are usually one of the following basic types or some modification of these types:

    1. Progressive dies
    2. Compound dies

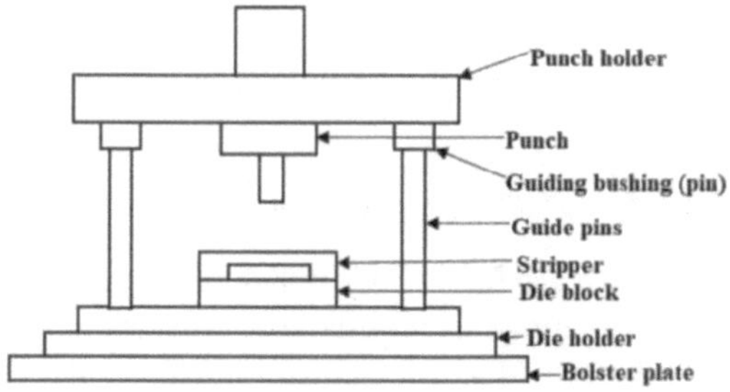

FIGURE 11.25 Details in dies.

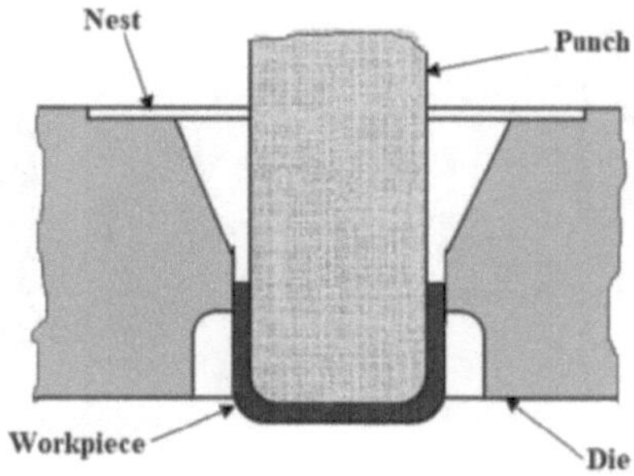

FIGURE 11.26 Single-action dies.

3. Single-action dies
4. Double-action dies
5. Multiple dies with transfer mechanism.

Selection of the die depends largely on the part size, severity of draw and quantity of parts to be produced.

(a) Single-action dies

Single-action dies are the simplest of all drawing dies and have only a punch and a die (Figure 11.26).

To position the blank, a nest or finder is supplied. The drawn component is forced through the die and is removed from the punch by the die's counterbore at the bottom. To make this feasible, the cup's rim expands slightly. Single-action dies should only be used when the forming limit allows for cupping without the use of a blankholder.

(b) Dies with two actions

A blank holder is included with double-action dies. This enables larger reductions as well as the drawing of flanged pieces. In this design, the die is attached to the bottom shoe, the punch to the inner or punch slide, and the blank holder to the outside slide. The pressure pad is used to keep the blank firmly against the punch nose when drawing and to extract the drawn cup from the die. In the absence of a die cushion, springs, air or hydraulic cylinders can be utilised; nevertheless, they are less effective than a die cushion, especially for deep draws.

Figure 11.27 depicts an inverted sort of double-action die that is used in single-action presses. The punch is positioned on the lower shoe, whereas the die is mounted on the top shoe in this arrangement.

(c) Compound dies

When the initial cost is justified by production demands, combining numerous procedures in a single die is feasible.

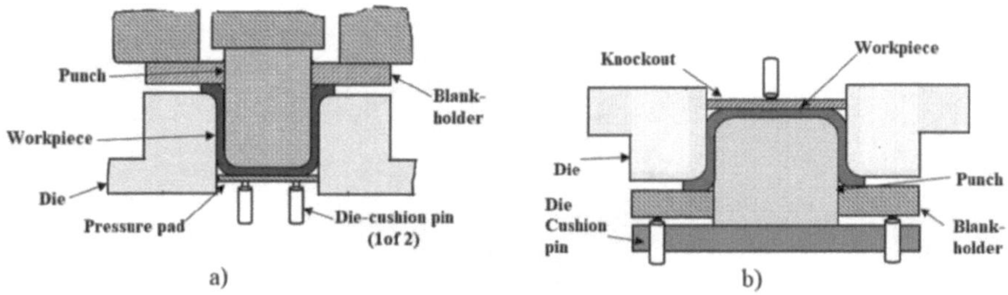

FIGURE 11.27 Double-action dies: (a) double-action die; (b) inverted type.

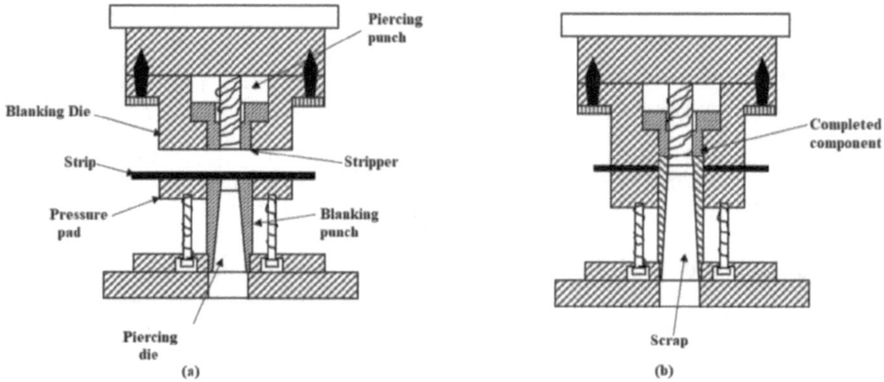

FIGURE 11.28 Compound dies.

Blanking and drawing are two processes that are frequently performed in compound dies. Work pieces can be made several times faster using compound dies than with simple dies shown in Figure 11.28.

(d) Progressive dies

The initial cost and length of bed required for progressive dies typically limit their applicability to relatively small workpieces. Figure 11.29 displays a typical six-station progression for mass producing small shell-like workpieces. Larger parts, such as liners for automotive headlights, have, on the other hand, been drawn in progressive dies.

11.10.2 Punches

The choice and type of the punch depend on the shape and size of the pierced or blanked contour and the work material. For example, large cutting perimeters require large punches which are inherently rigid and can be mounted directly. However, smaller size holes require punches which may have to be supported during the

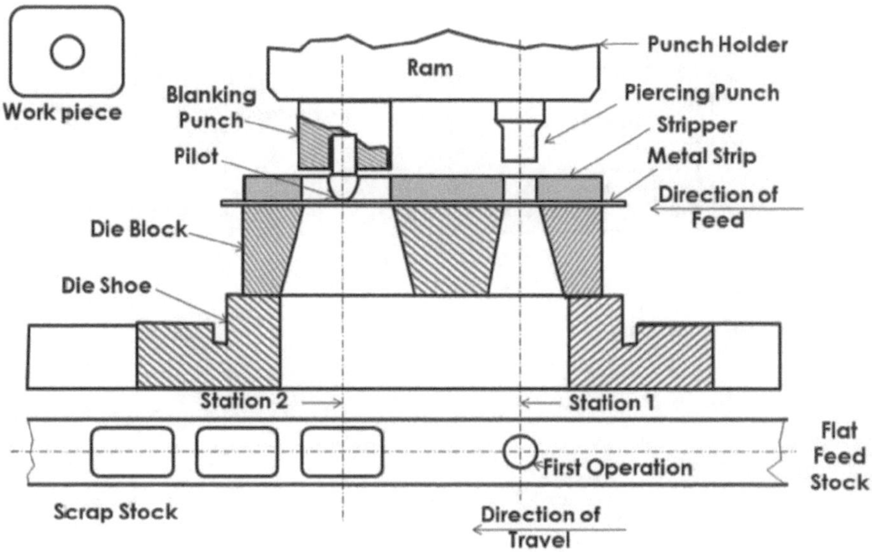

FIGURE 11.29 Progressive dies.

operation and therefore need to have another mechanism to join them to the punch holder. Punches are classified as follows:

- Plain punches
- Pedestal punches
- Punches mounted in punch plates
- Perforator punches
- Quill punches.

Descriptions of some of these follow.

(1) Plain punches

The most basic punches are composed of solid tool steel blocks and are immediately mounted to the punch holder. These punches are attached to the holder with dowels and screws. These must be large enough to accommodate the dowels while also being strong enough to sustain the punching force. The length and width of these punches must be greater than the punch's height. Figure 11.30 shows a plain punch.

The main advantage of plain punches is the cost-effectiveness of punch construction. For the sake of stability, the height of these punches is kept as low as possible. Because of its wide cross-sectional area, this saves a lot of tool material. Plain punches are very easy to mount to the punch holder. Rather than using a solid block, very large plain punches can be sectioned in the same way that die blocks are.

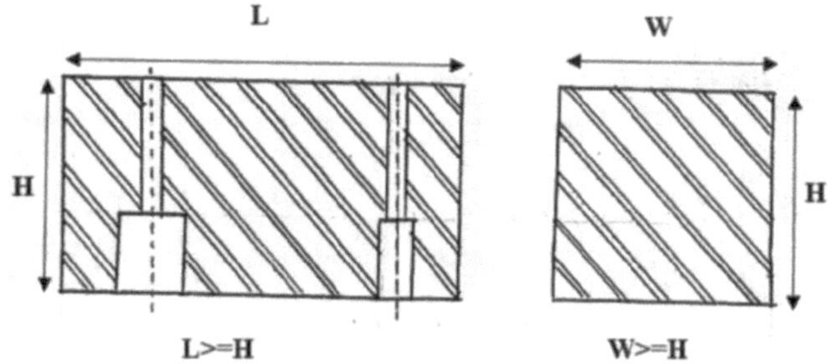

FIGURE 11.30 Plain punch.

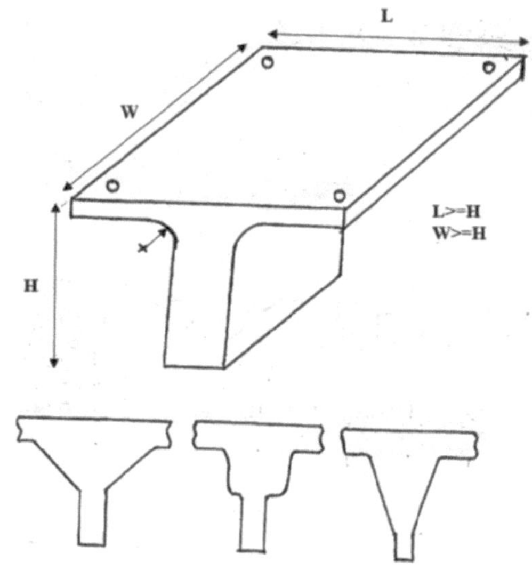

FIGURE 11.31 Pedestal punch.

(2) Pedestal punches

Also known as flanged punches or shoulder punches, these are distinguished by a broad base surface in comparison to the cutting face. The flanged part of the punch provides great punch stability. Because the vast base area disperses all of the cutting energy, pedestal punches are advantageous for heavy tasks. The base's length and width should be greater than or equal to the punch's height. The flange thickness and radius should be generously provided to withstand the considerable forces acting on the punch. Figure 11.31 shows a pedestal punch.

(3) Perforator type punches

Punches whose cutting face diameter is less than 25 mm are termed perforators.

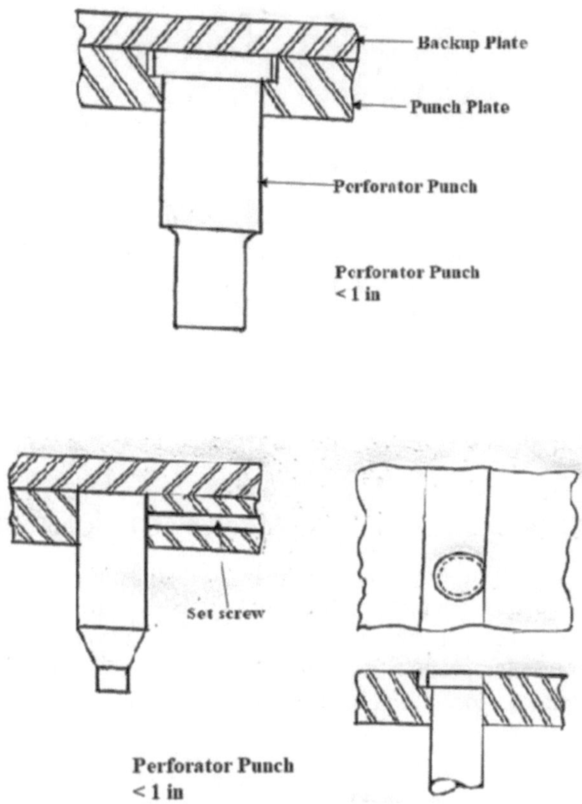

FIGURE 11.32 Perforator punch.

The punches do not have to be round, but the inserted circle of the punch must be smaller than 25 mm in diameter. Generally, all perforates are mounted in a punch plate. The simplest and most common perforator has a step head. The proportions of these punches are specified and are commercially available in a variety of sizes. Figure 11.32 shows details of perforator punch. If the cutting face is circular, the punch can be assembled in any orientation. However, for punches with other than round contours, some technique of limiting punch rotation is required.

(4) Quill punch

When piercing very small holes smaller than 6 mm in diameter, it is preferable to add extra support to the punch shank with a tightly fitted quill. Due to the close fit necessary between the punch and the quill sizes, quilled punches are more expensive when created separately. Figure 11.33 shows details of quill punch. As a result, they are mass produced in a variety of standard sizes.

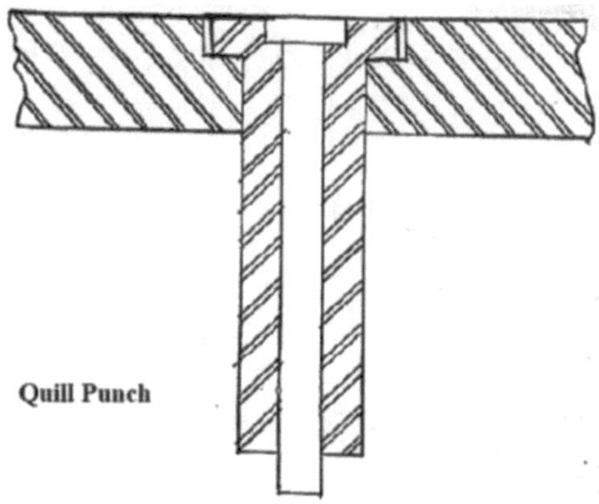

FIGURE 11.33 Quill punch.

11.11 AUTOMOTIVE STAMPING MANUFACTURING PROCESS

Metal stamping manufacturing processes may produce solid metal parts and forms in a speedy and clean manner. Specialised dies are used in the process to accurately shape metal sheets. Metal stamping is advantageous for auto manufacturers because the dies can be reused to produce consistently sized and shaped parts that meet stringent standards and tolerances. Among the many advantages of metal stamping are the following:

- **Cost-effectiveness**: After the dies are created, manufacturers can produce high-volume runs of car parts at a minimal cost. Metal stamping technologies can handle a wide range of metals, giving businesses a high degree of versatility without incurring additional costs. Metal stamping dies are very inexpensive to produce and maintain in excellent working order. The latter stages of production, such as plating, are similarly inexpensive.
- **Material efficiency**: Sheet metal is used to make parts in metal stamping procedures. Sheet metal is economical and does not add unnecessary bulk or thickness to the product. Sheet metal can be used for a variety of pieces, such as structural elements, chassis and mechanical components in engine or transmission systems.
- **Automation**: Metal stamping processes can be completely automated. This not only reduces the possibility of human error or inconsistency, but also reduces labour expenses and keeps personnel safe.

Technologies and techniques

With the introduction of new technologies, metal stamping processes continue to evolve and improve. Not only does enhanced end-to-end automation improve the manufacturing process, but stamping machinery is also becoming more advanced. The employment of electromagnetic tools is a common improvement. These devices can interact with sheet metal more effectively by precisely positioning the sheet metal using electromagnetic forces. By minimising the distance between stamps, electromagnetically assisted tools can create more elaborate, complicated structures while also reducing material waste.

Metal stamping presses for automotive and other industrial parts are classified into three types. They are as follows:

1. Mechanical presses with a mechanical flywheel driven by a motor.
2. Hydraulic presses, which use pressured fluids to manage pressure and force application on a workpiece.
3. Mechanical servo presses, which rely solely on motors and do not have mechanical flywheels.

To boost efficiency and manufacturing speed, all three press types can collaborate with automated systems such as automatic feeders. To achieve a quick run, automatic feeders feed coiled or blank sheet metal through the presses.

Key materials for automotive metal stamping

Metal stamping presses and dies can work with a variety of metals to produce a wide range of products. Aluminium, copper and steel are some of the most widely utilised metals in stamped vehicle parts. Each metal has distinct properties that make it appropriate for specific purposes.

- Aluminium: Because of its appealing appearance and light weight, manufacturers choose aluminium for visible parts and details. Companies can readily create alloys with aluminium and other metals to create stronger or more lasting variants.
- Copper: This metal is corrosion resistant, ductile, has a nice finish and is inexpensive. Copper is also a good choice for environmentally minded customers because it is easy to recycle and reuse.
  - Stainless steel: Because it contains at least 11% chromium, this metal resists corrosion and rust. It not only has greater durability, but also has a shiny, appealing appearance.
  - Steel and steel alloys: Steel alloys exist in a variety of forms, including mild steel, high-strength steel and specialty steel, which may be utilised to make a wide range of durable parts.

Stamping is a forming method that uses a number of stamping stations to create sheet metal. The stamping technique is used to mass manufacture finished goods in large quantities. Stamping sheet metals is described as the process of converting the shape of a sheet metal blank into a usable shape in the plastic deformation state utilising a die and a mechanical press; stamping is a net shaping process. In general, automobile stamping is divided into two types: deep drawing and stretch forming. The punch forms (draws) the sheet metal from the binder in the deep drawing mode.

When the clamping pressure of the holding down ring is extremely strong, the friction forces in the flange are high enough to allow the material to flow in. The metal flow patterns within the die cavity can be used to further assess the formability of sheet metal. Metal flow patterns in stamping are classified into three types: elastic flow, plastic flow and rigid movement.

Metal stamping is the process of producing metallic components by applying severe pressure to blank pieces or sheets of metal. The goal of this metal stamping method is to be able to create structures of any shape and size from metal sheet. It is closely related with the automobile industry because many pieces of a car are made of steel. Outer car panels, such as hoods and fenders, are common examples of metal stamping parts. Sheet metal stamping presses serve as carriages for transporting other machinery. Dies are simple machinery that fit within and are attached to a press. The steel sheet is reduced to size and bent and cut in a stamping machine to manufacture pieces such as car doors, roofs and hoods. Plastic forming processes 60–70% of metal parts in the automobile manufacturing process, and metal forming completes the stamping process, such as a variety of car body panels, car parts support, engine exhaust pipe and muffler, hollow camshaft, oil pan, engine cradles, frameworks, structures, horizontal beam, and so on. The stamping process is the so-called means of stamping components of the adaption process that are specified in size stampings, precision and base size, structure and shape with pressing processing technology needs. An automobile body may have more than 1500 stampings, which include the door, cover, chassis, floor, and so on, in addition to the large-scale panel.

11.12 IDENTIFICATION OF MAJOR AUTO COMPONENTS BY MATERIAL AND PROCESS

Over the past decade the automotive industry has seen massive changes in next-generation automotive manufacturing materials and processes. Vehicles of the past consisted of entirely steel-based products, while modern auto manufacturers are transitioning towards aluminium, magnesium and composite materials that deliver enhanced performance. To accommodate these new materials, new manufacturing techniques are also being adopted. Figure 11.34 focuses on upcoming processes and material changes that the automotive industry will be moving towards over the next 20 years.

Unsurprisingly, the study found that current vehicles are predominantly steel structures with some use of aluminium. Vehicle frames, including floors, doors, roofs, body side panels and fenders are all typically constructed from steel. As these are the components most responsible for driver safety, they are the most difficult to use other materials for. Materials used for other less-critical components such as a vehicle's hood, sunroof, bumper or engine cradle are often experimented with as they offer the opportunity to reduce overall vehicle weight.

11.13 SUMMARY

Press shop processes are also known as sheet-metal forming processes. These are among the most versatile of all metalworking operations. In sheet forming, sheet

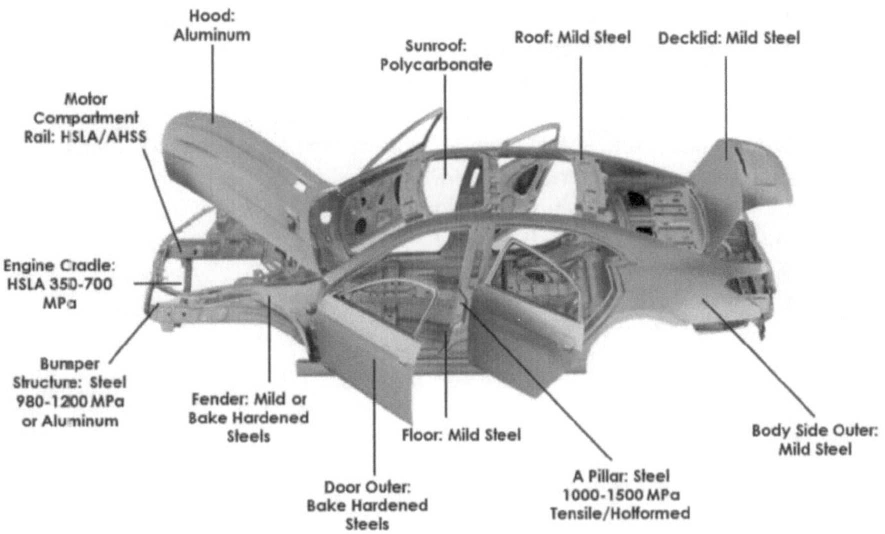

FIGURE 11.34 Materials used most commonly for major vehicle structures components in current fleets.

blanks are plastically deformed into a complex three-dimensional geometry, usually without any significant change in sheet thickness or surface characteristics. Bending is a procedure that deforms metal in such a way that the length and thickness before and after bending remain the same. It just alters the shape of the workpiece. Stamping sheet metals is described as the process of converting the shape of a sheet metal blank into a usable shape in the plastic deformation state utilising a die and a mechanical press. The tooling that performs sheet metalwork is called a punch-and-die; the term stamping die is also used. The sheet-metal products are called stampings. An automobile body may have more than 1500 stampings, which include the door, cover, chassis, floor, and so on, in addition to the large-scale panel.

11.14 REVIEW QUESTIONS

(1) List the various metal forming processes.
(2) List various types of dies used in the sheet metal forming process.
(3) State the advantages of a power press.
(4) State the advantages of the sheet metal forming processes.
(5) List the various types of punches used in the sheet metal forming process.
(6) Justify the roll of sheet metal operation in the automotive body manufacturing process.
(7) Describe the forming process with the help of a neat sketch.
(8) Describe the blanking process with the help of a neat sketch.
(9) Describe the piercing process with the help of a neat sketch.
(10) Describe the punching process with the help of a neat sketch.

(11) Describe the various of types of dies with the help of a neat sketch.
(12) Describe the construction and working of the power press.
(13) Give a detailed classification of presses.
(14) Give a detailed classification of dies.
(15) Draw a neat sketch of a single-action die, double-action die, compound die, progressive die, fly press and power press.
(16) Describe the construction and working of the fly press with a neat sketch.

12 Case Studies of Automotive Manufacturing Units

LEARNING OBJECTIVES

- To understand the practical cases of automotive manufacturing processes implemented in Industry.
- To apply knowledge of Heat Treatment, Casting, Forging, Welding, machining operation in vehicle manufacturing.
- To understand the applications of automotive plastic processing.

12.1 CASE STUDY 1: HEAT TREATMENT OF AUTOMOTIVE COMPONENTS – CURRENT STATUS AND FUTURE TRENDS

12.1.1 INTRODUCTION

The automobile is an important product of the manufacturing process, requiring a diverse range of materials as well as technologies. The current societal requirements require a reduction in vehicle weight by the use of comparatively lightweight materials. Despite growing usage of aluminium and plastics, a modern vehicle's iron and steel content remains as high as 70%. Metallic materials are well suited for use in high-stress components that demand a high level of durability. Functionality and component performance are inextricably linked to the efficient use of the processing technology used in a particular application.

Typically, parts of the body are made of steel sheets which are rolled and also thermally processed to get the appropriate qualities. The heavy body parts are made using processes known as stamping, welding as well as coating prior to assembly.

Gears used in automotive applications are another critical category of components that are subjected to extreme stress and require a high level of fatigue as well as wear-resistant performance. Effective and proper heat treatment and also surface processing technologies are used to maximise the qualities of almost all types of metal components, with a particular emphasis on endurance in a wide variety of applications.

DOI: 10.1201/9781003367321-12

Various forms of heat treatment and surface engineering procedures are used in the fabrication of automobile components, starting with raw metal products and ending with final component assembly. Heat treatment techniques add the needed strength or hardness attributes to a component based on its intended use. Additionally, metal processing operations such as forming, machining, quenching, tempering, carburising, hardening and nitriding may be used throughout production. When done appropriately, surface modification results in optimal surface qualities that increase corrosion and wear resistance while increasing frictional resistance.

12.1.2 MATERIALS AND THEIR APPLICATIONS IN AUTOMOBILES

Despite significant efforts to produce all-aluminium automobile bodies, the majority of automobiles on the road today are constructed of iron and steel (70%), aluminium (6%), plastics (9%), rubber (4%), glass (3%), and many other elements (8%). To meet fuel efficiency objectives, vehicle body weight must be reduced while engine and rolling energy losses are reduced. These advancements are being made by increasing the applications of steel sheets having high strength and/or by increasing the use of aluminium, magnesium and titanium alloys with lower specific weights than iron and steel. Weight reduction is possible through the use of ultra-high tensile strength sheet steels. These steel sheets with ultra-high tensile strength are manufactured using advanced technology of steel mills, including controlled rolling and cooling processes, as well as heat treatment. The challenges of body design and manufacturing technologies have been studied as part of a global development effort known as the Ultra Light All Steel Body (ULSAB).

12.1.2.1 Bake Hardening Steel Sheets

Steel sheets with a low yield strength and good formability are required for stamping and forming operations. Body panels, on the other hand, require rigidity and strength to improve crashworthiness and fatigue resistance. Initially, both of these requirements were met by bake hardening steel sheets. These sheets are formable during the stamping operation but must be baked after painting to solidify. The baking process can increase the strength of the material from less than 30 kg/mm^2 to approximately 45 kg/mm^2. This technique significantly increased productivity, increased the strength and longevity of the vehicle body, permitted the production of high-mileage automobiles and improved crashworthiness.

12.1.2.2 High-Tensile-Strength Steel Sheets

The chemical composition of sheet steels is optimised, as are the rolling and cooling processes. Increased use of high-tensile-strength sheets (Hiten) appears necessary to reinforce the body shell while reducing its weight in order to meet fuel economy requirements. Numerous research and development efforts have been made, as documented in the Ultra Light Steel Auto Body (ULSAB) and the Ultra Light Steel Auto Body (ULSAB) AVC (Advanced Vehicle Concept).

12.1.2.3 Corrosion-Resistant Coated Steel Sheets

Coating the body panel sheets will prevent perforation caused by salt splash during winter. Panels are coated in a variety of ways; including zinc dip coating, Zn-Fe alloy coating, Zn-Fe plating and Ni-Zn plating. Additionally, comparable to some types of stainless sheet and pipe, Al coating and Al-Zn coating are employed on exhaust line pipes and mufflers. Additionally, fuel tanks require corrosion-resistant coatings such as lead, however this is no longer necessary to reduce harmful leftovers.

12.1.2.4 Constructional Steels

Numerous construction steels are used to create high-performance components in engines, suspensions and power train systems that require strength and durability. Strength is introduced into components through the selection of the suitable steel grade and heat treatment to provide the required strength and fatigue durability. While weight reduction is required to enhance fuel economy, steel components must retain their strength, and various types of fasteners and screws needed to assemble the body, transmission, chassis and auxiliary sections are made of steel. However, a gradual switch to a lighter fastening mechanism will continue. The rivalry associated with such material transitions may force the creation of more effective component designs that are more durable through the effective usage of structural steel products that have undergone optimal heat treatment.

12.1.2.5 Case Hardening Steels

Case hardening procedures are critical for imparting hardness, static as well as dynamic strength, as well as wear and seizure resistance. Because the qualities of simple quenched and hardened steel are insufficient to withstand the bending, rotational stress and friction, automobile components are manufactured using a variety of surface hardening techniques. Carbon and/or nitrogen diffusion and quench hardening procedures such as carburising, nitriding, as well as induction hardening are used to increase case hardness. Case hardening steels of various grades are used to manufacture automotive components such as transmission shafts and gears.

12.1.2.6 Heat-Resistant Steels

Engine valves, particularly exhaust valves, require extreme temperature resistance and are manufactured with heat-resistant steels. The essential high-temperature strengths of the shaft area and valve seat part, on the other hand, are significantly different. Due to this, valves are constructed using two distinct types of heat-resistant steel. Additionally, valves have been applied nitriding to increase their wear as well as seizure qualities.

High-performance engine valve seats require high-temperature wear resistance during high-temperature operation, and high-energy cladding processes such as inert gas-arc or laser alloying are frequently used to deposit Stellite on the seat contact region.

Additionally, heat-resistant alloys are employed for the rotors of exhaust turbines in high-performance turbocharged engines. Although these components are manufactured using investment casting techniques, some are now constructed of sophisticated ceramics such as silicon nitride.

12.1.2.7 Copper Alloys

To produce plane bearings for engine crankshafts and connecting rods, bearing materials are required. Although copper alloys have historically been used as the primary material for aircraft bearings, the use of aluminium alloys has been progressively growing.

Copper alloys are used to make electrical wire products such as wire and thin ribbon wiring or cables. To optimise wear attributes, valve seats are made of Cu-based alloys.

The new trend toward hybrid and fuel cell vehicles with an electric motor drive train will result in an increase in the use of conductive materials in the near future. However, additional research and development are necessary to decrease the specific weight of magnetic and copper materials. These investigations are targeted at reducing the weight of components in aggregate.

12.1.2.8 Aluminium Alloys

Typical aluminium alloy components include transmission, differential and steering gearbox cases. Sheets of aluminium are utilised for panels such as the hood of the engine, body panels, trunk lid cover and suspension components in order to minimise the vehicle's weight. The same method used to harden steel sheets was employed to generate a new aluminium alloy for use in body panels (Figure 12.1).

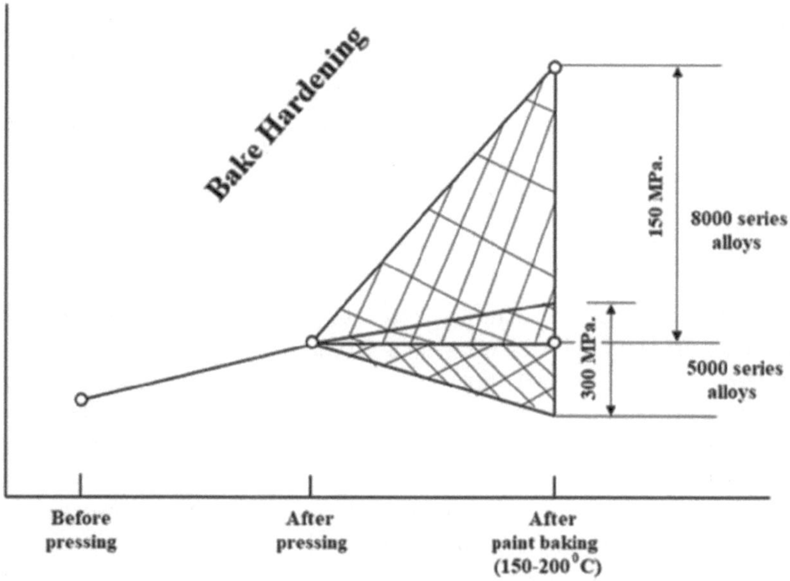

FIGURE 12.1 Change of yield strength by paint baking.

12.1.2.9 Magnesium Alloys

While magnesium's low weight makes it an attractive material for reducing component weight, its rigidity, strength and corrosion resistance make it unsuitable for automobile applications. However, magnesium alloys have been utilised in older Volkswagens since the 1950s because they are the lightest metal. Though magnesium alloys are expensive, their use is facilitated by their higher weight-to-strength ratio. Specifically, the corrosion resistance of commercial magnesium alloys is insufficient to meet application requirements, and hence a low-impurity alloy grade for automotive applications has been created.

12.1.2.10 Titanium Alloys

An alloy of titanium has a great specific rate – weight/strength ratio – but their high cost is the primary impediment to their widespread use in the automobile sector. Suspension coil springs, valve spring retainers, as well as connecting rods are all examples of components that are suitable for Ti alloy application.

The primary impediment to titanium alloy use has been overcome by the development of new manufacturing technologies utilising powder metallurgy procedures. Toyota developed a low-cost valve fabrication technology based on sponge titanium powder. These valves are less expensive than those made of heat-resistant steel. The primary production procedures depicted in Figure 12.2 permitted mass manufacture of lightweight titanium valves for passenger cars.

12.1.2.11 Composite Materials

Fibres or particles can be used to reinforce soft and weak materials. Glass fibre-reinforced polymers are utilised for roof panels and other plastic goods. Although a matrix of metal composites is not widely used in the automotive manufacturing industry, they can be used to boost local strength in areas where components require creep strength or stiffness.

Since 1981, Toyota has used composite engine pistons. It was developed successfully by the use of fairly basic alumina-silica fibres squeeze-cast in the top half of an Al alloy piston. From 1981, numerous technical advancements have been achieved, and the most modern MMC piston features a porous nickel performer surrounding the groove of ring.

MMC is used in automobiles to manufacture the following components: diesel engine pistons (Figure 12.3), engine cylinder block bore section, connecting rod, and crank dumper pulley shaft hole.

12.1.2.12 Plastics and Rubber

Plastics have steadily increased their share of the auto body. The primary objective is to develop environmentally friendly plastic materials that are recyclable in order to reduce waste. Aluminium and high-strength steel sheets compete with plastic panels. Plastics are increasingly being used as fasteners, connections and coverings.

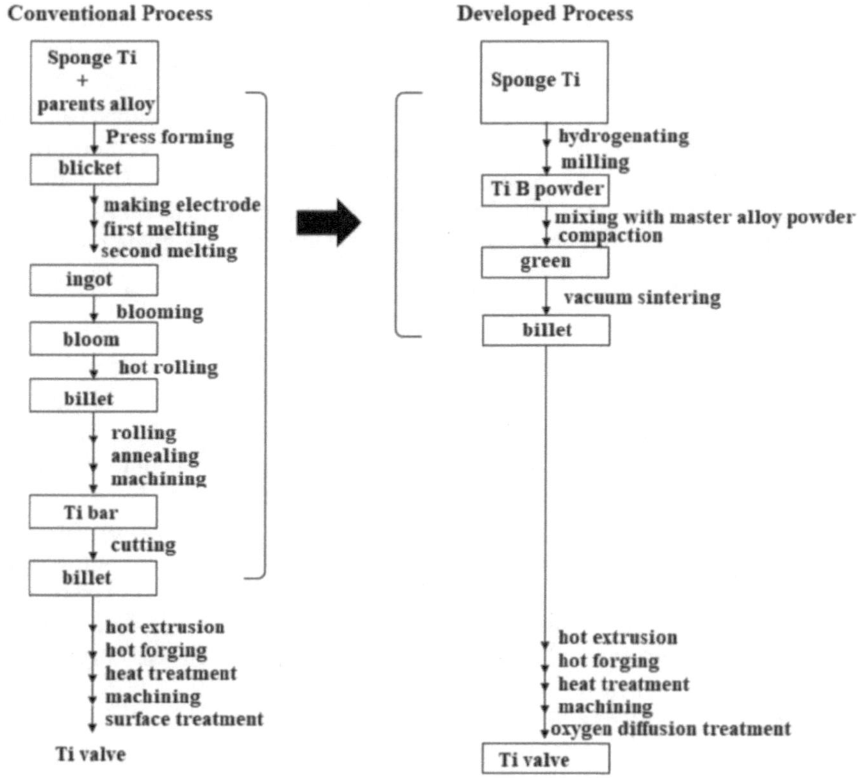

FIGURE 12.2 Flow charts for the valve making process.

Although the used volume or weight is modest, plastics as well as resins are used to manufacture metal-laminated products such that noise-absorbing sheets and light-weight laminated steel sheets with a specific strength compared to aluminium alloys.

Rubber has historically been utilised extensively in the manufacture of tyres and weather stripping. While rubber will continue to be the primary material used in tyres, several rubber uses such as weather strips and seals are under competition from new plastic grades.

12.1.2.13 Glass and Ceramics

While glass is still utilised for windows and covers for lighting, translucent polymers are being used for housing and covers for lights. Ceramics are used in a wide variety of applications, from classic spark plugs to modern ceramics such as turbocharged rotors, exhaust valves, as well as valve lifter shims. PZT actuators, diesel-specific filters and a variety of other ceramic items appear to be gradually expanding their uses in response to the rising use of electronic parts.

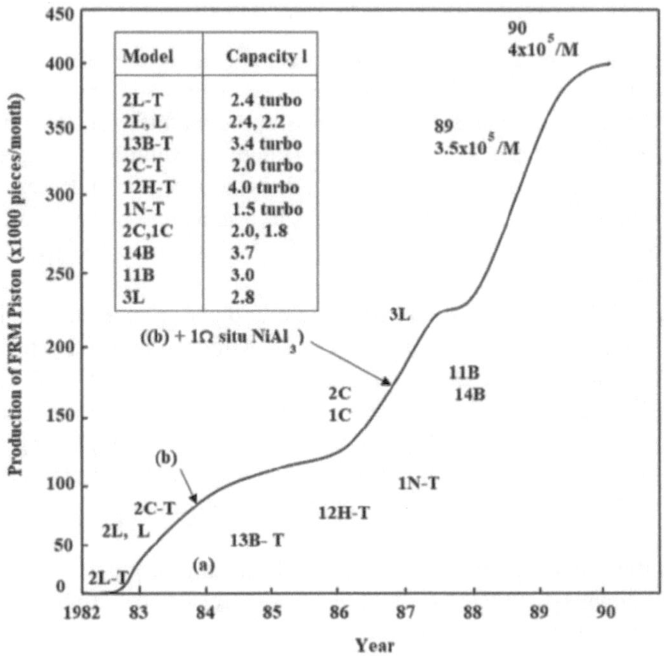

FIGURE 12.3 Production of the FRM piston.

12.1.3 HEAT TREATMENT

Heat treatment is the process of heating metal without letting it reach its molten, or melting, stage, and then cooling the metal in a controlled way to select the desired mechanical properties. Heat treatment is used to either make metal stronger or more malleable, more resistant to abrasion or more ductile.

12.1.3.1 Types of Heat Treatment

The choice of steel types as well as grades, and the use of specific heat treatment processes, is critical for producing components of consistent quality. Controlling the chemical composition of an alloy as well as the steel inclusion content has an effect on and can cause variation in an alloy's properties. Additionally, elements such as refining, casting, rolling and cooling processes all have an effect on the quality and durability of finished components. Additionally, strength, toughness, fatigue strength and wear qualities are substantially determined by the microstructure and hardness outcomes produced by the chosen heat treatment conditions and procedures. As a result, it is critical to be aware of these aspects and to employ proper solutions.

12.1.3.2 Processing Technology in Heat Treatment

From the introduction of gas carburising processes, numerous advancements in furnace, environment and control technologies have been made, and numbers of continuous heat treatment furnaces are now used in the manufacturing of automobiles. Improvements in furnace design have resulted in significant energy savings. Additionally, significant progress has been made in the adoption of decreased pressure-controlled processing technologies. Vacuum carburising and nitriding techniques are gaining popularity, and plasma-assisted technologies have significantly enhanced their capabilities.

Steel components are typically forged and after that can be quenched, tempered or isothermally annealed, machined, as well as case hardened to provide the appropriate strength, durability and fatigue life. These repeated heating and cooling operations require a significant quantity of thermal energy. Since the 1960s, forge as well as quench technologies have been used to conserve energy by eliminating the need for warming processes. This technology was initially applied to plain carbon steel in order to boost quench hardenability through the application of appropriate thermo-mechanical treatment.

Micro-alloyed steels permit heat treatment following forging. Controlled cooling technology, likewise, facilitated the fabrication of steel sheets with high strength. The forge as well as direct quench or direct control cool procedures eliminated the need for post heat treatment, resulting in a significant reduction in thermal energy consumption. When compared to conventional heat treatment techniques that required repeated heating and cooling, the forge quench approach can save roughly 80% of thermal energy. It is a very efficient heat treatment procedure that may be used on hot forged blanks and components. The heat energy of forged components is fully exploited to achieve the appropriate microstructure by the deigned controlled cooling methods such as water spray cooling air or quenching, and blasting. These alloys typically have the same strength as quenched and tempered products. Austempering procedures are used to increase the strength and ductility of cast products. Austempering cast iron has equivalent strength to quenched and tempered steels and is used in brackets and suspension.

12.1.3.3 Carburising and Carbonitriding

Carburising is an extensively used method of case hardening gears used in automotive applications. Carburising is a method that incorporates carbon into the surface which has undergone heat treatment in a controlled furnace followed by directly quenching the produced gears in a chosen quenchant, thereby introducing a high-carbon quenched and hard martensite case. Carbonitriding, on the other hand, diffuses both carbon and nitrogen concurrently.

12.1.3.4 Nitrocarburising

By enhancing wear and fatigue qualities, salt bath nitrocarburising techniques have made a significant contribution to component durability improvement. Because the procedure involved cyanide, environmental concerns impacted their use. As a result, various further methods of gas nitrocarburisation have been developed and their uses broadened.

Low-temperature nitriding, oxy-nitrocarburising and nitrocarburising, are gaining popularity as a result of benefits they provide in terms of avoiding thermal as well as transformation-related deformation. The layer of iron nitride at the component's surface provides good wear as well as seizure resistance, while nitrogen diffused into the component's sub-surface area boosts fatigue resistance by quenching to prevent iron nitride precipitation while retaining nitrogen within the ferrite matrix.

12.1.3.5 Induction Hardening

It is frequently used on automotive components which require localised hardening. By induction heating and quenching, ball joint studs, cam shafts and other components are hardened to a high surface hardness. Not only is induction heating an energy-efficient heat treatment technology for hardening, it also has the advantage of softening specific areas to increase the toughness of case hardened components. Recent technology has permitted precise control of the heating cycle and optimal coil design to enable hardening of the gear tooth profile. In comparison to conventional carburising and hardening techniques, this profile hardening process can impart an unusually high residual compressive stress on the surface layer. Combining case depth optimisation technologies with high residual compressive stresses appears to increase their use to transmission and machine gearing.

12.1.3.6 Powder Metallurgy and Sintering

Compaction and sintering in powder metallurgy are finding a broader range of uses. Transmission and engine components comprised of iron powder include the clutch hub, timing sprockets and gears. Powder processing technologies continue to advance, allowing for the introduction of new powder metallurgical products. Valve seats in the cylinder head are a great example of powder metallurgy products that permitted the introduction of specifically developed valve seat materials to withstand the extreme working conditions of an engine.

BMW developed and incorporated powder forged connecting rods into its engines. Toyota also pioneered the use of powder forged connecting rods in engine development. Additionally, the variable valve timing gears utilised in the engine are comprised of an iron-based powder alloy. The most advanced powder metallurgical application is in the manufacture of titanium valves, which are formed through compaction, sintering, extrusion and swaging.

12.1.4 KEY ISSUE IN HEAT TREATMENT: ATMOSPHERE CONTROL

Over the last decade, protective atmospheres utilised in a variety of heat treatment techniques have undergone significant alterations. While the generation of gas atmospheres began with charcoal, liquid petroleum gases (LPG) and natural gas have been widely used. While the manufacture of protective atmospheres has traditionally relied on generators to convert the source gas to a combination of N_2, CO and H_2, in situ gas generation methods have been developed and partially implemented in industry. These have resulted in a decrease in the consumption of source gas and associated processing expenses. Atmospheric control technologies have evolved from isolated measurements of dew points, CH_4, CO or CO_2 to continuous monitoring via oxygen sensors. However, a mismatch between air composition and oxygen sensor output current variations is unavoidable. This occurs as a result of the sensor degrading and also as a result of the sensor being subjected to significant changes in the atmosphere during processing. As a result, relying solely on an O_2 sensor is extremely risky. As a result, the need for accurate atmospheric control techniques like direct carbon potential measurement should be emphasised in order to strengthen the reliability of case hardening processes that do not require gas or low-pressure operations. Recent sophisticated measurement techniques that enable continuous as well as accurate monitoring (in situ and sample gas monitoring using laser spectroscopy and a high-speed gas chromatograph approach) have been developed and are being used in part for heat treatment processes. These novel techniques enabled precise control on the heat treatment process, resulting in high-quality products with minimal waste.

12.1.4.1 Carbon Potential Control

The carbon potential control loop increases the carbon potential by opening a solenoid valve which allows a carburising or enriching gas (e.g. propane) to enter the furnace. Conversely, to decrease the carbon potential, dilution air or nitrogen is introduced into the furnace.

12.1.4.1.1 Gas Carburising Processes

Controlling the heat treatment environment is critical for producing high-quality goods, regardless of whether they are produced using classic gas or vacuum techniques. However, despite a long history, controlling the carbon potential (CP) of protective atmospheres is not well done, despite the fact that quenching, carburising and carbonitriding all require precise atmospheric controls to avoid decarburisation and over-carburisation and to introduce optimal compressive residual stresses. The diffusion depth and hardened case depth are both affected by the atmosphere's carbon potential. CP is influenced by a variety of parameters, including the material of the furnace insulator, the state of sooting, the composition of the gas, the pressure and the operating temperature, and is not stable at any point during the carburising process, as previously assumed.

Appropriate control of the carbon phase offers time savings and ensures the surface carbon concentration and diffusion pattern, which have a direct effect on the state of the introduced hardness, microstructure and strength of treated parts. The updated technology that offers more trustworthy results is a multi-element infrared or laser gas analyser, or an in situ carbon potential measurement method.

12.1.4.1.2 Reduced Pressure Carburising (Vacuum Carburising)

The technology of vacuum carburising has improved significantly. Continuous monitoring of the carburising condition under reduced pressure is enabled by newly developed gas sampling methods (similar to gas carburising) and permits accurate control on carburising process.

Carbon diffusion speed is directly related to the temperature of the heat treatment, with higher temperatures increasing the diffusion speed. Due to the fact that the carburising condition with high temperature frequently results in grain development during treatment, it is necessary to employ an enhanced case hardening steel that allows for some grain growth prevention. Steel of vacuum carburising grade having less than 0.1% Ti and Nb raises the grain growth temperature limit to 1050°C.

12.1.4.1.3 High-Pressure Gas Quenching

Recent advancements in cooling technologies have utilised pressured gas in the case of quench hardening. However, the technology is not designed to impart an adequate amount of hardness while minimising distortion. Generally, the cooling power of industrially suitable pressure used for gas quenching is substantially lower than the cooling power of oil quench and aqueous solutions used for quench hardening. Additionally, distortions arise when cooling power is increased to achieve almost the same hardness as with standard quench procedures. As a result, cooling technology optimisation should occur concurrently.

12.1.4.1.4 Carbonitriding

Nitrogen is a critical alloying element that contributes to the hardenability of steels. Additionally, it is effective at lowering the requirement for additional alloying elements. Since the mid-1960s, carbonitriding processes have been utilised to case-harden automobile components, primarily to impart high case hardness to low-alloy and plain-carbon steels. Components such as bushing clutch release forks and tiny universal joint cups are typical examples. Nitrogen concentrations of up to around 1% are extremely effective at increasing case hardenability and resistance to softening caused by heat generated during operation due to sliding or rolling fatigue phenomena. It is also excellent at recovering reduced surface hardenability caused by grain boundary oxidation, which occurs frequently in steels containing Cr and Mn. However, caution should be exercised to avoid excessive nitrogen concentrations to avoid pore development induced by nitrogen gas escape from the component matrix. Recent developments in carbonitriding technology have resulted in the development of gears for automated transmissions that contain finely dispersed carbide and are reinforced by a fine martensite matrix.

12.1.4.1.5 Low-Temperature Nitrocarburising and Oxy-Nitrocarburising

For more than three decades, the automotive industry has relied heavily on the low-temperature (570–580°C) cyanide salt bath nitriding process. Not only does the salt bath nitriding procedure deposit a coating of nitride substance on the component surface. Nitrogen diffuses into the matrix, resulting in an increase in fatigue strength due to nitrogen saturation in the ferrite matrix structure. Thousands of vehicle components composed of low-carbon iron basis materials have been manufactured using this technology.

Salt bath nitriding, which employs cyanate rather than cyanide, was developed in the late 1960s to address toxic as well as hazardous waste disposal issues. Additionally, operating temperatures have been increased from 400^{0}C to 630^{0}C to accommodate application requirements and enhance component attributes. Nitriding of stainless steel avoids compromising their corrosion resistance while improving their surface hardness to micro-Vickers hardness (HV) 1000 or above.

The amounts of N and C in the salt bath nitriding and gas procedures vary depending on the technique utilised. Additionally, the treatment periods for these processes vary according to the N, C and O concentrations. Throughout the last decade, numerous low-temperature nitriding techniques utilising gas and specifically formulated liquids have been created. Recent developments in Europe have resulted in the development of a new plasma nitriding technology. The novel "through cage" or "active screen plasma" (ASP) nitriding techniques have demonstrated great results and may create a new industry for low-temperature nitriding.

12.1.5 SURFACE MODIFICATION AND TREATMENT

There are many surface modification approaches available today for enhancing the characteristics of components. Additionally, it requires the adoption of hybrid procedures to meet the requirements of this component group. To address unique property requirements and applications, heat treatment or dual heat treatment in combination with an additional surface modification process is used. More precisely, there are numerous popular combinations of modifications of surface techniques like phosphate treatment, vapour deposition and/or solid lubricant coatings that have the potential to meet unique combined property requirements.

A coat quench technique, such as the one that can be used to increase the surface friction coefficient for more than 30 years, as well as other methods to resist rusting by producing a surface oxide film after cooling into aqueous solutions, is a unique cost-effective coating with heat treatment.

12.1.5.1 Coatings of Solid Lubricant

Numerous varieties of metallic, polymeric and composite coating films are used to increase the friction and corrosion resistance of various automotive components. The thickness and kind of coating on materials are chosen to meet the requirements of the individual components and design at hand. Although this coating is not a heat treatment, it significantly lowers surface friction and is a widely utilised method of surface modification. Coatings require specialised technologies which have been processed in designated plant facilities to ensure the component's quality is maintained. Various process control measures become necessary depending on the methods used. Throughout all phases of processing, methods and processes for monitoring and controlling accurate compositions and combinations should be standardised within ranges of specified limits to ensure that processed components exhibit the desired qualities.

12.1.6 EMERGING TECHNOLOGIES IN MATERIALS, HEAT TREATMENT AND SURFACE PROCESSING OPERATIONS

12.1.6.1 Materials

Today, social concerns about climate change, the global environment and conservation of energy are the primary factors influencing technological choices. To reduce fossil fuel use, innovative vehicles such as fuel cell, hybrid and electric vehicles have been available on the market since 2001, thanks to the advancements made by mass manufacturing of Toyota Hybrid automobiles. This tendency will alter the composition of the materials used in the manufacture of automobiles, as well as the processing processes. Research and development efforts will be directed on developing lighter, stronger materials like light metals and polymers, while specific steels will become the material of choice for applications requiring durability and comfort.

Heat treatment necessitates fundamental study in order to find methods for enhancing the components' stiffness, strength, durability and wear resistance. Simultaneous energy saving is feasible by maximising the usage of all available technologies, such as minimising heating energy, eliminating re-heating processes, improving the design of burners/furnaces, lowering processing temperatures and reducing the processing time. Collaboration between material makers, automobile engineers, as well as heat treatment engineers will yield greater results.

Surface processing technologies will become increasingly significant as an optimal mix of base material properties, heat treatment properties and final surface processing operations resulting in superior products. Ferrous, plastics, light metals, as well as glass-ceramic materials should be combined optimally to develop new grades of specified weight-to-strength ratios to meet the demands of the twenty-first century.

12.1.6.2 Carburising and Carbonitriding

- A reduction in carburising pressure requires extra technologies to shorten the time required for heating as well as cooling, which should be referred to as "controlled pressure processing".
- Increased cooling power without increasing distortion is required for high-temperature and reduced-pressure carburising with gas quench.
- The creation of grain growth regulated steel using scattered TiN, AlN and NbN precipitates, as well as thermomechanical process control, are required technologies.
- The role of rare earths appears to be critical in reducing carburisation under reduced pressure, and additional research is required.
- Even with decreased pressure methods, the processing time for a particular effective case depth is closely related to the carbon potential of the carburising

atmosphere, and standardisation of process parameters must take carbon potential into account to provide accurate and efficient carburising. As a result, it is required to create more accurate carbon potential control systems.

- To improve the efficiency of processing, it is necessary to boost surface activity by the use of catalytic promoter. Controlling surface activity is critical for increasing nitriding power, but it is not well understood and requires further exploration. The fact that reduction and oxidation phenomena appear to be tied to specific results observed in some circumstances will be critical in breaking through conventional processes. Additional approaches for increasing surface reactivity include atmospheric and plasma control.

12.1.7 CONCLUSION

This case study provides an overview of the materials, heat treatment and surface treatments currently employed in automotive applications. Additional research and development efforts are required to produce eco-friendly technologies capable of meeting industry objectives for increased fuel efficiency, safety, comfort, cost, durability and compliance with emission standards. International collaboration along the lines of ULSAB-AVC is critical to addressing the automotive industry's difficulties.

12.2 CASE STUDY 2: CASTING IN THE AUTOMOTIVE INDUSTRY

12.2.1 INTRODUCTION

The vehicle industry is a significant market for high-pressure die casting. Since the recession, the car industry's revenues have been steadily increasing. The graph depicts the auto sales trend since 1990, which has maintained yearly sales of roughly 17 million vehicles over the last three years. Although experts assert that sales above 16 million vehicles are healthy, not everyone believes they are sustainable. Manufacturers of electric vehicles were optimistic about the likelihood of an upward trend in EV sales, but underestimated the market for these vehicles. With less than 1% of overall auto sales, industry analysts anticipated that EV sales would plateau in 2018. The Chevrolet Bolt, Nissan Leaf and Tesla Model 3 have become the most successful mass-market automobiles, owing to their surprising success or pre-ordered potential. Analysts assert that sales of electric vehicles could improve if more economical models become accessible, which has yet to occur. Technology is a significant element in sales growth. Cameras, sensors and electronic communications innovation have resulted in important safety and technological advancements. Federal legislation addressing distracted driving continues to provide engineering teams with a challenge in terms of creating a safe, smartphone-friendly workplace.

12.2.2 AUTOMOTIVE MANUFACTURING AND DIE CASTING

In the automobile component sector, die castings are commonly employed. For four primary reasons, the HPDC technology is appropriate for automotive applications:

1) Production at a high rate. Die casting enables the rapid and high-volume production of complex part shapes in near net shape. Tooling is capable of producing hundreds of thousands of castings before it has been replaced. Die casting cycle times are significantly shorter than those of other metal casting processes and are comparable to those of stamping and injection moulding.
2) Accuracy and stability of dimensions. The high-pressure die casting method enables the production of extremely robust and dimensionally stable products with tight tolerances.
3) Strength and weight. A well-balanced combination of physical and mechanical qualities, combined with near net shape design and thin wall die casting techniques, results in strong and lightweight components.
4) Numerous choices for concluding. Die cast items' surfaces often require little to no surface preparation.

Aluminium and magnesium have been increasingly significant in the rising automobile business during the last few decades. There are several significant advantages to transitioning from steel to aluminium or magnesium. Aluminium is a very effective material for light weighting since it weighs less than half as much as steel. Magnesium is the most effective material for light weighting at 14 times less than the weight of steel and two-thirds the weight of aluminium. Weight reduction promotes performance, safety and the environment. Aluminium and magnesium both exhibit a stable set of mechanical and physical properties. The figure below depicts the current and planned aluminium components for a certain vehicle. The darker blue colour denotes the existing aluminium components of a car. The sky blue tint represents metal components that will be constructed in the near future.

Along with the numerous power train and structural applications for high-pressure die casting, the method is used to manufacture a variety of non-structural components within the vehicle. Several of these components include electronic housings, ECU encasements, mirror mounts, headlamp assemblies, vehicle camera systems, GPS systems, and gearbox housings, among others. Annually, Chicago White Metal (CWM) produces millions of automotive components.

12.2.3 HID LIGHT BALLAST UNIT HOUSING

Material: Magnesium AZ91D

Figure 12.4 shows HID light ballast unit housing. These are the cover and base of the high-intensity discharge (HID) light ballast unit, which are found on a variety of automotive makes and models. Chicago White Metal offered three installation options: a basic cover, one with four mounting holes and one with three mounting holes. CWM worked with the customer to design pieces that were suitable for die casting, ensuring that strict tolerances and key points were satisfied.

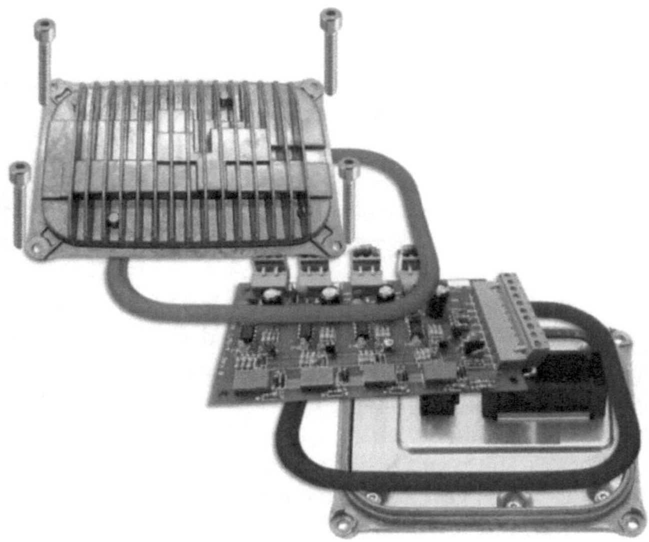

FIGURE 12.4 HID light ballast unit housing.

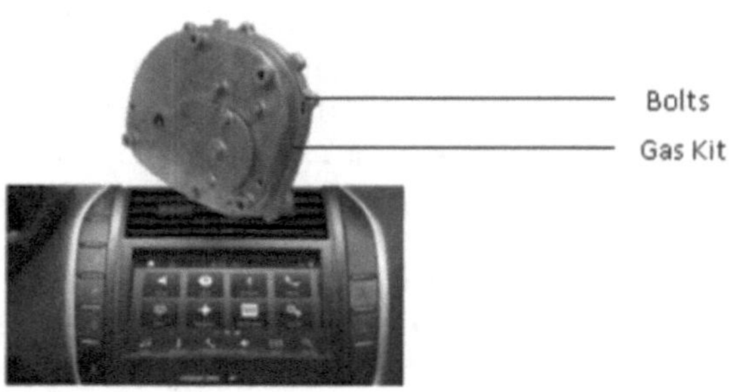

FIGURE 12.5 Transmission Range Control Module (TRCM) housing.

12.2.4 TRANSMISSION RANGE CONTROL MODULE (TRCM) HOUSING

Material: Aluminium A380

This component is the aluminium enclosure for the Transmission Range Control Module (often referred to as the "shift-by-wire" technology. Figure12.5 shows transmission range control module housing.). In various different GM, Buick and Cadillac automobiles, the control module is mounted externally to the transmission casing. CWM assisted in the design of the part, which adhered to the automobile industry's stringent criteria. The team offered engineering help to ensure crucial tolerances were maintained and to eliminate unnecessary procedures by casting rather than machining the mounting holes.

12.2.5 FORD HYDRA ECU HOUSING

Material: Aluminium A380

Figure 12.6 shows ford hydra ECU housing. The cover and base plate castings, seen in blue in the above figure, serve as the outer housing for the printed circuit board and associated components of a vehicle camera system. There are two distinct cover options to accommodate two distinct connector sizes. As a new design, CWM and the customer concentrated on discussing feature and tolerance adjustments that would improve the components' manufacturability.

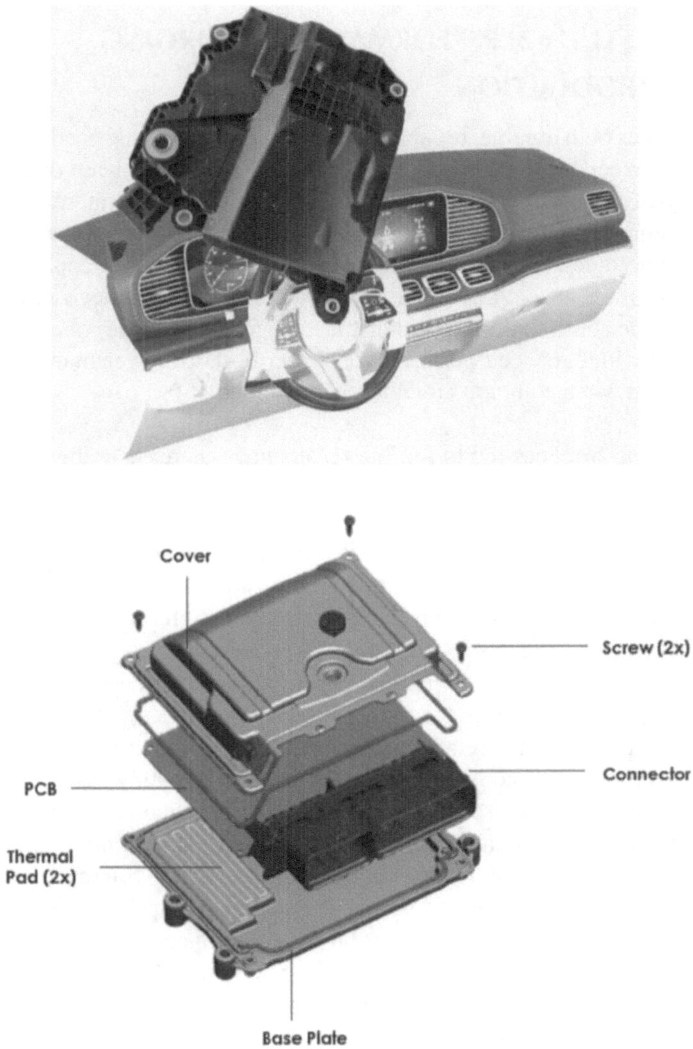

FIGURE 12.6 Ford Hydra ECU housing.

12.2.6 CONCLUSION

Die casting has great potential to produce different components with the features like stiff but lightweight, long lasting, ductile but strong, consistent and versatile. With the semi-solid die-casting methods, non-ferrous metal alloys like magnesium alloys and aluminium alloys are most frequently used in industry. A reduction in the vehicular weight is an important characteristic feature of the material used in the advanced die casting process for vehicular applications. Reduction in weight of the vehicle drastically reduces fuel consumption, leading to minimising the environmental pollution.

12.3 CASE STUDY 3: PERFORMANCE OF FORGING

12.3.1 INTRODUCTION

Forging achieves both durable, reliable component shapes and the need for engineered metallurgy to meet specific product requirements. They have been chosen to show the majority of the forging methods, a variety of alloys, and a number of the prior chapters' themes. While some of the goods were easily created using industry-standard technologies, others necessitated the use of specialised capabilities. Some of the products are new, while others date back more than a decade. All are contemporary in the sense that they demonstrate the forging industry's capacity to provide products with enhanced performance, lower costs or fewer manufacturing and assembly issues when compared to other technologies.

(A) Changing from casting to forging results in an increase in the life of massive connecting rods

Field failures, which are costly as far as downtime is concerned and with significant damage to the pump, demonstrated that the cast connecting rods applied to drive the pumps which maintain the movement of coal slurry throughout pipelines were insufficiently strong. The issue was resolved by reimagining the cast connecting rod as a forging. Additional strength was provided to the points of greatest stress by refinement of grain flow during the closed-die forging process. Not only did forging increase the tensile strength relative to the casting, but it also greatly enhanced the transverse strength of the connecting rods. With the alternating loads that the rod is subjected to throughout its service, a longer life due to increased fatigue strength became a possibility. Critical straightness and thickness tolerances were obtained on the forged 4140 steel connecting rod (see Figure 12.7). Tolerances are vital, as as-forged surfaces must adhere to stringent dimensional specifications in order to assure trouble-free installation. Due to the tight tolerances, a considerable section could be used as forged without any finishing machining. As they are larger in size and capacity, high-power pumps are conceived and built to transport greater quantities of slurry via bigger pipelines for longer distances; the increased effectiveness of forged connecting rods can help boost the pumps' reliability. A steel connecting rod delivers improved transverse strength, permitted forging to succeed where casting failed.

FIGURE 12.7 Refined grain flow in a 2450 lb.

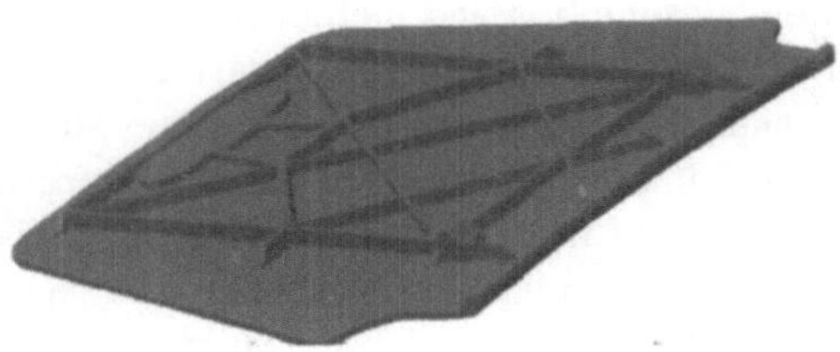

FIGURE 12.8 Precision forged 2014-T6 aluminium to reduce cost.

(B) Compared to fabricated parts, forging blocker doors were utilised to decelerate planes while landing resulted in cost reduction

Previously manufactured as a critical fabricated assembly, precision and forged aluminium blocker doors used in jet engines resulted in anticipated cost savings of several fold. Previously, the blocker doors were a built-up structure consisting of a honeycomb and aluminium sheet that were both mechanically fastened and adhesively bonded and were activated by the thrust reversers to reverse the jet stream and slow down a plane during landing. The redesigned blocker door was forged in one piece and included integral reinforcing ribs and a hinged attachment; both were made in a net shape. Figure 12.8 shows precision forged 2014-T6 aluminium to reduce cost. By the development of a single database for programming tool travel paths for both the EDM electrodes utilised for manufacture the dies used in forging and the fixture used in machine/straightening, the forger was able to achieve tighter tolerances. Along with manufacturing and assembly cost benefits, the forged doors were much more damage-resistant compared than the sheet/honeycomb sandwich structure. As a result, the increased life resulted in additional cost reduction. A blocker door for a jet engine also outperformed its labour-intensive fabricated predecessor. Shown is the detailed rib side and hinged attachment area, which was a forged "net".

(C) Precision forging provided critical properties and cost savings over a machined block

The trailing-edge flap retraction mechanism supporting the flap track actuator bracket on the Airbus A-320 was a "natural" application for high-precision forged aluminium due to its fracture-critical nature. Not only did the design necessitate the safety as well as qualities of forging, but it also proved to be incredibly cost-effective, as alternatives would have required significant, difficult machining to achieve the complicated shape, as shown in Figure 12.9. Refer to the illustration on the right. Castings were ruled out due to performance requirements such as mechanical qualities in thick sections, stress-corrosion resistance and high fracture toughness. Furthermore, the item was too large to be machined from plate. As a result, the only other viable choice was a "hogout". The severe internal radii that machining processes would have needed to cut with required cutting tools have the diameter of a pencil and a length of 15 inches. Hence these were quickly abandoned in favour of precision forging. Braking of this tool leads to more time consumption and inefficient machining. Because a component failure could have impacted the flaps' proper raising or lowering, 7050-T74 was chosen above more standard aerospace aluminium alloys such as 7075-T73 for its high fracture toughness and ability to retain mechanical properties over portions up to 6 inches thick. At the end of the day, precision forging proved to be more cost-effective than anticipated. The customer saved around 25% on the cost of machining a rectangular block by using the net-forged item, which required just minor machine finishing on some tight tolerance parameters. Material savings also played a significant role in cost savings. The actuator brackets possessed high strength in thick section. If the parts were machined from a rectangular block, the cost would have increased by four times as compared to the precision forging.

(D) Forged microalloyed steel crankshafts

Crankshafts made of vanadium microalloyed steel are possible. Of course, economics is the driving force. Cost reductions of 10% or more can be realised by omitting heat treatments that are customary for quenched-and-tempered steels. Refer to Figure 12.10. Additionally, increased machinability may result in cost savings. Microalloyed steel is appropriate for forging applications requiring moderate strength, such as crankshafts, that do not encounter significant impact loads in service. The vanadium-modified microalloy is a low-carbon, high-manganese variant that has 0.3% carbon, 1.50%

FIGURE 12.9 Made from a precision forged aluminium alloy, 7050-T74.

FIGURE 12.10 Starting with a 6 in. round cornered billets.

manganese and 0.11% vanadium. It exhibits increased strength, hardness and induction hardening properties due to the high Mn content and the microalloying element. Forging technique development that optimises the hot working, heating and constant volume cooling, cylindrical steel billets is critical for the crankshaft application. For instance, decreasing the forging temperature while increasing the reduction leads to a very fine size of austenite grain, optimising combination of the properties. Forged crankshafts of microalloyed steel can provide strength-to-toughness ratios for high-volume applications in an acceptable range. Most importantly, strength as well as hardness ratings are nearly comparable from the crankshaft's surface to its centre. Due to the finer grain size, the surface has slightly higher ductility and toughness qualities. The fatigue strength is predicted to be similar to that of plain carbon steel quenched and tempered. Billets were induction heated, the microalloyed steel crankshaft (5 in. main bearing diameter) was press forged then fan cooled.

12.3.2 Conclusion

Forging is a manufacturing process associated for the shaping of metals by using localised compressive stresses. There are various advantages to the forging process over the other manufacturing processes used in the automobile manufacturing process such as strength, material consumption, flexibility as well as product life. The demands for the forging increasing day by day from all fields of manufacturing processes as an effect of advancements in precision level and ease of forging of complex shapes with the help of ultra-high-temperature forging.

12.4 CASE STUDY 4: AUTOMOTIVE APPLICATIONS OF WELDING TECHNOLOGY

12.4.1 Introduction

In general, welding is used to create a permanent junction between two metals by applying a proper mix of pressure, temperature and metallurgical approaches. Different sets of temperature and pressure have resulted in the development of a diverse range of welding methods. Welding is the primary method for producing and repairing metal items and is utilised across all industries. Among its numerous uses, welding is heavily utilised in the automotive industry. Metal inert gas (MIG) welding, resistance spot welding (RSW), resistance seam welding (RSEW), tungsten inert gas (TIG) welding, friction welding (FW), laser beam welding (LBW) and plasma arc welding are the most frequently utilised welding processes for automotive components. Advanced welding methods have been developed for automotive applications with the goal of reducing vehicle weight as well as increasing fuel efficiency. In old welding, an extra material is always used to the welding that flows into the components being connected, resulting in an incredibly strong bond. The extra metal of every welding joint increases the vehicle's weight, reducing fuel economy. This section discusses welding techniques applied in automotive manufacturing, including RSW, FW, RSEM and LBW. Additionally, magnetic pulse welding (MPW), the advanced technique of welding, is discussed, which is used to make lighter weld components. This study also discusses medium-frequency welding (MFW), which is employed in the automotive industry.

12.4.2 Automotive Applications of Welding

Welding is used to unite a wide variety of automotive components. The demand for innovative welding procedures for automotive applications continues to grow as new material preparations for auto body parts are introduced. The demand for new welding methods has increased significantly in recent years, as car manufacturers have shifted their focus to lighter yet sturdy and fuel-efficient vehicles made of lighter alternative materials. The following sections discuss the most frequently used welding processes in automobile applications.

12.4.2.1 Resistance Spot Welding

On average, a car's standard steel body comprises 4500 spot weld joints. For many years, resistance spot welding has been the primary joining method utilised in the automotive industry. The joint is formed in this approach by the heat created by the resistance of the workpieces to current flow and pressure application. Because the weld is limited to the overlapped work parts, it is not continuous. As shown in Figure 12.11, the pointed copper electrodes conduct the welding current to the work area and also serve to apply pressure to build a strong junction.

Automated RSW is utilised in the automobile industry with robotic spot welding to join sheet metals to construct the car body. An actual photograph of industrial robots spot welding the automobile body on a production line is displayed in Figure 12.12.

12.4.2.2 Resistance Seam Welding

The junction is formed gradually over the entire weld length in this type of resistance welding. This gives a continuous and leak-tight sheet metal junction. The weld may be constructed from continuous or overlapping work components. In the automobile industry, to create leak-proof gasoline tanks ,welding methodology has been adopted. The RSEW idea is illustrated in Figure 12.13. While this procedure allows for rapid

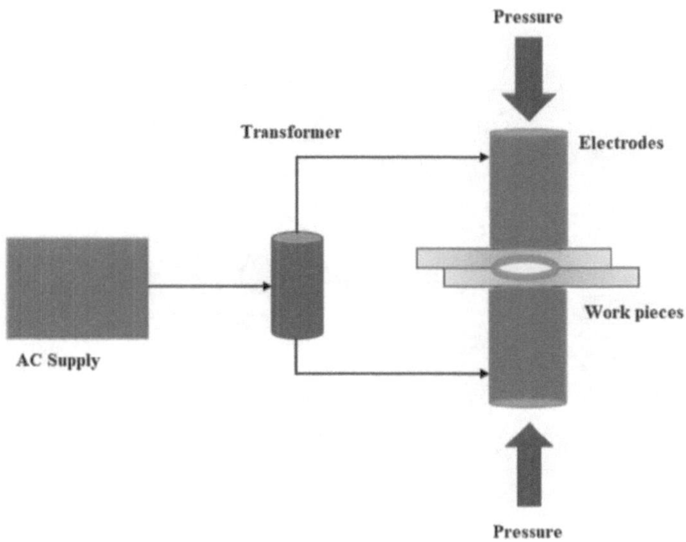

FIGURE 12.11 Resistance spot welding.

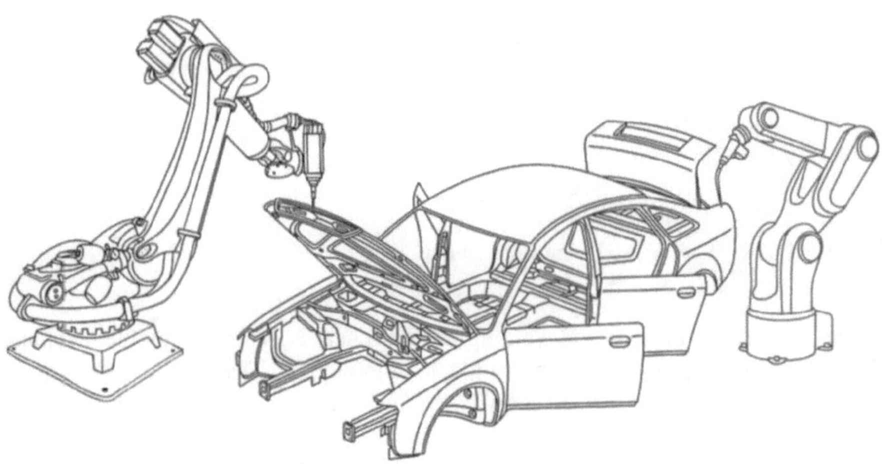

FIGURE 12.12 Industrial robots welding a car body.

welding, its applicability is limited by the geometry of the component and wheel access. Figure 12.14 illustrates the seam welding process in action.

12.4.2.3 Friction Welding

In the case of solid-state welding, development of the weld joint can be carried out by applying pressure to the workpieces without significantly melting any of them. Friction

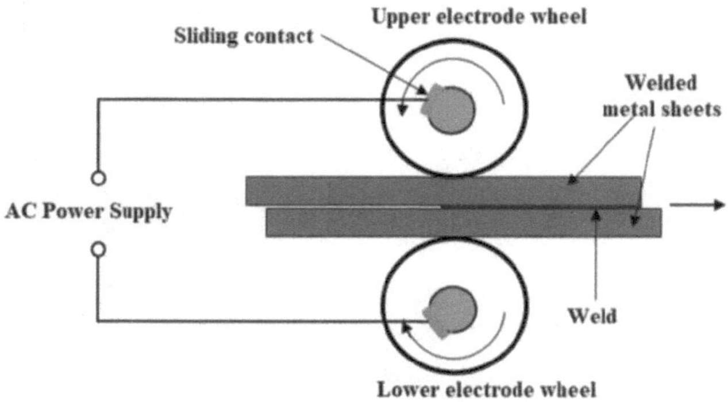

FIGURE 12.13 Principle of resistance seam welding.

FIGURE 12.14 RSEW producing a continuous joint.

welding is a type of solid-state welding in which heat is generated by the mechanically induced sliding motion of the weldable components. The weld components are held together under pressure. Generally, frictional heat is generated when one part rotates against another. When a predetermined temperature is attained, the circular motion is halted and the components are welded together by the applied pressure. Figure 12.15 illustrates the two shafts connected by the FW method. Controlling this welding process entails adjusting the rotational speed, time and pressure.

Friction welding is a simple process for metallic materials that exhibit a certain degree of flexibility at elevated temperatures and thermal stability. While many typical technical alloys are capable of being friction welded, cast iron does not come under this category. The FW is used to create a broad variety of components in the automobile industry, including half shafts, steering columns, axle cases, piston rods, hydraulic cylinders and engine valves. Figure 12.16 illustrates a few vehicle components that have been friction welded.

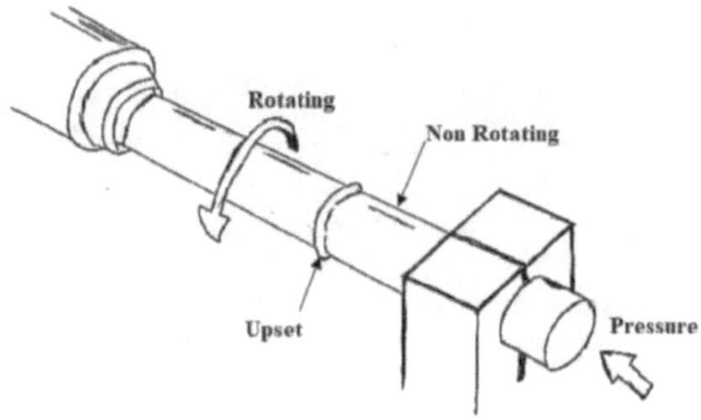

FIGURE 12.15 Shafts joined by the friction welding process.

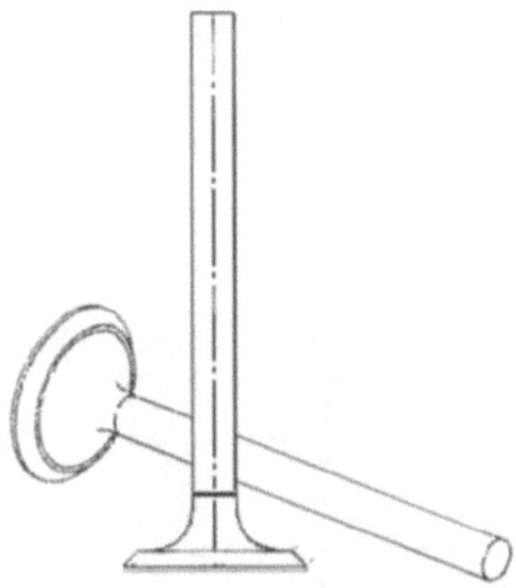

FIGURE 12.16 Automotive components manufactured by friction welding.

- **Laser beam welding**

Laser technology has gained favour for welding of high-volume automobile parts due to its specific advantages. The primary benefits include increased productivity, significant cost savings in terms of maintenance and energy, and the production of a robust weld. The LBW procedure makes use of the heat created by the impingement of a concentrated laser beam on the joint. Laser welding is possible on metal sheets with a thickness of 0.2–6 mm. The majority of automobile industries utilise cross-flow CO_2 laser systems with a power output of between 3 and 5 kW. Figure 12.17 illustrates the schematic organisation of the components of a laser welding system.

Using copper mirrors, the laser beam directed at the workstation is focused on the components to be welded. This is a non-contact procedure that lends itself nicely to automated applications. The weld depth as well as width are determined by process parameters such as speed, laser power and concentrated spot size. A beam from a single laser source can simply be switched between many workstations, which allows

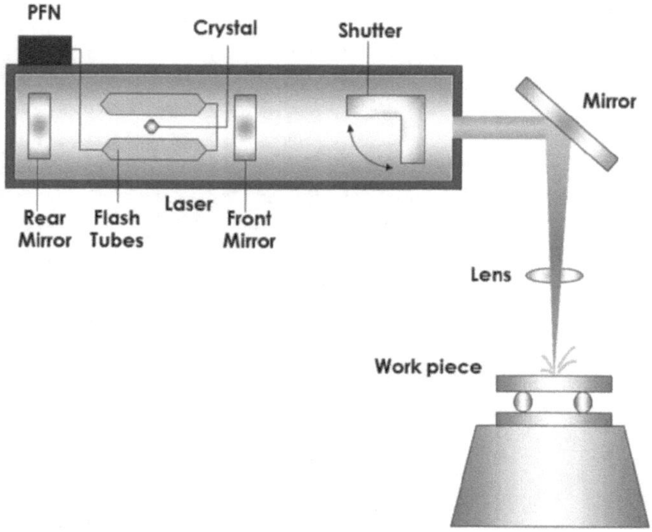

FIGURE 12.17 Laser beam welding.

FIGURE 12.18 Laser beam welding of low-carbon steel sheet.

for optimal laser utilisation for a variety of welding purposes. This intrinsic flexibility of LBW enables it to meet the requirements of higher volume of automotive production for a variety of part geometries without requiring considerable setup. Figure 12.18 illustrates the welding of a low-carbon steel sheet with a CO_2 laser system.

Combining automatic transmission components has been identified as being particularly well-suited for LBW. Powder metallurgy (PM) components are increasingly being employed in automobiles. The performance of PM components is determined by a combination of the finished density, the alloying element composition and the microstructure of the part. These parameters increase the degree of freedom in determining the weldability of PM components. Density is said to have a significant effect on performance and weldability. LBW is also used in various automotive applications, such as welding the roof to the side panels of the vehicle's body structure. As less material is displaced, a continuous watertight joint with greater accuracy can be created. Additionally, post-weld treatment is not required. The laser beam procedure is used for connecting hinges to the strengthening framework of an automobile door. The evaluation utilised CO_2 laser equipment with a 6 kW output power. Because the surface finish of the parts was harmed by LBW porosity creation, an improved surface finish could resolve this issue.

12.4.2.4 Medium-Frequency Welding

The medium-frequency welding process is a variation on the resistance welding process and is the latest product from Bosch-Rexroth Ltd in the United Kingdom. Figure 12.19 illustrates the fundamental operation of MFW. Rectified three-phase 50 Hz alternating current (AC) is delivered to an inverter during this welding operation. The inverter converts the current to a 1000 Hz (medium) frequency, which is then delivered to a transformer, which is typically built into the welding gun.

12.4.2.5 Magnetic Pulse Welding

Recently, demand for lighter, more fuel-efficient vehicles has surged. Automobile manufacturers make a concerted effort to develop lighter components. This helps improve the fuel efficiency of existing cars and meets the criteria for alternative fuel vehicles, such as fuel cell cars and hybrid gas/electric vehicles. Utilising lighter materials such as aluminium and developing new production procedures that require less steel in the weld can assist in achieving the stated aim. As with conventional welding, the additional material deposited on the weld junction increases the weight of the welded component. The MPW is a revolutionary method created by Dana Corporation in the United States of America that enables the bonding of aluminium and steel (dissimilar metals) without the use of additional metal at the weld junction. Using accurately machined die cavities, pre-shaped aluminium and steel tube stock is exposed to high pressure in this procedure. This procedure, referred to as hydroforming, leads to more accurate fitting of various structural components with minimal fill material required in following welding operations.

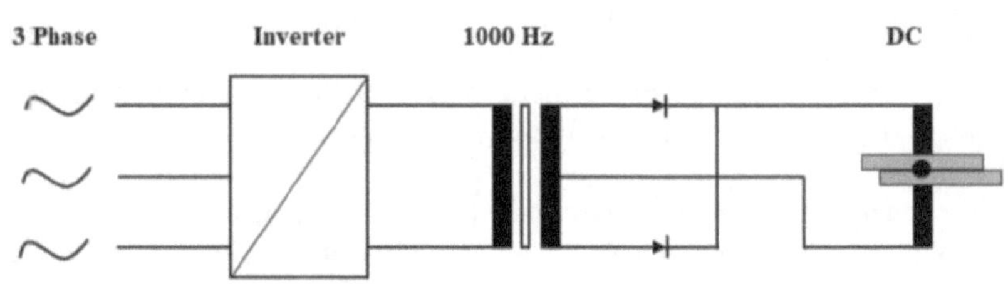

FIGURE 12.19 Working principle of MFW.

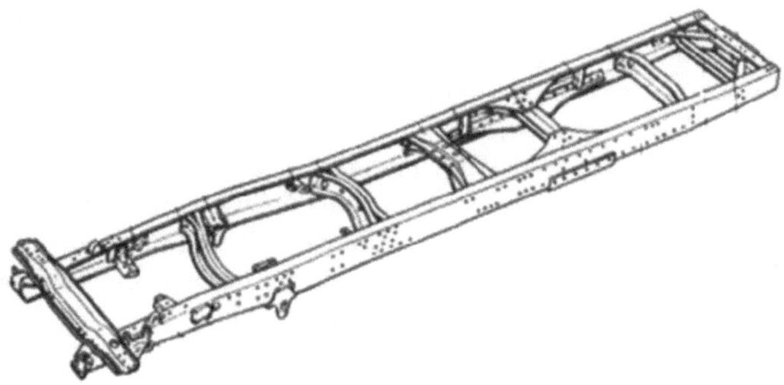

FIGURE 12.20 Application of the MPW process.

The hydroformed components are then loosely assembled. A rapidly switching magnetic field generated by an inductor (which may be either internal or external) allows one metallic component to develop fast and contact the other stationary metal component with enough velocity and force to make a weld. This requires sophisticated machinery capable of producing the correct welds in intricate geometric shapes. Due to the elimination of production processes, materials, equipment and people costs, the MPW process increases manufacturing productivity. The MPW-welded car structure is said to be two-thirds lighter, resulting in an 8–10% increase in fuel efficiency. This contributes to further reductions in air pollution. MPW is a more efficient method of manufacturing vehicle components such as frames, cradles, side rails, stampings, space frames and bumper reinforcements (Figure 12.20).

As commercialisation of the MPW process is concerned, it is now possible to most successfully weld dissimilar metals (bimetallic welding). This enables the development of novel geometries for automobile transmissions and undercarriage systems by combining lightweight materials in novel ways to improve fuel economy and cost savings.

12.4.3 CONCLUSION

The following conclusions were drawn from the aforementioned study.

- Automotive makers concentrated on developing lighter, stronger and more fuel-efficient automobiles through the use of new and improved alternative materials. This necessitates the use of the most efficient welding processes possible without adding additional material.
- The commercialisation of magnetic pulse welding may benefit the automotive industry by allowing for the development of lighter, more fuel-efficient vehicles.
- The recent advancement of laser beam medium-frequency welding enables the evolution of new geometries for automobile transmissions as well as undercarriage systems using novel material combinations, particularly those derived from the powder metallurgy process.

12.5 CASE STUDY 5: MACHINING OPERATION IN THE AUTOMOTIVE INDUSTRY

- Application of SMED to electron-beam machining in the automotive industry

12.5.1 INTRODUCTION

Lean manufacturing processes and tools have been widely employed to eliminate waste, to satisfy customer needs in the desired quantity and on schedule, resulting in competitive advantages over competitors. Due to the fact that lean manufacturing environments are often characterised by small batch sizes and substantial product variation, a new strategy for minimising setup time's needs to be devised. Automobile manufacturing firms are currently enjoying prosperous times, as seen by a significant global increase in the number of automobiles manufactured. This obviously results in an increase in component manufacturing. As a result, the industrial unit where this case study was conducted decided that an increase in production was necessary to remain competitive and expand into new markets. The flexibility and agility required of component manufacturing lines are a result of the market's increasing demand for customised and diverse goods. The industrial unit's primary activity is manufacturing electrical wires as well as cables for the car industry, where it is a market leader. This report details the efforts made by the industrial unit's cross-linked wires production department to attain the required target of production. Due to the capacity of the spools, which vary in cross-section as well as diameter, it is necessary to change the old spool with a new one frequently in each shift. Due to the fact that the amount of new loads cannot be reduced due to various constraints, machine setup times are critical. This case study provides a SMED implementation study and examines the findings from an analysis of a real-world problem in an industrial setting based on a review of the literature on lean manufacturing as well as SMED methodology. This research is structured around a SMED implementation methodology and model, which the authors have validated in several firms, and in which all stages necessary to implement and control various SMED improvements are explicitly stated.

12.5.2 STUDY FRAMEWORK

Three electron beam machines are owned by the industrial unit where this case study was developed (EBM). Due to the fact that each machine is unique, the providers of electron beam machines gave them unique names.

To facilitate their identification in the paper, the nomenclatures used by their supplier were also utilised, as they are placed in separate manufacturing units with their own irradiation area. The Tuna machine is located in Manufacturing Unit 1, the Farrusco machine is located in Manufacturing Unit 2 and the Cora machine is placed in Manufacturing Unit 3.

Tuna is an inline electron beam machine, as extrusion and irradiation occur on the same manufacturing line, in contrast to Farrusco and Cora (see Figure 12.21), which do irradiation offline. In this situation, the extrusion takes place prior to the injection, on other production lines. As a result, a workgroup was formed to examine

FIGURE 12.21 Offline electron beam machine.

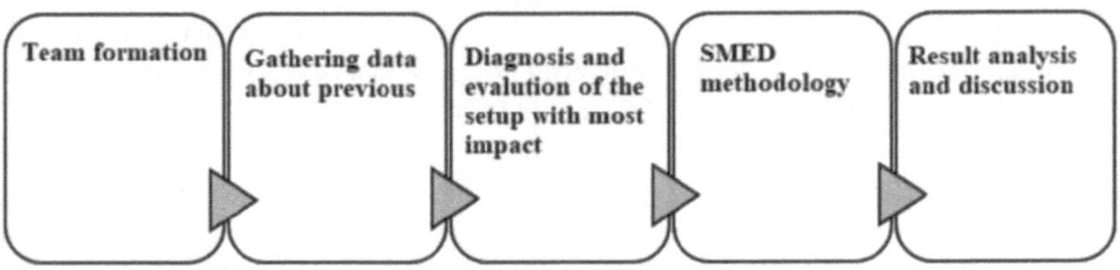

FIGURE 12.22 Different steps applied.

and implement the SMED technique in the industrial unit's offline machines (Cora and Farrusco). This multidisciplinary group was formed by the Departments of Production, Maintenance, Industrial Performance, EHS (Environmental, Health, and Safety), Corporate Process Engineer and the industrial unit's Plant Manager. As a result, this study involved a total of six individuals. When assistance was required, the supplier of the electron beam machines provided it.

Following the team's initial meeting, a schedule and action diagram were approved, as illustrated in Figure 12.22.

During the irradiation process, two primary setup types must be performed, each of which can be subdivided into several subtypes. Reel loading with the same recipe indicates that the operator will load a fresh reel using the previous one's reference. In comparison, the operator may be required to load a new recipe in order to cross-link a new reference that may or may not have a different diameter. As will be seen later in this study, changing the diameter has a significant effect on the setup effort. Cables coiled in reels during extrusion are supplied into the radiation field via a pay-off, which makes numerous passes.

EBM threading is typically performed when a wire breaks in the manufacturing process. There is no need to switch the rolls inside the conveyor in this instance. There are various rolls available depending on the wire diameter to cross-link. If the difference between the two product kinds is sufficiently large, operators may be required to load additional rolls into the EBM. When the operators needed to change the rolls, a production instruction was prepared. Only the Farrusco machine utilises a variety of rolls.

During production, the most frequent activity is reel loading, which is performed more frequently on Farrusco machines than on Cora machines. This distinction between Cora and Farrusco machines is due to the Farrusco machine's ability to handle a greater variety of wires and cables. Due to the Farrusco machine's increased capacity in comparison to the Cora machine, it is capable of handling battery cable. Due to the larger diameter of the battery cable compared to ordinary wire, the length of the used spool is reduced, requiring operators to load additional spools per shift.

Loading reels (regardless of the wire diameter) account for a significant portion of the setup time during the irradiation process. Due to the magnitude of this impact, the SMED study focused on reducing the time required to load reels. Simultaneously, some enhancements were suggested for future SMED investigations.

12.5.3 SMED IMPLEMENTATION PROGRAM

Stage 0 (preliminary)

With the assistance of the Industrial Performance Department, a Corporate Process Engineer member diagnosed the actual setup times. The results are illustrated in Figures 12.23 and 12.24. The average time period used in this analysis is the time interval between the last spool generated at production speed (ramp down) and the next spool produced at production speed (ramp up). According to the analysis in Figure 12.24, the setup durations on the Farrusco machine were much longer than those on the Cora machine (regardless of the wire diameter).

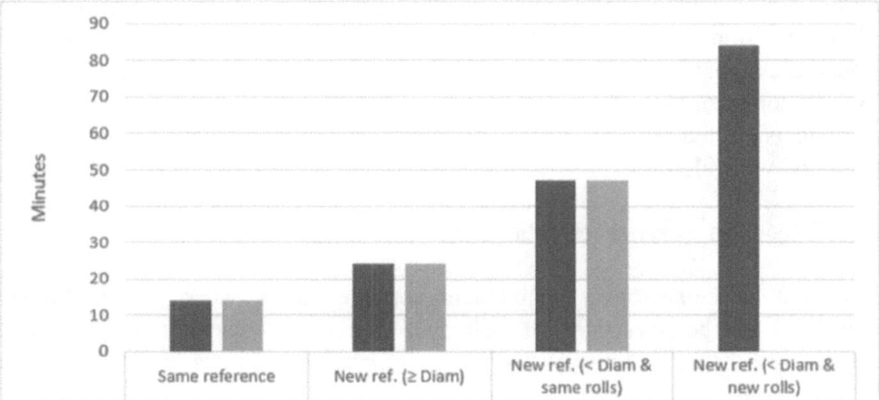

	Same reference	New ref. (≥ Diam)	New ref. (< Diam & same rolls)	New ref. (< Diam & new rolls)
Farrusco	14	24	47	84
Cora	14	24	47	0

FIGURE 12.23 Setup before the SMED workshop.

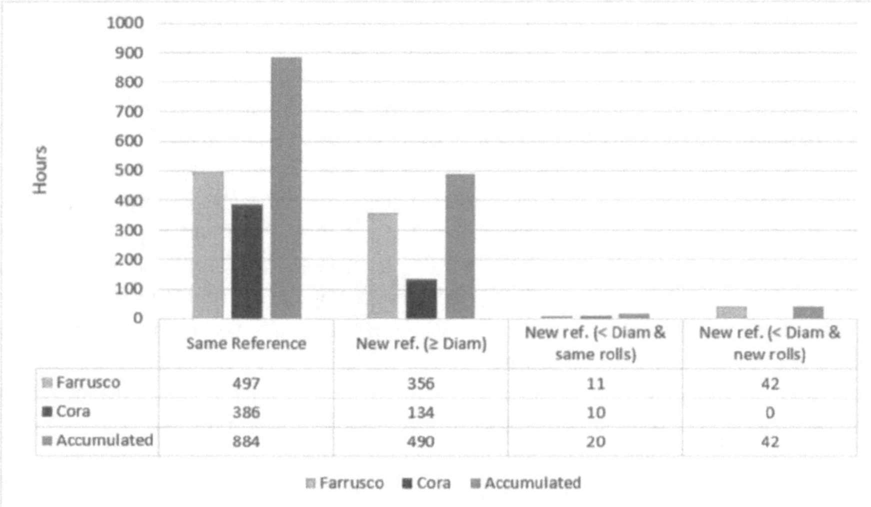

FIGURE 12.24 Setup time analysis.

Additionally, at this stage, all setup tasks were recognised, along with their durations and setup classes. Separation of internal as well as external activities was not accomplished at this stage, according to the Kusar model's suggestions.

12.5.3.1 Separate Internal and External Setups

On the basis of the data gathered, it can be stated that equipment has a significant impact on setup chores regardless of the setup type. Since, as a first step, the operators' setup duties (manual tasks) were divided into internal as well as external setup, and a thorough examination was conducted to discover potential to convert internal setup chores to external ones, 41 internal and two external tasks were recognised in this case. To complete manual setup jobs precisely, operators must employ a variety of instruments, including wire cutters and duct tape, which are always carried in the operators' work clothing.

Stage 2 (convert internal setup tasks into external setup tasks)

This stage can be seen as a complement to the operators' work, as they establish procedures and best practices on their own initiative to meet productivity targets and simplify their own jobs. At that moment, only actions/tasks involving real reel changes and/or recipe modifications were evaluated.

Five internal setup tasks were transformed to external tasks (stage 0). Converting internal configurations to external configurations has little effect on the global configuration, since it was lowered by only 35 seconds. From the operators' perspective, it is evident that those changes benefitted them and did not result in increased or additional labour. They viewed this move as a means to make better use of their own time.

12.5.3.2 STAGE 3 (STREAMLINE OPERATIONS)

At this point, the emphasis was on reengineering the arrangement, with a particular emphasis on two primary questions:

- Are the existing setup chores truly necessary?
- Because equipment plays a significant role in setup, may it be adjusted to reduce setup time?

Each action and job was analysed critically to determine if they were indeed necessary for the setup function or if they could be avoided. Two questions were asked for each of them: "Is this task truly necessary?" and "Is it possible to complete the assignment more quickly?" The Corporate Process Engineer and Industrial Performance conducted several trials under the supervision of the Production, Health and Safety Departments. The jobs performed in the EBM as well as pay-off were separated and analysed to identify their relative value, which was determined by taking into account operator and equipment safety, and whether any negative influence is conveyed to the product. Each activity deemed relevant was denoted by an X.

In terms of pay-off, about 40% of the time can be regarded optional and can be removed if the equipment is upgraded.

Previously, the reel could be loaded or emptied only with the pay-off door closed. If the door stays open, neither the reel nor movement from the platform is permitted. A software upgrade could rectify this scenario, releasing 24 seconds every configuration regardless of setup type.

Grounding the reel is required for this process, as the EBM's beam may reach the conductor of the wire. The reel becomes a giant capacitor as a result of the generated electricity, and the electricity on it should be discharged to avoid quality concerns. This objective necessitated the development of a unique tool. Grounding the reel takes 40 seconds, which was allotted for external setup. Thus, the upgrades and new tools used in pay-off reduced setup time by 64 seconds, accounting for more than 40% of the overall setup time associated with pay-off.

Concerning the EBM itself, the operators slowed production in the final spool in order to avoid, allegedly, wire fractures. Numerous testings were conducted in Cora and Farrusco machines with varying diameters, cross-sections and production speeds, and a break was verified in each case. This modification shortened the setup time to five minutes (or as little as one minute, depending on the wire diameter). Additionally, as the pace changes, the ramping up of the EBM was divided into two steps. The operators were operating the machine at a reduced speed due to the node connecting the two spools (completed and new), since the operators assumed the node might break. The SMED team could demonstrate that, if done correctly, the node cannot fail, allowing them to ramp up the machine from zero to full performance in less than a minute.

Due to the high volume of product types generated by the industrial unit and the operators' requirement to decondition and condition the EBM each time a new reference is produced, unloading and loading a new recipe has a significant effect on productivity. These tasks were omitted during the software upgrade of the EBM. This enhancement enabled the approximate loading of a reel with the same reference and loading a reel with larger diameter settings to be performed in the same duration. This second configuration decreased by more than 55%.

Additionally, as previously stated, the wire width has a significant impact on how operators load new product kinds. If the fresh wire has a smaller diameter than the last wire in production, it will break inside the EBM because of the stretch caused by the rollers' many passing. While this stretch is unavoidable, it can be absorbed by a more elastic material. Between the spools, a typical elastic rope was placed with great results, since no additional breakages were verified. As a result of this enhancement, the action node between spools is increased to 120 seconds. However, because this new tool prevents the wire from breaking within the EBM when threading it and all associated tasks (reconditioning EBM, prepping rope for threading, opening and threading, shutting and conditioning EBM) are omitted, almost 23 minutes are saved. Additionally, loading a reel with a smaller diameter took 47 minutes prior to the SMED adoption. It now takes around 75% of the time it did previously.

12.5.4 CONCLUSION

Since its inception, the SMED technique has consistently produced positive results. Converting internal settings to exterior setups had a negligible effect in this case study. However, the investigation resulted in significant adjustments to the equipment, allowing for a more than 50% reduction in setup time. In this way, it was feasible to demonstrate that Shingo methodology remains popular due to its effectiveness in lowering setup times, which is critical in highly competitive sectors such as the automotive industry.

12.6 CASE STUDY 6: AUTOMOTIVE PLASTIC PROCESSING

Assisting in the quality control of plastics processes for automobile dashboard materials.

12.6.1 INTRODUCTION

Metals are being phased out of the automobile industry in favour of plastics. As a result of the lowered vehicle weight, component performance and corrosion resistance are increased. Increased emphasis on pollution control and vehicle weight will result in increased fuel market demand in the approaching period. The automobile industry is seeing an increasing trend toward vehicle weight reduction in order to improve fuel efficiency. Plastics demand is increasing as automobiles become lighter.

The advantages of plastic components in the automotive industry:

- Low corrosion, resulting in a longer vehicle life
- Significant design freedom, which fosters creativity and innovation
- Adaptability in terms of component integration

- Security, convenience and economy
- Recyclability.

- Automotive plastics process control case study

Automotive polymer characterisation and process control involved testing against national and international OEM standards. Styrene maleic anhydride copolymer materials from Polyscope are used in a variety of applications, including dashboards. This material not only gives the desired appearance, but also the necessary qualities to ensure safety in all conditions. Numerous OEMs have designated SMA as their primary material for crucial structural automotive components.

A piping pollution incident resulting from a compounding process was seen. Contamination was identified through the use of FTIR (attenuated total reflection). The results indicated that the majority of the ingredients were organic, consisting of ABS and wax, rather than SMA:

- GC-FID analysis of monomers, oligomers, and residual solvents
- Viscosity, acidity, opacity, yellowness and percentage of water
- Support for specific tests during the creation of new product types
- Physical characterisation of SMA and its derivatives by a number of laboratory testing methodologies, including Charpy, IZOD, Tensile, HDT, Vicat, MFI, VEM, Density, and DSC.
  According to a recent study, every 10% reduction in vehicle weight results in a 6–8% reduction in fuel consumption. Automobile makers are now introducing more modern plastic materials into their vehicles in order to reduce weight and increase fuel efficiency. These sophisticated plastics have the following advantages when used in transportation vehicles:
- Increased security and comfort
- Fuel economy
- Adaptability in the integration of components
- Design liberty
- Weight loss.

We consider eight innovative plastics that are utilised in the automotive industry.

12.6.1.1 Polyvinyl Chloride

PVC is well-known for its flexibility, thermal stability and flame-retardant properties, as well as its extremely low lead concentration. It can be injection moulded, compression moulded or blow moulded to produce a wide variety of stiff or flexible plastic goods, depending on the type and quantity of plasticisers employed. It is utilised in the manufacture of automotive doors, electrical cable sheathing and instrument panels.

12.6.1.2 Polypropylene

Polypropylene is a saturated polymer derived from propylene as the monomer. One of the most significant advantages of polypropylene is its resistance to acids, bases and chemical solvents. It is used in gasoline cans, carpet fibres, bumpers, cable insulation and chemical storage tanks.

12.6.1.3 ABS

ABS is popular because of its low production cost and the simplicity with which plastic manufacturers can machine the material. It is synthesised through the polymerisation of styrene and acrylonitrile. Additionally, the styrene imparts a glossy, impermeable surface to the plastic. It performs exceptionally well at high and low temperatures, has excellent insulating capabilities and is easy to paint and glue. It is used in dashboards, wheel covers and other vehicle body elements.

12.6.1.4 Polyamide

Polyamide is a highly absorbent material with excellent mechanical qualities and a stiff nature. It is a multipurpose polymer that is extrudable and mouldable. Polyamides provide a number of advantages, including high strength, abrasion resistance and resilience. They are used in gears, cams, bearings and waterproof coatings in automobiles.

12.6.1.5 Polystyrene

Polystyrene is incredibly resistant to electrical and chemical attack. It is simple to create, has a high degree of elasticity, and softens when heated above the glass transition temperature. Strength, elongation, impact strength, toughness and modulus are just a few of its mechanical qualities. It is utilised in automobile accessories, display bases and buttons.

12.6.1.6 Polyoxymethylene (POM)

POM is a thermoplastic material that is well-known for its dimensional stability, creep resistance and strong heat resistance, as well as its electrical and dielectric properties. It is a technical material that is utilised in precision-machined components. POM is used in a variety of high-performance components, including gears, interior and exterior trim and fuel systems.

12.6.1.7 Polycarbonate

Polycarbonate is a thermoplastic material with exceptional thermal, electrical, impact, optical and weathering qualities. It possesses an unusual balance of hardness, stiffness and toughness. It is a popular choice for headlight lenses, bumpers, helmets and bullet-proof glass due to its exceptional impact strength.

12.6.1.8 Polyethylene

Polyethylene has great resistance to impact, low density, and excellent toughness. It is particularly cost-effective, moisture-resistant and adaptable to a wide variety of thermoplastic processing processes. Electrical insulation and reinforced glass are two areas where it is used. Due to the great benefits they provide, high-performance polymers will continue to play a critical role in the automobile sector. If you work in the automobile industry, Plastivision is the platform for you because it will simplify your operation tremendously well.

12.6.2 CONCLUSION

Plastics are a most versatile category of material with thousands of the polymer options having a variety of mechanical properties. Another important quality is that plastics can be processed in any shape easily. Weight reduction is an important aspect in vehicle design which will enhance the fuel efficiency and finally reduce emissions. Advanced plastic processing technology with recent-generation plastics technology will fulfil the changed scenario of automobile manufacturing processes.

12.6.3 REVIEW QUESTIONS

(1) Describe any five materials used in vehicle with their justifications.
(2) State the purpose of using metal matrix composites in industry.
(3) Explain the process of heat treatment and list the vehicle components where heat treatment is absolutely essential.
(4) Describe the various types of heat treatment processes normally used in automotive industry.
(5) List the vehicle components produced by casting process.
(6) Explain with neat sketches Resistance Spot Welding, Resistance seam welding and Friction welding.
(7) Demonstrate the stages used in SMED implementation.
(8) State the advantages of Plastic components in the automotive industry.

Multiple Choice Questions (MCQs) with Answers and Necessary Explanation

CHAPTER 1: AUTOMOTIVE MATERIALS

(1) The normal carbon percentage in mild steel is______________________
(a) 0.51%; (b) 0.15; (c) 0.87%; (d) 1.80%

(2) The hardness of brass material can be increased by
(a) Pack carburising; (b) Cold working; (c) Induction hardening; (d) Nitriding

(3) Which of the following materials has maximum ductility?
(a) Mild steel; (b) Nickel; (c) Copper; (d) Aluminium

(4) The property of a material essential for spring material is
(a) Ductility; (b) Stiffness; (c) Resilience; (d) Plasticity

(5) The steel widely used for motor car crankshafts is
(a) Chrome steel; (b) Nickel steel; (c) Silicon steel; (d) Nickel-chrome steel

(6) The metal suitable for a bearing subjected to heavy load is
(a) White metal; (b) Silicon bronze; (c) Phosphor bronze; (d) Monel metal

(7) An alloy of copper and zinc is known as __________
(a) Brass; (b) Bronze; (c) Nichrome; (d) Duralumin

(8) An alloy of Ni and Fe is termed as ____
(a) Brass; (b) Bronze; (c) Invar; (d) Duralumin

(9) The major constituent of gun metal is ____________
(a) Nickel; (b) Copper; (c) Iron; (d) Zinc

(10) The major constituent of Muntz metal is ____________
(a) Zinc; (b) Nickel; (c) Iron; (d) Copper

Answers:
1. (b) 2. (b) 3. (a) 4. (c) 5. (a) 6. (a) 7. (a) 8. (c) 9. (b) 10. (d)

CHAPTER 2: NONFERROUS MATERIALS

(1) The major constituent of Nichrome is ____________
(a) Copper; (b) Iron; (c) Nickel; (d) Zinc

(2) The major constituent of Constantan alloy is ____________
(a) Nickel; (b) Copper; (c) Iron; (d) Zinc

(3) The major constituent of Elektron alloy is ____________
(a) Copper; (b) Nickel; (c) Zinc; (d) Magnesium

(4) The major constituent of gun metal is ____________
(a) Copper; (b) Nickel; (c) Iron; (d) Zinc

(5) Which of the following alloys is widely used in thermocouples?
(a) Brass; (b) Bronze; (c) Duralumin; (d) Nichrome

(6) What is the approximate percentage of lead in soft solder?
(a) 60; (b) 50; (c) 90; (d) 99.02

(7) Thermosetting plastics have ____________
(a) 1-Degree bond; (b) 2-Degree bond; (c) 3-Degree bond; (d) 0-Degree bond

(8) Which of the following is a primary bond network of thermosetting plastics?
(a) 1-Dimensional; (b) 3-Dimensional; (c) 2-Dimensional; (d) 0-Dimensional

(9) Thermoplastics have ____________
(a) 1-Degree bond; (b) 2-Degree bond; (c) 3-Degree bond; (d) 0-Degree bond

Answers:
1. (C) 2. (b) 3. (d) 4. (a) 5. (d) 6. (b) 7. (a) 8. (b) 9. (b)

CHAPTER 3: HEAT TREATMENT

(1) Which of the following is the hardest constituent of steel?
(a) Ledeburite; (b) Austenite; (c) Bainite; (d) Martensite

(2) Which of the following form of iron is magnetic in nature?
(a) α; (b) δ; (c) γ; (d) λ

(3) For steel, which one of the following properties can be enhanced upon annealing?
(a) Hardness; (b) Toughness; (c) Ductility; (d) Resilience

(4) In annealing, cooling is done in which of the following media?
(a) Air; (b) Water; (c) Oil; (d) Furnace

(5) In normalising, cooling is done in which of the following media?
(a) Air; (b) Water; (c) Oil; (d) Furnace

(6) Mild steel can be converted into high-carbon steel by which of the following heat treatment processes?
(a) Annealing; (b) Normalising; (c) Case hardening; (d) Nitriding

(7) Upon annealing, eutectoid steel converts to which of the following?
(a) Perlite; (b) Cementite; (c) Austenite; (d) Martensite

(8) Fastest cooling is obtainable by cooling in
(a) Water; b) Air; c) Brine; d) None

(9) The hardness of steel increases with the
(a) Increase of carbon %; (b) Decrease of carbon %; (c) By slow cooling; (d) None

(10) Hypo-eutectoid steel contains carbon
(a) <0.022%; (b) < 0.770%; (c) < 6.77%; (d) None

Answers:
1. (d) 2. (a) 3. (c) 4. (d) 5. (a) 6. (c) 7. (a) 8. (c) 9. (a) 10. (b)

CHAPTER 4: MOULDING AND CASTING

(1) The casting process is preferred for parts having
(a) A few details; (b) Many details; (c) No details; (d) Non-symmetrical shape

(2) Cores are used to
(a) Support loose pieces; (b) Strengthen moulding sand; (c) Make desired recess in castings; (d) Remove pattern easily

(3) The purpose of a gate is
(a) Act as reservoir for molten metal; (b) Feed the casting at a rate consistent with the rate of solidification; (c) Help feed the casting until all solidification takes place; (d) Feed molten metal from pouring basin to gate

(4) The purpose of a pouring basin is to
(a) Feed the casting at a rate consistent with the rate of solidification; (b) Act as a reservoir for molten metal; (c) Help feed the casting until all solidification takes place; (d) Feed molten metal from pouring basin to gate

(5) The purpose of a riser is to
(a) Feed the casting at a rate consistent with the rate of solidification; (b) Act as a reservoir for molten metal; (c) Feed molten metal from pouring basin to gate; (d) Help feed the casting until all solidification takes place

(6) The purpose of a chaplet is to
(a) Provide bending; (b) Induce directional solidification; (c) Compensate shrinkage; (d) Support the core

(7) Cope in foundry practice refers to
(a) Bottom of moulding box; (b) Top of moulding box; (c) Middle portion of moulding box; (d) Heavy weight kept on moulding box to overcome buoyant effect of molten metal

(8) Cast iron pipes are cast by
(a) Loam moulding process; (b) Die casting process; (c) Centrifugal casting process; (d) Investment casting process

(9) True centrifugal casting is used to
(a) Cast objects symmetrical about an axis; (b) Get accurate castings; (c) Get dynamically balanced castings; (d) Get statistically balanced castings

(10) In the die casting process
(a) Any metal can be cast; (b) Any size of casting can be prepared; (c) The cost of dies is generally insignificant; (d) Very high production rates are possible

Answers:
1. (a) 2. (c) 3. (b) 4. (a) 5. (d) 6. (d) 7. (b) 8. (c) 9. (a) 10. (d)

CHAPTER 5: FORGING PROCESS

(1) Shaping of metal by squeezing it between two or more dies in order to obtain the desired shape is done by?
(a) Forming; (b) Forging; (c) Welding;(d) Grinding

(2) Forging is carried out at which temperature?
(a) Below recrystallisation temperature; (b) Above recrystallisation temperature; (c) Below or above recrystallisation temperature; (d) Above melting point

(3) Which of the following is a type of forging?
(a) Open die; (b) Closed die; (c) Impression dies; (d) Hold dies

(4) Which of the following forging metals is kept in the lower die?
(a) Open die; (b) Closed die; (c) Impression dies; (d) Hold dies

(5) In which of the forging processes is metal kept between a pair of dies and a gutter is provided in the lower die?
(a) Open die; (b) Closed die; (c) Impression dies; (d) Hold dies

(6) In which of the forging processes is metal kept between a pair of dies and no gutter is provided in the lower die?
(a) Open die; (b) Closed die; (c) Impression dies; (d) Hold dies

(7) The extra metal which settles down in the gutter is known as?
(a) Flash; (b) Slag; (c) Flux; (d) Barrelling

(8) In which of the following forging processes is no flash formed?
(a) Open die; (b) Closed die; (c) Impression dies; (d) Hold dies

(9) In which of the following forging processes does poor material utilisation occur?
(a) Open die; (b) Closed die; (c) Impression dies; (d) Hold dies

(10) Cogging, which is also called drawing out, is basically?
(a) Open die forging operation; (b) Closed die forging operation; (c) Impression dies forging operation; (d) Hold die forging operation

Answers:
1. (b) 2. (c) 3. (d) 4. (a) 5. (b) 6.(c) 7. a) 8. (b) 9. (a) 10. (a)

CHAPTER 6: WELDING PROCESS

(1) Which of the following can be easily welded using a flash butt welding process?
(a) Tin; (b) Lead; (c) Cast irons; (d) Carbon steel

(2) Industries using spot welding are
(a) Vehicles and electronics; (b) Pressure vessels; (c) Both (a) and (b); (d) None

(3) Electrodes used in spot welding are made up of which material?
(a) Only copper; (b) Copper and tungsten; (c) Copper and chromium; (d) Copper and aluminium

(4) How are the metals to be welded connected to each other in spot welding?
(a) Electric contact; (b) Magnetic field; (c) Mechanical pressure; (d) Direct contact

(5) In which of the following welding processes are heat and pressure applied on the joint but no filler material or flux is added?
(a) Arc welding; (b) Resistance welding; (c) Gas welding; (d) Thermite welding

(6) Up to what thickness can steel be welded using the spot welding process?
(a) 10 mm; (b) 11 mm; (c) 12 mm; (d) 13 mm

(7) In which of the following gas welding processes is a non-consumable electrode used?
(a) Submerged arc welding; (b) Tungsten inert gas welding; (c) Stud welding; (d) Gas metal arc welding

(8) What is the only difference between plasma arc welding and TIG welding?
(a) Flux is not used; (b) Construction of torch is different; (c) Gas is not used; (d) Tungsten electrode is not used

(9) Which of the following gas welding processes uses constant voltage?
(a) Submerged arc welding; (b) Tungsten inert gas welding; (c) Stud welding; (d) Gas metal arc welding

(10) In plasma arc welding the gas is?
(a) Ionised; (b) Heated; (c) Magnetised; (d) Vaporised

Answers:
1. (d) 2. (a) 3. (d) 4. (c) 5. (b) 6. (c) 7. (b) 8. (b) 9. (d) 10. (a)

CHAPTER 7: MATERIAL REMOVAL PROCESSES

(1) Why is the metal removal process costly?
(a) More energy is required; (b) Some of the material is wasted; (c) Both more energy is required and some of the material is wasted; (d) None of the mentioned

(2) In which machining process is removed metal negligible?
(a) Surface finishing; (b) Metal removal; (c) Both surface finishing and metal removal; (d) None of the mentioned

(3) Which of the following processes is not grouped under metal removal processes?
(a) Boring; (b) Milling; (c) Tumbling; (d) Rolling

(4) Which of the following is not grouped under the surface finishing process?
(a) Sawing; (b) Tapping; (c) Buffing; (d) Polishing

(5) In how many groups can the metal removal process be classified?
(a) 2; (b) 3; (c) 4; (d) 5

(6) In which type of metal removal process is grinding included?
(a) Conventional machining; (b) Abrasive process; (c) Non-traditional machining; (d) None of the mentioned

(7) ______ metal removal process includes milling.
(a) Conventional machining; (b) Abrasive process; (c) Non-traditional machining; (d) None of the mentioned

(8) In which type of metal removal process is thermal energy included?
(a) Conventional machining;(b) Abrasive process; (c) Non-traditional machining; (d) None of the mentioned

(9) Which of the following is a type of non-traditional machining?
(a) Turning;(b) Drilling; (c) Milling; (d) None of the mentioned

(10) In which metal removal process is material removed by particles?
(a) Conventional machining; (b) Abrasive process; (c) Non-traditional machining; (d) None of the mentioned

Answers:
1. (c) 2. (a) 3. (c) 4. (a) 5. (b) 6. (b) 7. (a) 8. (c) 9. (d) 10. (b)

CHAPTER 8: PLASTIC PROCESSING IN THE AUTOMOTIVE INDUSTRY

(1) Which of the following materials is not used in extrusion?
(a) Wax; (b) Granules; (c) Powder; (d) Pellets

(2) How are extruded materials cooled?
(a) Water; (b) Contact with chilled surface; (c) Air; (d) Oil

(3) The film extrusion process best involves film having a thickness below what length?
(a) 0.2 mm; (b) 0.3 mm; (c) 0.4 mm; (d) 0.5 mm

(4) Which of the following materials is not made by injection moulding?
(a) Nuts; (b) Tubes; (c) Car handles; (d) Electrical fittings

(5) What is the minimum temperature allowed to be given to the injection moulding process?
(a) 120°C; (b) 130°C; (c) 140°C; (d) 150°C

(6) Which of the following processes of moulding is widely used for the manufacturing of bottle caps and automotive dashboards?
(a) Compression moulding; (b) Transfer moulding; (c) Injection moulding; (d) Jet moulding

(7) What is the minimum air pressure required in the blow moulding process?
(a) 350 kPa; (b) 400 kPa; (c) 450 kPa; (d) 500 kPa

(8) Which of the following cooling systems is used in the injection moulding process to increase the solidification rate of components made?
(a) Air jet cooling system; (b) Water cooling system; (c) Cooling with free convection; (d) Cooling with fins

(9) Which of the following factors is not considered in a ram type injection moulding?
(a) Inner pressure of material;(b) Outer pressure of material; (c) Volume of material; (d) Temperature of material

(10) What is the maximum temperature allowed to be used in the injection moulding process?
(a) 300°C; (b) 320°C; (c) 350°C; (d) 400°C

Answers:
1. (a) 2. (d) 3. (d) 4. (b) 5. (d) 6. (c) 7. (b) 8. (b) 9. (c) 10. (a)

CHAPTER 9: POWDER METALLURGY

(1) What is the dimensional accuracy in powder metallurgy?
(a) High; (b) Medium; (c) Low; (d) Sometimes high and sometimes low

(2) Wastage of material in powder metallurgy as scrap is ____________
(a) Large; (b) Small; (c) Depends on other factors; (d) medium

(3) Which of the following tools is manufactured by powder metallurgy?
(a) High-speed steel; (b) Sintered carbides; (c) High-carbon steel; (d) Low-carbon steel

(4) For powder of aluminium and its alloys, the sintering temperature and time are
(a) 370–500°C, up to 24 hrs; (b) 250–350°C, up to 18 hrs; (c) 400–600°C, up to 20 hrs; (d) 550–700°C, up to 22 hrs

(5) A part produced by powder metallurgy is described as
(a) Welded part; (b) Cast part; (c) Forging part; (d) Sintered part

(6) Which of the following methods is used to make powder for brittle metals?
(a) Mechanical pulverisation; (b) Electrolytic process; (c) Chemical reduction; (d) Atomisation

(7) Complex shapes can be formed effectively using?
(a) Powder metallurgy; (b) Turning; (c) Sand casting; (d) Metal casting

(8) The process of forming metal powder by directing molten metal through an orifice after which it is broken into small particles using high-pressure fluid is known as?
(a) Atomisation; (b) Reduction; (c) Crushing; (d) Electrolysis

(9) Production of pure powder of iron and copper can be effectively done using?
(a) Atomisation; (b) Reduction; (c) Crushing; (d) Electrolysis

(10) Powder of various non-ferrous materials which becomes brittle on heating can be formed using?
(a) Atomisation; (b) Reduction; (c) Crushing; (d) Electrolysis

Answers:
1. (a) 2. (b) 3. (b) 4. (a) 5. (d) 6. (a) 7. (a) 8. (a) 9. (d) 10. (c)

CHAPTER 10: SURFACE TREATMENT

(1) Which of the following coatings has a glass composition?
(a) Paint; (b) Galvanised; (c) Enamel; (d) Anodised

(2) Which of the following is not a type of protective coating?
(a) Metallic; (b) Non-metallic; (c) Organic; (d) Inorganic

(3) An example of an anodic coating is _____
(a) Zinc; (b) Copper; (c) Nickel; (d) Chromium

(4) Which of these methods uses a filler wire at a high-temperature flame?
(a) Hot dipping; (b) Metal spraying; (c) Vapour plating; (d) Cementation

(5) Alclad is the cladding method where _____ is coated with pure aluminium.
(a) Duralumin; (b) Molybdenum; (c) Tin; (d) Silver

(6) Phosphate coating and chromate coating are classifications of ______ coatings.
(a) Anodic; (b) Cathodic; (c) Chemical; (d) Vitreous

(7) A varnish is a mixture of _____ and oil.
(a) Resin; (b) Pigment; (c) Turpentine; (d) Soybean

(8) Which common application do anodising and galvanising serve?
(a) Corrosion resistance; (b) Improved surface; (c) Zinc coating; (d) Increased strength

(9) With the help of brushing, which weld surface can be finished?
(a) Internal only; (b) External only; (c) both a and b; (d) Neither

(10) The precision surface finishing operations, employed for producing extremely high surface finish, are called
(a) Macro finishing operations; (b) Micro finishing operations; (c) Precise finishing operations;
(d) All of the above

(11) In super finishing operations
(a) The work rotates, the abrasive block reciprocates;
(b) The abrasive block rotates, the work reciprocates;
(c) Both abrasive block and work rotate;
(d) Both abrasive block and work reciprocate

(12) The wheels used in polishing are disc-shaped and termed as
(a) Pops; (b) Bobs; (c) Tops; (d) Lops

(13) In which of the following processes are highly polished steel balls used instead of abrasive
(a) Honing; (b) Lapping; (c) Polishing; (d) Burnishing

Answers:
1. (c) 2. (b) 3. (a) 4. (b) 5. (a) 6. (c) 7. (a) 8. (a) 9. (c) 10. (b) 11. (a) 12. (b) 13. (d)

CHAPTER 11: PRESS SHOP PROCESS

(1) Which of the machine components is designed under bending stress?
(a) Shaft; (b) Arm of a lever; (c) Key; (d) Belts and ropes

(2) For bending equations to be valid, the radius of curvature of the beam after bending should be
(a) Equal to its transverse dimensions; (b) Infinity; (c) Very large compared to its transverse dimensions; (d) Double its transverse dimensions

(3) The neutral axis of a beam always coincides with
(a) Axis passing through bottom of beam; (b) Axis passing through height h/2 from bottom; (c) Axis passing through height h/3 from bottom; (d) Axis passing through centroid

(4) If the depth of a beam is doubled then changes in its section modulus
(a) Will remain the same; (b) Will decrease; (c) Will be doubled; (d) Will increase by four times

(5) RCC beams are designed assuming
(a) Concrete can take no compressive load; (b) Concrete can take no compressive stress; (c) Concrete can take no tensile stress; (d) Concrete can take no tensile load

(6) Bending stress is
(a) Neither tensile nor compressive stress; (b) Tensile or compressive but cannot be added algebraically with direct tensile stress; (c) Tensile or compressive and can also be added algebraically with direct tensile stress; (d) None of the above

(7) Which of the following is not a shearing operation?
(a) Blanking; (b) Piercing; (c) Punching; (d) Forming

(8) Which of the following processes is not the type of bulk forming process in metal forming?
(a) Bending; (b) Rolling; (c) Forging; (d) Extrusion

(9) Which of the following manufacturing processes is mainly considered for producing components of very high strength?
(a) Casting; (b) Forging; (c) Extrusion; (d) Rolling

(10) Which of the following metal forming processes is most suitable for making wires?
(a) Forging; (b) Extrusion; (c) Drawing; (d) Rolling

(11) Which of the following components are manufactured by the sheet metal forming process?
(a) Engine blocks; (b) Connecting rods; (c) Electric wires; (d) Car bodies

(12) Which of the following processes is not a type of metal forming process?
(a) Extrusion; (b) Injection moulding; (c) Forging; (d) Drawing

(13) Which of the following can help in determining the behaviour of the material in metal forming?
(a) Size of material; (b) Shape of material; (c) Stress–strain curve; (d) Colour of material

(14) The shear angle in the piercing operation is provided on.......
(a) Die; (b) Punch; (c) Half on die and half on punch; (d) Die or punch depending on material and thickness of sheet

(15) The clearance in a blanking operation is provided on.........
(a) Die; (b) Punch; (c) Half on die and half on punch; (d) Die or punch depending on material and thickness of sheet

Answers:
1. (b) 2. (c) 3. (d) 4. (d) 5. (c) 6. (c) 7. (d) 8. (a) 9. (b) 10. (c) 11. (d) 12. (b) 13. (c) 14. (b) 15. (b)

CHAPTER 12: CASE STUDIES OF AUTOMOTIVE MANUFACTURING

(1) For stamping and forming operations of steel sheets with ________________ and ________________ are required.
(a) Low Yield Strength and good formability; (b) High Yield Strength and good formability; (c) Strength and Strength; (d) None of the above

(2) Case hardening procedures are critical for imparting________________
(a) Hardness; (b) Static as well as Dynamic strength; (c) Wear and Seizure resistance; (d) Hardness, Static as well as Dynamic strength, Wear and Seizure resistance

(3) Metal Matrix Composites is used in automobiles to manufacture the following components:
(a) Vehicle Dashboard; (b) Diesel engine pistons, engine; (c) Head Lamps and Tail Lamps; (d) Battery

(4) Method of Pressure die casting is applied in automotive industry to manufacture a variety of non-structural components within the vehicle includes_ ________________
(a) Electronic housings; (b) ECU encasements; (c) Mirror mounts; headlamp (d) All of the above

(5) The majority of automobile industries utilise cross-flow CO_2 laser systems with a power output of between________________
(a) 1 and 2 KW; (b) 100 and 150 KW; (c) 3 and 5 kW; (d) 0.1 and 100 KW

(6) Metals are being phased out of the automobile industry in favour of plastics and Plastics are used due to ________________
(a) Increased corrosion resistance; (b) Unable to recycle; (c) Higher vehicle weight; (d) None of the above

(7) Which one the following is not the advantages of sophisticated plastics used in transportation vehicles.
(a) Increased security and comfort; (b) Fuel economy; (c) Adaptability in the integration of components; (d) Design limitations

Answers:
(a) 2. (d) 3. (b) 4. (d) 5. (c) 6. (a) 7. (d)

Bibliography

Adedayo, A. V., Ibitoye, S. A., & Oyetoyan, O. A. (2010). Annealing heat treatment effects on steel welds. *Journal of Mineral, Materials Characterization and Engineering*, *9*(6), 547–557.

Adefuye, O. A., Raji, N. A., Adedeji, K. A., Fadipe, O. L., & Olowu, B. (2019). Additive manufacturing and sand-casting foundries practices in Nigeria. *Journal of Engineering and Technology Research, 4*, 55–63.

Adeleke, A. A., Oki, M., Anyim, I. K., Ikubanni, P. P., Adediran, A. A., Balogun, A. A., ... & Akinlabi, E. T. (2022). Recent development in casting technology: A pragmatic review. *Revue des Composites et des Matériaux Avancés*, *32*(2).

Adetunji, O. R., Adegbola, A. O., & Afolalu, S. A. (2015). Comparative study of case-hardening and water-quenching of mild steel rod on its mechanical properties. *International Journal of Advance Research*, *3*(6), 1–9.

Agarwal, J., Sahoo, S., Mohanty, S., & Nayak, S. K. (2020). Progress of novel techniques for lightweight automobile applications through innovative eco-friendly composite materials: A review. *Journal of Thermoplastic Composite Materials*, *33*(7), 978–1013.

Agrawal, A. R., Pandelidis, I. O., & Pecht, M. (1987). Injection-molding process control—A review. *Polymer Engineering & Science*, *27*(18), 1345–1357.

Ahmadein, M., Elsheikh, A. H., & Alsaleh, N. A. (2022). Modeling of cooling and heat conduction in permanent mold casting process. *Alexandria Engineering Journal*, *61*(2), 1757–1768.

Akafuah, N. K., Poozesh, S., Salaimeh, A., Patrick, G., Lawler, K., & Saito, K. (2016). Evolution of the automotive body coating process—A review. *Coatings*, *6*(2), 24.

Akhtar, S., Saad, M., Misbah, M. R., & Sati, M. C. (2018). Recent advancements in powder metallurgy: A review. *Materials Today: Proceedings*, *5*(9), 18649–18655.

Alden Abd Al Al, S., Gáspár, M., & Meilinger, Á. (2022). Properties of Hybrid Aluminium-Steel Joints Made by Resistance Spot Welding. In *Defect and Diffusion Forum* (Vol. 416, pp. 131–138). Trans Tech Publications Ltd.

Ali, N. I. M., Ali, M. A. M., Maidin, S., Sulaiman, M. A., Salleh, M. S., & Bakar, M. H. A. (2021, November). Review on Experimental Design, Process Parameters and Responses of Compression Moulding Process. In *Symposium on Intelligent Manufacturing and Mechatronics* (pp. 407–414). Springer, Singapore.

Altan, T., & Tekkaya, A. E. (Eds.). (2012). *Sheet Metal Forming: Fundamentals*. ASM International.

Altan, T., Ngaile, G., & Shen, G. (Eds.). (2004). *Cold and Hot Forging: Fundamentals And Applications* (Vol. 1). ASM international.

Ambadekar, P. K., & Choudhari, C. M. (2020). CNN based tool monitoring system to predict life of cutting tool. *SN Applied Sciences*, *2*(5), 1–11.

Aneiba, A., & Brown, S. Smart Die Casting: A New Approach. *Innovation in manufacturing through digital technologies and applications: Thoughts and Reflections on Industry 4.0.*, 80.

Anjo, V., & Khan, R. (2013). Gating system design for casting thin aluminium alloy (Al-Si) plates. *Leonardo Electronic Journal of Practices and Technologies*, *23*, 51–62.

Ariharan, N., Sriram, C. G., Radhika, N., Aswin, S., & Haridas, S. (2022). A comprehensive review of vapour deposited coatings for cutting tools: Properties and recent advances. *Transactions of the IMF*, 1–14.

Asghari, S. (2019). *Innovative cutting tool design for machining advanced engineering materials* (Doctoral dissertation, University of Bath).

Ashby, M. F., Shercliff, H., & Cebon, D. (2018). *Materials: Engineering, Science, Processing and Design*. Butterworth-Heinemann.

Askeland, D. R., & Wright, W. J. (2018). *Essentials of Materials Science and Engineering*. Cengage Learning.

Asthana, R., Kumar, A., & Dahotre, N. B. (2006). *Materials Processing and Manufacturing Science*. Elsevier.

Azevedo, J. M., CabreraSerrenho, A., & Allwood, J. M. (2018). Energy and material efficiency of steel powder metallurgy. *Powder Technology*, *328*, 329–336.

Baisukhan, A., Nakkiew, W., & Wisittipanit, N. (2022). Optimization of Tungsten Inert Gas Welding Process Parameters for AISI 304 Stainless Steel. In *Defect and Diffusion Forum* (Vol. 417, pp. 23–28). Trans Tech Publications Ltd.

Banabic, D. (2007). *Advanced methods in material forming*. Springer-Verlag Berlin Heidelberg.

Barletta, M., Aversa, C., Puopolo, M., & Vesco, S. (2019). Extrusion blow molding of environmentally friendly bottles in biodegradable polyesters blends. *Polymer Testing*, *77*, 105885.

Bayramoglu, M., Polat, H., & Geren, N. (2008). Cost and performance evaluation of different surface treated dies for hot forging process. *Journal of Materials Processing Technology*, *205*(1-3), 394–403.

Bein, T., Bös, J., Mayer, D., & Melz, T. (2012). Advanced Materials and Technologies for Reducing Noise, Vibration and Harshness (NVH) in Automobiles. In *Advanced Materials in Automotive Engineering* (pp. 254–298). Woodhead Publishing.

Bektemirov, B. S., Ulashov, J. Z., Akhmedov, A. K., & Gopirov, M. M. (2021, June). Types of Advanced Cutting Tool Materials and Their Properties. In *Euro-Asia Conferences* (Vol. 5, No. 1, pp. 260–262).

Belcher, S. L. (2011). Blow Molding. In *Applied Plastics Engineering Handbook* (pp. 267–288). William Andrew Publishing.

Bell, G. (1997). Advances in powder coating technology for automotive applications. *Pigment & Resin Technology*.

Bell, T. (1991). Surface treatment and coating of PM components. *Powder Metallurgy*, *34*(4), 253–258.

Bewilogua, K., Bräuer, G., Dietz, A., Gäbler, J., Goch, G., Karpuschewski, B., & Szyszka, B. (2009). Surface technology for automotive engineering. *CIRP Annals*, *58*(2), 608–627.

Bhalekar, A., Ingale, A. N., & Lekurwale, R. A. (2019). Experimental Analysis and Effect of Coating on HSS Cutting Tool. *Wear*, *4*(65), 915–8.

Bickerton, S., & Abdullah, M. Z. (2003). Modeling and evaluation of the filling stage of injection/compression moulding. *Composites Science and technology*, *63*(10), 1359–1375.

Black, J. T., & Kohser, R. A. (2017). *DeGarmo's materials and processes in manufacturing*. John Wiley & Sons.

Boljanovic, V. (2004). *Sheet metal forming processes and die design*. Industrial Press Inc..

Bolton, W. (2016). *Manufacturing technology for higher technicians*. Elsevier.

Bonollo, F., Gramegna, N., & Timelli, G. (2015). High-pressure die-casting: contradictions and challenges. *JOM*, *67*(5), 901–908.

Boothroyd, G. (1988). *Fundamentals of metal machining and machine tools* (Vol. 28). CRC Press.

Brinson, H. F., & Brinson, L. C. (2015). Characteristics, applications and properties of polymers. In *Polymer Engineering Science and Viscoelasticity* (pp. 57–100). Springer, Boston, MA.

Brosius, A., Hermes, M., Ben Khalifa, N., Trompeter, M., & Tekkaya, A. E. (2009). Innovation by forming technology: motivation for research. *International Journal of Material Forming*, *2*(1), 29–38.

Bruschi, S., Altan, T., Banabic, D., Bariani, P. F., Brosius, A., Cao, J., ...& Tekkaya, A. E. (2014). Testing and modelling of material behaviour and formability in sheet metal forming. *CIRP Annals*, *63*(2), 727–749.

Budkin, Y. V. (2011). Welding joints in dissimilar metals. *Welding International*, *25*(7), 523–525.

Buekens, A., & Letcher, T. M. (2020). The treatment of plastic in automobile shredder residue. In *Plastic Waste and Recycling* (pp. 401–414). Academic Press.

Burke, P., Kipouros, G. J., Fancelli, D., & Laverdiere, V. (2009). Sintering fundamentals of magnesium powders. *Canadian Metallurgical Quarterly*, *48*(2), 123–132.

Buryak, V. G., Buryak, A. V., & Buryak, V. V. (2019). Formation of the characteristics of the energy state in the local contact of the processing and tool materials in the cutting process.

Byrne, G., Dornfeld, D., & Denkena, B. (2003). Advancing cutting technology. *CIRP Annals*, *52*(2), 483–507.

Campbell, J. (2015). *Complete casting handbook: metal casting processes, metallurgy, techniques and design*. Butterworth-Heinemann.

Campos, J., Seatzu, C., & Xie, X. (Eds.). (2018). *Formal methods in manufacturing*. CRC press.

Čapek, J., & Vojtěch, D. (2013). Properties of porous magnesium prepared by powder metallurgy. *Materials Science and Engineering: C*, *33*(1), 564–569.

Cha, S. C., & Erdemir, A. (Eds.). (2015). *Coating technology for vehicle applications* (pp. 25–35). Basel, Switzerland: Springer.

Chandrasekharaiah, M. N. (1995). Weldability and weld joint failures. *Sadhana*, *20*(1), 213–232.

ChandraYadaw, R., Singh, S. K., Chattopadhyaya, S., Kumar, S., & Singh, R. C. (2018). Comparative study of surface coating for automotive application—a review. *International Journal of Applied Engineering Research*, *13*(6), 17–22.

Chao, B. T., & Bisacre, G. H. (1951). The effect of speed and feed on the mechanics of metal cutting. *Proceedings of the Institution of Mechanical Engineers*, *165*(1), 1–13.

Chastain, S. (2004). *Metal casting: a sand casting manual for the small foundry* (Vol. 1). Stephen Chastain.

Chaudhary, K. (2017). A Review on Advance Welding Processes. *Int. J. Eng. Tech*, *3*, 86–96.

Chen, J. J., Li, M. Z., Liu, W., & Wang, C. T. (2005). Sectional multipoint forming technology for large-size sheet metal. *The International Journal of Advanced Manufacturing Technology*, *25*(9), 935–939.

Chen, J., & Chang, Z. (2021). Recent Investigations on Incremental Sheet Forming: From Fundamentals to Industrial Application Technologies. In *Forming the Future* (pp. 369–382). Springer, Cham.

Chen, M., Fan, G., Tan, Z., Xiong, D., Guo, Q., Su, Y., ...& Zhang, D. (2018). Design of an efficient flake powder metallurgy route to fabricate CNT/6061Al composites. *Materials & Design*, *142*, 288–296.

Chen, S. B., & Lv, N. (2014). Research evolution on intelligentized technologies for arc welding process. *Journal of Manufacturing Processes*, *16*(1), 109–122.

Chen, X., Tang, J., Ding, H., & Liu, A. (2021). Experimental study on the evolution of chip morphology, chip formation, and surface topography with cutting parameters, and their relationships in dry milling of cast aluminum alloy with PCD inserter. *Journal of Mechanical Science and Technology*, *35*(4), 1651–1662.

Chen, Z., & Turng, L. S. (2005). A review of current developments in process and quality control for injection molding. *Advances in Polymer Technology: Journal of the Polymer Processing Institute*, *24*(3), 165–182.

Choi, J., Choi, B., Heo, S., Oh, Y., & Shin, S. (2018). Numerical modeling of the thermal deformation during stamping process of an automotive body part. *Applied Thermal Engineering*, *128*, 159–172.

Chubb, J. P., & Billingham, J. (1980). Coated cutting tools—a study of wear mechanisms in high speed machining. *Wear*, *61*(2), 283–293.

Conti, M. C., Mallia, B., Sinagra, E., Wismayer, P. S., Buhagiar, J., &Vella, D. (2019). The effect of alloying elements on the properties of pressed and non-pressed biodegradable Fe–Mn–Ag powder metallurgy alloys. *Heliyon*, *5*(9), e02522.

Creese, R. (2017). *Introduction to manufacturing processes and materials*. CRC Press.

Cselle, T., & Barimani, A. (1995). Today's applications and future developments of coatings for drills and rotating cutting tools. *Surface and Coatings Technology*, *76*, 712–718.

da Cunha, T. V., & Bohórquez, C. E. N. (2015). Ultrasound in arc welding: a review. *Ultrasonics*, *56*, 201–209.

Das, S. (2001). *The cost of automotive polymer composites: a review and assessment of DOE's lightweight materials composites research* (p. 47). Oak Ridge, TN, USA: Oak Ridge National Laboratory.

Davim, J. P. (2013). Materials Forming, Machining and Tribology.

Davim, J. P. (Ed.). (2008). *Machining: fundamentals and recent advances*, Springer.

Davim, J. P. (Ed.). (2012). *Dynamic methods and process advancements in mechanical, manufacturing, and materials engineering*. IGI Global.

Davim, J. P. (Ed.). (2012). *Tribology in manufacturing technology*. Springer Science & Business Media.

Davis, J. R. (Ed.). (2002). *Surface hardening of steels: understanding the basics*. ASM International.

De Chiffre, L. (1977). Mechanics of metal cutting and cutting fluid action. *International Journal of Machine Tool Design and Research*, *17*(4), 225–234.

Deng, L., Xia, J., & Wang, X. (2022). Fundamental of Precision Forging Technology for Aluminum Alloy. In *Precision Forging Technology and Equipment for Aluminum Alloy* (pp. 11–40). Springer, Singapore.

Deng, L., Xia, J., & Wang, X. (2022). Precision Forging Technology for Long Shaft Parts. In *Precision Forging Technology and Equipment for Aluminum Alloy* (pp. 63–119). Springer, Singapore.

Dobrzański, L. A. (Ed.). (2017). *Powder Metallurgy: Fundamentals and Case Studies*. BoD–Books on Demand.

Doege, E., & Bohnsack, R. (2000). Closed die technologies for hot forging. *Journal of Materials Processing Technology*, *98*(2), 165–170.

Dolinšek, S., Šuštaršič, B., & Kopač, J. (2001). Wear mechanisms of cutting tools in high-speed cutting processes. *Wear*, *250*(1-12), 349–356.

Douglas, R., & Kuhlmann, D. (2000). Guidelines for precision hot forging with applications. *Journal of Materials Processing Technology*, *98*(2), 182–188.

Drucker, D. C. (1949). An analysis of the mechanics of metal cutting. *Journal of Applied Physics*, *20*(11), 1013–1021.

Duarte, J. M., Sales, N. G. S., Sousa, M. H., Bridge, C., Maric, M., & de Andrade Gomes, J. (2020). Automotive paint analysis: How far has science advanced in the last ten years?. *TrAC Trends in Analytical Chemistry*, *132*, 116061.

DuPont, J. N., &Marder, A. R. (1995). Thermal efficiency of arc welding processes. *Welding Journal – Including Welding Research Supplement*, *74*(12), 406s.

Dwivedi, D. K. (2022). Classification of Joining Processes. In *Fundamentals of Metal Joining* (pp. 33–45). Springer, Singapore.

Dwivedi, D. K. (2022). Design of Welded Joints: Weld Bead Geometry: Selection, Welding Parameters. In *Fundamentals of Metal Joining* (pp. 343–351). Springer, Singapore.

Edwards, J. (1997). Coating and surface treatment systems for metals: a comprehensive guide to selection.

El-Eskandarany, M. S. (2015). *Mechanical alloying: nanotechnology, materials science and powder metallurgy*. Elsevier.

Elmarakbi, A. (2013). *Advanced composite materials for automotive applications: Structural integrity and crashworthiness*. John Wiley & Sons.

Enahoro, H. E., & Welsh, M. J. M. (2022). The relevance of the mechanics of metal cutting to machinability.

Enahoro, H. E., & Welsh, M. J. M. (2022). The relevance of the mechanics of metal cutting to machinability.

Erofeev, V. V., Troyanovskaya, I., Ignatiev, A. G., Sharafiev, R., & Gilmanshin, R. (2022). Welding Mode Justification for a T-Joint. In *Key Engineering Materials* (Vol. 910, pp. 73–81). Trans Tech Publications Ltd.

Fande, A. W., Taiwade, R. V., & Raut, L. (2022). Development of activated tungsten inert gas welding and its current status: A review. *Materials and Manufacturing Processes*, *37*(8), 841–876.

Faoro, G. (2001). UV Protection and Coating for Plastics in the Automobile Industry. *Paint & Coatings Industry*, *17*(10), 34–38.

Faruk, O., Tjong, J., & Sain, M. (Eds.). (2017). *Lightweight and sustainable materials for automotive applications*. Boca Raton: CRC Press.

Fathi, M. S. A., Ismael, Q. A. Y. S. H. A. Z. I. M., & Saleh, K. A. (2019). An effect of welding type on the mechanical properties of welding joints. *International Journal of Mechanical and Production Engineering Research and Development*, *9*(4), 699–707.

Fentahun, M. A., & Savas, M. A. (2018). Materials used in automotive manufacture and material selection using Ashby charts. *Int. J. Mater. Eng*, *8*(3), 40–54.

Fernandes, C., Pontes, A. J., Viana, J. C., & Gaspar-Cunha, A. (2018). Modeling and Optimization of the Injection-Molding Process: A Review. *Advances in Polymer Technology*, *37*(2), 429–449.

Fredriksson, H., & Akerlind, U. (2006). *Materials processing during casting* (Vol. 210). Chichester: Wiley.

Friedrich, K. (2016, May). Carbon fiber reinforced thermoplastic composites for future automotive applications. In *AIP Conference Proceedings* (Vol. 1736, No. 1, p. 020001). AIP Publishing LLC.

Friedrich, K., & Almajid, A. A. (2013). Manufacturing aspects of advanced polymer composites for automotive applications. *Applied Composite Materials*, *20*(2), 107–128.

Froes, F. H. S. (2015). A historical perspective of titanium powder metallurgy. *Titanium Powder Metallurgy*, 1–19.

Froes, F. H., & Eylon, D. (1990). Powder metallurgy of titanium alloys. *International Materials Reviews*, *35*(1), 162–184.

Froes, F. H., Mashl, S. J., Hebeisen, J. C., Moxson, V. S., & Duz, V. A. (2004). The technologies of titanium powder metallurgy. *JOM*, *56*(11), 46–48.

Fu, Q., Wu, S., Li, C., Xu, J., & Wang, D. (2022). Delamination and chip breaking mechanism of orthogonal cutting CFRP/Ti6Al4V composite. *Journal of Manufacturing Processes*, *73*, 183–196.

Fuchs, E. R., Field, F. R., Roth, R., &Kirchain, R. E. (2008). Strategic materials selection in the automobile body: Economic opportunities for polymer composite design. *Composites science and technology*, *68*(9), 1989–2002.

Gangopadhyay, A. (2015). Coating technologies for automotive engine applications. In *Coating Technology for Vehicle Applications* (pp. 63–79). Springer, Cham.

Garcia Filho, F. C., & Monteiro, S. N. (2020). Welding joints in high entropy alloys: a short-review on recent trends. *Materials*, *13*(6), 1411.

Gaspar-Cunha, A., Covas, J. A., & Sikora, J. (2022). Optimization of Polymer Processing: A Review (Part II-Molding Technologies). *Materials*, *15*(3), 1138.

Gengel, P. (2010). Foundry technologies and their influence on environment. *International Multidisciplinary Scientific GeoConference: SGEM*, *2*, 581.

Ghassemali, E. (2022). Forging of metallic parts and structures.

Ghassemieh, E. (2011). Materials in automotive application, state of the art and prospects. *New Trends and Developments in Automotive Industry*, *20*, 365–394.

Ghosh, A., & Mallik, A. K. (2010). *Manufacturing science*. Affiliated East-West Press Private Limited.

Giampieri, A., Ling-Chin, J., Ma, Z., Smallbone, A., & Roskilly, A. P. (2020). A review of the current automotive manufacturing practice from an energy perspective. *Applied Energy*, *261*, 114074.

Glownia, J. (1993). Metal casting and moulding processes. In *Manufacturing Engineer's Reference Book* (pp. 3-1). Butterworth-Heinemann.

Gronostajski, Z., Kaszuba, M., Hawryluk, M., & Zwierzchowski, M. (2014). A review of the degradation mechanisms of the hot forging tools. *Archives of Civil and Mechanical Engineering*, *14*(4), 528–539.

Groover, M. P. (2020). *Fundamentals of modern manufacturing: materials, processes, and systems*. John Wiley & Sons.

Grzesik, W. (2008). *Advanced machining processes of metallic materials: theory, modelling and applications*. Elsevier.

Gu, F., Guo, J., Zhang, W., Summers, P. A., & Hall, P. (2017). From waste plastics to industrial raw materials: A life cycle assessment of mechanical plastic recycling practice based on a real-world case study. *Science of the Total Environment*, *601*, 1192–1207.

Guma, T., & Uche, O. L. (2019). A typification of foundry practices for correct artisanal sand casting of aluminum pots. *Int. J. Eng. Appl. Sci. Technol*, *4*, 169–178.

Güney, B., Mutlu, İ., & Küçüksariyildiz, H. (2019). The effect of flame spray coating on the tribological properties of brake disc. *PoliteknikDergisi*, *23*(3), 755–761.

Gupta, K. (Ed.). (2020). *Materials forming, machining and post processing*. Springer.

Gupta, K. M., Ramdev, K., Dharmateja, S., & Sivarajan, S. (2018). Cutting characteristics of PVD coated cutting tools. *Materials Today: Proceedings*, *5*(5), 11260–11267.

Haga, T., Sugishita, T., Fuse, H., Watari, H., & Nishida, S. (2022). Hot Forging of Gravity Casted Al-Mg Alloy. In *Key Engineering Materials* (Vol. 918, pp. 3–8). Trans Tech Publications Ltd.

Hartlieb, M. (2013). Aluminum alloys for structural die casting. *Die Casting Engineer*, *57*(3), 40–43.

Hattel, J. H., Pryds, N., Thorborg, J., Lipinski, M., & Schneider, M. (2005). *Fundamentals of numerical modelling of casting processes*. PolytekniskForlag.

Hausner, H. H. (2012). *Modern developments in powder metallurgy: volume 1: fundamentals and methods*. Springer Science & Business Media.

Hawryluk, M., & Jakubik, J. (2016). Analysis of forging defects for selected industrial die forging processes. *Engineering Failure Analysis*, *59*, 396–409.

Higgins, R. A. (2010). *Materials for engineers and technicians*. Routledge.

Holbery, J., & Houston, D. (2006). Natural-fiber-reinforced polymer composites in automotive applications. *JOM*, *58*(11), 80–86.

Hong, K. M., & Shin, Y. C. (2017). Prospects of laser welding technology in the automotive industry: A review. *Journal of Materials Processing Technology*, *245*, 46–69.

Hosford, W. F., &Caddell, R. M. (2011). *Metal forming: mechanics and metallurgy*. Cambridge University Press.

Hovorun, T. P., Berladir, K. V., Pererva, V. I., Rudenko, S. G., & Martynov, A. I. (2017). Modern materials for automotive industry.

Hsu, T. Y., & Xu, Z. Y. (2007). Design of structure, composition and heat treatment process for high strength steel. In *Materials Science Forum* (Vol. 561, pp. 2283–2286). Trans Tech Publications Ltd.

Htun, M. S., Kyaw, S. T., & Lwin, K. T. (2008). Effect of heat treatment on microstructures and mechanical properties of spring steel. *Journal of Metals, Materials and Minerals*, *18*(2), 191–197.

Huang, H. X., Li, J. C., Li, D., & Huang, G. Q. (2011). New strategies for predicting parison dimensions in extrusion blow molding. *Polymer-Plastics Technology and Engineering*, *50*(13), 1329–1337.

Huang, R., Sone, M., Ma, W., & Fukanuma, H. (2015). The effects of heat treatment on the mechanical properties of cold-sprayed coatings. *Surface and Coatings Technology*, *261*, 278–288.

Hutmann, P. (2006). The application of mechanical surface treatment in the passenger car industry. *Shot Peening*, 3.

Ibrahim, I. A., Mohamat, S. A., Amir, A., & Ghalib, A. (2012). The Effect of Gas Metal Arc Welding (GMAW) processes on different welding parameters. *Procedia Engineering*, *41*, 1502–1506.

Ibrahim, I. A., Mohamat, S. A., Amir, A., & Ghalib, A. (2012). The Effect of Gas Metal Arc Welding (GMAW) processes on different welding parameters. *Procedia Engineering*, *41*, 1502–1506.

Ingarao, G., Di Lorenzo, R., & Micari, F. (2011). Sustainability issues in sheet metal forming processes: an overview. *Journal of Cleaner Production*, *19*(4), 337–347.

Iqbal, M. (2014). Gating design criteria for sound casting. *International Journal of Mechanical Engineering and Robotics Research*, *3*(3), 675.

Isik, K., Silva, M. B., Tekkaya, A. E., & Martins, P. A. F. (2014). Formability limits by fracture in sheet metal forming. *Journal of Materials Processing Technology*, *214*(8), 1557–1565.

Ivasishin, O. M., Anokhin, V. M., Demidik, A. N., & Savvakin, D. G. (2000). Cost-effective blended elemental powder metallurgy of titanium alloys for transportation application. In *Key Engineering Materials* (Vol. 188, pp. 55–62). Trans Tech Publications Ltd.

Ivasishin, O. M., Sawakin, D. G., Moxson, V. S., Bondareval, K. A., & Froes, F. H. (2002). Titanium powder metallurgy for automotive components. *Materials Technology*, *17*(1), 20–25.

Jain, P. L. (2003). *Principles of foundry technology*. Tata McGraw-Hill Education.

Jain, V. K. (2009). *Advanced machining processes*. Allied Publishers.

Jaspart, J. P., & Weynand, K. (2016). Design of joints in steel and composite structures.

Ji, Y., Wu, S., & Zhao, D. (2016). Microstructure and mechanical properties of friction welding joints with dissimilar titanium alloys. *Metals*, *6*(5), 108.

Jindal, U. C. (2012). *Material Science and Metallurgy*. Pearson Education India.

Joao, D., Milliken, N., Tutunea-Fatan, O. R., & Bordatchev, E. V. (2019). One-Side Cutting Strategy for Ultraprecise Single Point Cutting of V-grooves Case 1: Constant Chip Thickness. *IFAC-PapersOnLine*, *52*(10), 306–310.

Jović, D., &Milićević, J. (2017). Influence of Application of New Material in Automotive Industry on Improving Quality of Life. *Center For Quality*, 331–332.

Kah, P., Suoranta, R., & Martikainen, J. (2013). Advanced gas metal arc welding processes. *The International Journal of Advanced Manufacturing Technology*, *67*(1), 655–674.

Kah, P., Suoranta, R., & Martikainen, J. (2013). Advanced gas metal arc welding processes. *The International Journal of Advanced Manufacturing Technology*, *67*(1), 655–674.

Kalpakjian, S., & Schmid, S. R. (2009). Manufacturing Engineering. *Technology, Addison-Wesley, Reading, MA (1989)*, 661–663.

Kalyon, D., Tan, V., & Kamal, M. R. (1980). The dynamics of parison development in blow molding. *Polymer Engineering & Science*, *20*(12), 773–777.

Kano, M. (2015). Overview of DLC-coated engine components. In *Coating technology for vehicle applications* (pp. 37–62). Springer, Cham.

Karpat, Y., & Özel, T. (2008). Mechanics of high speed cutting with curvilinear edge tools. *International Journal of Machine Tools and Manufacture*, *48*(2), 195–208.

Katuku, K. (2022). Tool-chip contact and chip formation in dry finish-machining of austempered ductile iron. *Journal of Manufacturing Processes*, *78*, 288–307.

Kaushish, J. P. (2010). *Manufacturing processes*. PHI Learning Pvt. Ltd..

Kik, T. (2014). Numerical analysis of MIG welding of butt joints in aluminium alloy. *BiuletynInstytutuSpawalnictwa w Gliwicach*, *58*(3), 37–49.

Kim, C., Choi, W., Kim, J., & Rhee, S. (2008). Relationship between the weldability and the process parameters for laser-TIG hybrid welding of galvanized steel sheets. *Materials transactions*, *49*(1), 179–186.

Kleiner, M., Chatti, S., & Klaus, A. (2006). Metal forming techniques for lightweight construction. *Journal of Materials Processing Technology*, *177*(1-3), 2–7.

Koç, M., & Özel, T. (Eds.). (2019). *Modern manufacturing processes*. John Wiley & Sons.

Kodama, S., Ishida, Y., Furusako, S., Saito, M., Miyazaki, Y., & Nose, T. (2013). Arc welding technology for automotive steel sheets. *Nippon Steel Technical Report*, *103*(103), 83–90.

Kolnes, M., Kübarsepp, J., Sergejev, F., Kolnes, M., Tarraste, M., & Viljus, M. (2020). Performance of ceramic-metal composites as potential tool materials for friction stir welding of aluminium, copper and stainless steel. *Materials*, *13*(8), 1994.

Kopp, R. (1996). Some current development trends in metal-forming technology. *Journal of Materials Processing Technology*, *60*(1-4), 1–9.

Kopp, R., Wiedner, C., & Meyer, A. (2005). Flexibly rolled sheet metal and its use in sheet metal forming. In *Advanced Materials Research* (Vol. 6, pp. 81–92). Trans Tech Publications Ltd.

Kot, P. (2008). Computer aided design of plastic components used in automobile industry. *CzasopismoTechniczne. Mechanika*, *105*(3-M), 111–120.

Krishna, B. N., Tech, N. N. M., & Tech, K. G. M. (2021). Modelling and analysis of single point cutting tool using FEM method. *Turkish Journal of Computer and Mathematics Education (TURCOMAT)*, *12*(13), 7402–7409.

Kulekci, M. K. (2008). Magnesium and its alloys applications in automotive industry. *The International Journal of Advanced Manufacturing Technology*, *39*(9), 851–865.

Kulkarni, H., & Dabhade, V. V. (2019). Green machining of powder-metallurgy-steels (PMS): an overview. *Journal of Manufacturing Processes*, *44*, 1–18.

Kumar, K., & Davim, P. J. (Eds.). (2020). *Modern Manufacturing Processes*. Woodhead Publishing.

Kumar, N., Bharti, A., Dixit, M., & Nigam, A. (2020). Effect of powder metallurgy process and its parameters on the mechanical and electrical properties of copper-based materials: Literature review. *Powder Metallurgy and Metal Ceramics*, *59*(7), 401–410.

Kumar, R., Singh, R., Ahuja, I. P. S., Penna, R., & Feo, L. (2018). Weldability of thermoplastic materials for friction stir welding-A state of art review and future applications. *Composites Part B: Engineering*, *137*, 1–15.

Kumaravel, G., &Jabarin, S. A. (1996). Extrusion blow molding of high-density polyethylene–poly (ethylene terephthalate) blends. *Advances in Polymer Technology: Journal of the Polymer Processing Institute*, *15*(3), 191–204.

Kurka, P., & Schieck, F. (2014). Innovations in forming technologies. *Achieves of Materials Science and Engineering*, *65*, 72–76.

Kuzin, V. V., Dos' ko, S. I., Popov, V. F., Fedorov, S. Y., & Fedorov, M. Y. (2005). Tooling for high-speed cutting. *Russian Engineering Research*, *25*(9), 20–25.

Lal, R. K., Choubey, V. K., Dwivedi, J. P., & Kumar, S. (2018). Study of factors affecting Springback in sheet metal forming and deep drawing process. *Materials Today: Proceedings*, *5*(2), 4353–4358.

Landel, R. F., & Nielsen, L. E. (1993). *Mechanical properties of polymers and composites*. CRC Press.

Laschet, G., Jakumeit, J., & Benke, S. (2004). Thermo-mechanical analysis of cast/mould interaction in casting processes. *International Journal of Materials Research*, *95*(12), 1087–1096.

Laughlin, D. E., & Hono, K. (Eds.). (2014). *Physical metallurgy*. Newnes.

Li, K., & Zhang, Y. (2007). Metal transfer in double-electrode gas metal arc welding.

Li, L., Pratihar, D. K., Chakrabarty, S., & Mishra, P. C. (2020). *Advances in Materials and Manufacturing Engineering* (Vol. 119, No. 125, p. 0). Springer.

Li, M. Z., Cai, Z. Y., Sui, Z., & Yan, Q. G. (2002). Multi-point forming technology for sheet metal. *Journal of Materials Processing Technology*, *129*(1–3), 333–338.

Li, Y., Zou, W., Lee, B., Babkin, A., & Chang, Y. (2020). Research progress of aluminum alloy welding technology. *The International Journal of Advanced Manufacturing Technology*, *109*(5), 1207–1218.

Lim, S. C., Lee, S. H., Liu, Y. B., & Seah, K. H. W. (1993). Wear maps for uncoated high-speed steel cutting tools. *Wear*, *170*(1), 137–144.

Liou, S. Y., & Miller, R. A. (1991). Design for die casting. *International Journal of Computer Integrated Manufacturing*, *4*(2), 83–96.

Lippold, J. C. (2014). *Welding metallurgy and weldability*. John Wiley & Sons.

Liu, C., Wan, M., Zhang, W., & Yang, Y. (2021). Chip formation mechanism of Inconel 718: a review of models and approaches. *Chinese Journal of Mechanical Engineering*, *34*(1), 1–16.

Liu, Y., Chen, L. F., Tang, H. P., Liu, C. T., Liu, B., & Huang, B. Y. (2006). Design of powder metallurgy titanium alloys and composites. *Materials Science and Engineering: A*, *418*(1-2), 25–35.

Lokesh, M. T. S., & Kumar, M. (2022). Effect of Filler Rod on Microstructure and Mechanical Properties of Bimetallic Weld Joint.

Louda, P. (2007). Applications of thin coatings in automotive industry. *Journal of Achievements in Materials and Manufacturing Engineering*, *24*(1), 50–56.

Lovano, S. (1996). Automotive applications for powder coating. *Metal Finishing*, *94*(9), 20–24.

Madan, J., Mani, M., & Lyons, K. W. (2013, June). Characterizing energy consumption of the injection molding process. In *International Manufacturing Science and Engineering Conference* (Vol. 55461, p. V002T04A015). American Society of Mechanical Engineers.

Madavi, K. R., Jogi, B. F., & Lohar, G. S. (2022). Metal inert gas (MIG) welding process: a study of effect of welding parameters. *Materials Today: Proceedings*, *51*, 690–698.

Mae, H., Teng, X., Bai, Y., & Wierzbicki, T. (2008). Comparison of ductile fracture properties of aluminum castings: Sand mold vs. metal mold. *International Journal of Solids and Structures*, *45*(5), 1430–1444.

Mahajan, S., & Chakraborty, K. (2022). The Effect of HSS and Carbide Tools on the Mechanism of Chip Formation During Turning of Medium Carbon Steel (AISI1030). In *Recent Advances in Manufacturing, Automation, Design and Energy Technologies* (pp. 165–172). Springer, Singapore.

Makhlouf, M. M., & Apelian, D. (2002). *Casting characteristics of aluminum die casting alloys* (No. DOE/ID/13716). Worcester Polytechnic Institute (US).

Mallick, P. K. (2012). Advanced materials for automotive applications: An overview. *Advanced Materials in Automotive Engineering*, 5–27.

Malloy, R. A. (2012). *Plastic part design for injection molding: an introduction.* Carl HanserVerlag GmbH Co KG.

Marco, Y., & Chevalier, L. (2009). Blow moulding of polymers. *Advances in Polymer Processing. From macro to nano scales*, 226–286.

Mason, A., Romanov, D., Cordova-Lopez, L. E., Ross, S., & Korostynska, O. (2022). Smart knife: technological advances towards smart cutting tools in meat industry automation. *Sensor Review*.

Mayer, E. J. (1924). *Die-casting machine* (Doctoral dissertation, Colorado Agricultural College).

McKie, M. G. (2015). *Research and Development of a Sustainable Technology Improvement Model for the Foundry Industry* (Doctoral dissertation, Doctoral Dissertation, Teesside University, Middlesbrough, England, 10.13140/RG. 2.2. 23201.25443).

Mechraoui, A., Barzegari, M. R., Riedl, B., & Rodrigue, D. (2008). Compression moulding of polypropylene foams and their properties. *Cellular Polymers*, *27*(4), 217–233.

Mehaboob, S. (2021). Impact of thermal treatment process over carbon based single point cutting tool. *Journal of Scholastic Engineering Science and Management*, *1*(1), 29–37.

Mehrabi, H., Salonitis, K., & Jolly, M. (2016, April). Sustainable Investment Casting. In *14th World Conference in Investment Casting*.

Menary, G. H., & Armstrong, C. G. (2006). Experimental study and numerical modelling of injection stretch blow moulding of angioplasty balloons. *Plastics, rubber and composites*, *35*(8), 348–354.

Merchant, M. E. (1945). Mechanics of the metal cutting process. I. Orthogonal cutting and a type 2 chip. *Journal of Applied Physics*, *16*(5), 267–275.

Michaeli, W., Spennemann, A., &Gärtner, R. (2002). New plastification concepts for micro injection moulding. *Microsystem Technologies*, *8*(1), 55–57.

Mills, B. (1996). Recent developments in cutting tool materials. *Journal of Materials Processing Technology*, *56*(1-4), 16–23.

Ming, C., Fanghong, S., Haili, W., Renwei, Y., Zhenghong, Q., & Shuqiao, Z. (2003). Experimental research on the dynamic characteristics of the cutting temperature in the process of high-speed milling. *Journal of materials processing technology*, *138*(1-3), 468–471.

Ming, C., Fanghong, S., Haili, W., Renwei, Y., Zhenghong, Q., & Shuqiao, Z. (2003). Experimental research on the dynamic characteristics of the cutting temperature in the process of high-speed milling. *Journal of Materials Processing Technology*, *138*(1-3), 468–471.

Mishra, B., Panda, R. R., & Mohanta, D. K. (2014). Metal Inert Gas (MIG) welding parameters optimization. *International Journal of Multidisciplinary and Current Research*, *2*(1), 637–639.

Modi, S. (2016). Material Qualification in the Automotive Industry. *Center for Automotive Research*.

Modi, S. (2016). Material Qualification in the Automotive Industry. *Center for Automotive Research*.

Mohl, W., Kunstmann, R., Weichert, J., & Gonschor, R. (2001). Coating Technology and Materials for Lighting Applications. *Proceedings of PAL 2001-Progress in automobile lighting, Held Laboratory of Lighting Technology, September 2001. VOL 9.*

Muley, S. V., & McCarthy, J. J. (2022). Case study: Investigation of shorter lifetime of upper Cu weld wheels in an industrial mash seam welding setup. *Engineering Failure Analysis*, *135*, 106113.

Murphy, A. B. (2015). A perspective on arc welding research: the importance of the arc, unresolved questions and future directions. *Plasma Chemistry and Plasma Processing*, *35*(3), 471–489.

Murray, P. E., & Scotti, A. (1999). Depth of penetration in gas metal arc welding. *Science and Technology of Welding and Joining*, *4*(2), 112–117.

Mvola, B., Kah, P., & Martikainen, J. (2014). Dissimilar ferrous metal welding using advanced gas metal arc welding processes. *Reviews on Advanced Materials Science*, *38*(2).

Nakagawa, T. (1993). Recent developments in auto body panel forming technology. *CIRP Annals*, *42*(2), 717–722.

Nakata, K., Kim, Y. G., Ushio, M., Hashimoto, T., & Jyogan, S. (2000). Weldability of high strength aluminum alloys by friction stir welding. *ISIJ International*, *40*(Suppl), S15–S19.

Nam, H., Park, C., Kim, C., Kim, H., & Kang, N. (2018). Effect of post weld heat treatment on weldability of high entropy alloy welds. *Science and Technology of Welding and Joining*, *23*(5), 420–427.

Narasimhan, K. S. (2001). Sintering of powder mixtures and the growth of ferrous powder metallurgy. *Materials Chemistry and Physics*, *67*(1-3), 56–65.

Nee, A. Y. C. (2014). *Handbook of manufacturing engineering and technology*. Springer Publishing Company, Incorporated.

Neville, B. P., & Rabiei, A. (2008). Composite metal foams processed through powder metallurgy. *Materials & Design*, *29*(2), 388–396.

Nguyen, Q. T., Luu, V. T., & Le, T. K. (2022). Study on the Process and Formation of Fiber Structure During Forming of an Automobile Joint Part in Closed-Die Forging. In *Regional Conference in Mechanical Manufacturing Engineering* (pp. 49–55). Springer, Singapore.

Niederhofer, P., Richrath, L., Huth, S., & Theisen, W. (2016). Influence of conventional and powder-metallurgical manufacturing on the cavitation erosion and corrosion of high interstitial CrMnCN austenitic stainless steels. *Wear*, *360*, 67–76.

Ning, J., & Liang, S. Y. (2019). A comparative study of analytical thermal models to predict the orthogonal cutting temperature of AISI 1045 steel. *The International Journal of Advanced Manufacturing Technology*, *102*(9), 3109–3119.

Niu, J., Huang, C., Li, C., Zou, B., Xu, L., Wang, J., & Liu, Z. (2020). A comprehensive method for selecting cutting tool materials. *The International Journal of Advanced Manufacturing Technology*, *110*(1), 229–240.

Novák, P. (2020). Advanced powder metallurgy technologies. *Materials*, *13*(7), 1742.

Ocheri, C., Mbah, A. C., & Mbah, C. N. (2017). Focusing Foundry Engineering for Economic Recovery: A Case Study of the Ajaokuta Foundry Shop and Making Shop. *International Journal of Swarm Intelligence and Evolutionary Computation*, *6*(1).

Odebiyi, O. S., Adedayo, S. M., Tunji, L. A., &Onuorah, M. O. (2019). A review of weldability of carbon steel in arc-based welding processes. *Cogent Engineering*, *6*(1), 1609180.

Omar, M. A. (2011). *The automotive body manufacturing systems and processes*. John Wiley & Sons.

Orban, R. L. (2004). New research directions in powder metallurgy. *Romanian Reports in Physics*, *56*(3), 505–516.

Osborne, J. (2013). Automotive composites–in touch with lighter and more flexible solutions. *Metal Finishing*, *111*(2), 26–30.

Özel, T., & Altan, T. (2000). Determination of workpiece flow stress and friction at the chip–tool contact for high-speed cutting. *International Journal of Machine Tools and Manufacture*, *40*(1), 133–152.

Panda, A., Dobránsky, J., Jančík, M., Pandová, I., & Kačalová, M. (2018). Advantages and effectiveness of the powder metallurgy in manufacturing technologies. *Metalurgija*, *57*(4), 353–356.

Pantani, R., Coccorullo, I., Speranza, V., & Titomanlio, G. (2005). Modeling of morphology evolution in the injection molding process of thermoplastic polymers. *Progress in polymer science*, *30*(12), 1185–1222.

Parashivamurthy, K. I. (2012). *Material Science and Metallurgy* (p. 147). Pearson.

Park, J. J., & Kim, Y. H. (2003). Fundamental studies on the incremental sheet metal forming technique. *Journal of Materials Processing Technology*, *140*(1-3), 447–453.

Patel, A. V., Patel, V. P., & Patel, C. K. A Review On Optimization of Process Parameters For PVC Pipe Using Extrusion Process.

Patil, A., Patel, A., & Purohit, R. (2017). An overview of polymeric materials for automotive applications. *Materials Today: Proceedings*, *4*(2), 3807–3815.

Patil, R. T., Metri, V. S., & Tambore, S. S. (2015). Causes of casting defects with remedies. *International Journal of Engineering Research & Technology*, *4*(11), 639–644.

Patnaik, L., Saravanan, I., & Kumar, S. (2020). Die casting parameters and simulations for crankcase of automobile using MAGMAsoft. *Materials Today: Proceedings*, *22*, 563–571.

Pattnaik, S., Karunakar, D. B., & Jha, P. K. (2012). Developments in investment casting process—a review. *Journal of Materials Processing Technology*, *212*(11), 2332–2348.

Pickens, J. R. (1981). Aluminium powder metallurgy technology for high-strength applications. *Journal of Materials Science*, *16*(6), 1437–1457.

Pittner, A., & Rethmeier, M. (2022). Life Cycle Assessment of Fusion Welding Processes—A Case Study of Resistance Spot Welding Versus Laser Beam Welding. *Advanced Engineering Materials*, 2101343.

Prengel, H. G., Pfouts, W. R., & Santhanam, A. T. (1998). State of the art in hard coatings for carbide cutting tools. *Surface and Coatings Technology*, *102*(3), 183–190.

Pritchard, P. (1923). Modern Automobile Foundry Practice. *Proceedings of the Institution of Automobile Engineers*, *18*(1), 381–433.

Puglia, D., Biagiotti, J., & Kenny, J. M. (2005). A review on natural fibre-based composites—Part II: Application of natural reinforcements in composite materials for automotive industry. *Journal of Natural Fibers*, *1*(3), 23–65.

Qian, M., & Froes, F. H. (Eds.). (2015). Titanium powder metallurgy: science, technology and applications.

Rajan, T. S., Sharma, C. P., & Sharma, A. K. (2011). *Heat treatment: principles and techniques*. PHI Learning Pvt. Ltd..

Rajan, T. S., Sharma, C. P., & Sharma, A. K. (2011). *Heat treatment: principles and techniques*. PHI Learning Pvt. Ltd..

Rajput, R. K. (2007). *A textbook of manufacturing technology: Manufacturing processes*. Firewall Media.

Ramakers-van Dorp, E., Blume, C., Haedecke, T., Pata, V., Reith, D., Bruch, O., ...& Hausnerova, B. (2019). Process-dependent structural and deformation properties of extrusion blow molding parts. *Polymer Testing*, *77*, 105903.

Ramakrishnan, P. (2013). Automotive applications of powder metallurgy. In *Advances in powder metallurgy* (pp. 493–519). Woodhead Publishing.

Randhawa, H. (1986). TiN-coated high-speed steel cutting tools. *Journal of Vacuum Science & Technology A: Vacuum, Surfaces, and Films*, *4*(6), 2755–2758.

Rao, P. N. (2001). *Manufacturing Technology, Foundry, Forming, and Welding*. Tata McGraw-Hill Education

Rao, P. N. (2013). *Manufacturing technology: metal cutting and machine tools* (Vol. 2). Tata McGraw-Hill Education.

Rao, R. V. (2008). A decision making methodology for material selection using an improved compromise ranking method. *Materials & Design*, *29*(10), 1949–1954.

Ravi, B. (1997). Intelligent design of gating channels for casting. *Materials Science and Technology*, *13*(9), 785–790.

Ray, B. C., Prusty, R. K., & Nayak, D. (2020). *Phase Transformations and Heat Treatments of Steels*. CRC Press.

Rizzo, A., Goel, S., Luisa Grilli, M., Iglesias, R., Jaworska, L., Lapkovskis, V., ...& Valerini, D. (2020). The critical raw materials in cutting tools for machining applications: A review. *Materials*, *13*(6), 1377.

Rosenthal, S., Maaß, F., Kamaliev, M., Hahn, M., Gies, S., & Tekkaya, A. E. (2020). Lightweight in automotive components by forming technology. *Automotive Innovation*, *3*(3), 195–209.

Ruan, B., Li, G. Q., Chen, Y., Yuan, X., & Li, H. B. (2014). Research on mechanical characteristics of plastic materials in automobile manufacturing. In *Advanced Materials Research* (Vol. 1039, pp. 25–29). Trans Tech Publications Ltd.

Ružbarský, J., Paško, J., & Gašpár, Š. (2014). *Techniques of Die casting*. RAM-Verlag.

Sabroff, A. M., Boulger, F. W., & Henning, H. J. (1968). Forging Materials and Practices.

Saha, D. C., Han, S., Chin, K. G., Choi, I., & Park, Y. D. (2012). Weldability Evaluation and Microstructure Analysis of Resistance-Spot-Welded High-Mn Steel in Automotive Application. *Steel Research International*, *83*(4), 352–357.

Sahoo, M. (2014). *Principles of metal casting*. McGraw-Hill Education.

Saikaew, C., & Wiengwiset, S. (2012). Optimization of molding sand composition for quality improvement of iron castings. *Applied Clay Science*, *67*, 26–31.

Samal, P., & Newkirk, J. (2015). Powder metallurgy methods and applications. *ASM Handbook of Powder Metallurgy*, *7*.

Sands, M., & Shivkumar, S. (2003). Influence of coating thickness and sand fineness on mold filling in the lost foam casting process. *Journal of Materials Science*, *38*(4), 667–673.

Sar, M. H., Ridha, M. H., Husain, I. M., Barrak, O. S., & Hussein, S. K. (2022). Influence of Welding Parameters of Resistance Spot Welding on Joining Aluminum with Copper. *International Journal of Applied Mechanics and Engineering*, *27*(2), 217–225.

Sarma, B., & Ravi Chandran, K. S. (2011). Recent advances in surface hardening of titanium. *Jom*, *63*(2), 85–92.

Satoh, H., & Nakayama, T. (2004). Nonferrous Materials. *Zairyo-to-Kankyo*, *53*(4), 178–184.

Saxena, P., Papanikolaou, M., Pagone, E., Salonitis, K., & Jolly, M. R. (2020). Digital manufacturing for foundries 4.0. In *Light Metals 2020* (pp. 1019–1025). Springer, Cham.

Schulz, H., & Moriwaki, T. (1992). High-speed machining. *Cirp Annals*, *41*(2), 637–643.

Seetharaman, S. (Ed.). (2005). *Fundamentals of metallurgy*. Taylor & Francis.

Sen, A., & Boriwal, L. A Review on Resistance Spot Welding process for Steel Sheets. *Journal homepage:* www. ijrpr. com *ISSN*, *2582*, 7421.

Sen, R., Choudhury, S. P., Kumar, R., & Panda, A. (2018). A comprehensive review on the feasibility study of metal inert gas welding. *Materials Today: Proceedings*, *5*(9), 17792–17801.

Senthilkumar, T., & Ajiboye, T. K. (2012). Effect of heat treatment processes on the mechanical properties of medium carbon steel. *Journal of Minerals & Materials Characterization & Engineering*, *11*(2), 143–152.

Sethuramalingam, P., & Ambigai, R. (2021). The performance analysis of functionally graded material (FGM) based single point cutting tool. *International Journal on Interactive Design and Manufacturing (IJIDeM)*, *15*(2), 259–270.

Shackelford, J. F. (2005). *Introduction to materials science for engineers*. Pearson Education India.

Shafiee, M. R. M., Hashim, M. Y. B., & Said, M. N. B. (2009). Effects of gating design on the mechanical strength of thin section castings. *Proceeding of MUCEET. Pahang: MUCEET*, 1–4.

Sharma, S R (1994) *Forging Technology – As Applied to Automobile Industry.* In: Tailoring of Properties through Mechanical Processing, 23–24, September 1994, NML, Jamshedpur.

Shi, H. (2018). Metal Cutting Theory. *New Perspectives and New Approaches.*

Silberschmidt, V. V. (Ed.). (2020). *Mechanics of materials in modern manufacturing methods and processing techniques*. Elsevier.

Smith, B., Spulber, A., Modi, S., & Fiorelli, T. (2017). Technology roadmaps: Intelligent mobility technology, materials and manufacturing processes, and light duty vehicle propulsion. *Center for Automotive Research, Ann Arbor, MI.*

Song, M. C., Liu, Z., Wang, M. J., Yu, T. M., & Zhao, D. Y. (2007). Research on effects of injection process parameters on the molding process for ultra-thin wall plastic parts. *Journal of Materials Processing Technology*, *187*, 668–671.

Speich, G. R., & Leslie, W. C. (1972). Tempering of steel. *Metallurgical Transactions*, *3*(5), 1043–1054.

Srivastava, D., Chang, I. T. H., & Loretto, M. H. (2001). The effect of process parameters and heat treatment on the microstructure of direct laser fabricated TiAl alloy samples. *Intermetallics*, *9*(12), 1003–1013.

Staněk, M., Maňas, D., Maňas, M., & Šuba, O. (2011). Optimization of injection molding process. *International Journal of Mathematics and Computers in Simulation.*

Stephenson, D. A., & Agapiou, J. S. (2018). *Metal cutting theory and practice*. CRC Press.

Stojanović, B., &Ivanović, L. (2015). Application of aluminium hybrid composites in automotive industry. *Tehničkivjesnik*, *22*(1), 247–251.

Stroetmann, R., Kästner, T., Rust, B., & Schmidt, J. (2022). Welded connections at high-strength steel hollow section joints. *Steel Construction*, *15*(S1), 10–21.

Subramonian, S., Altan, T., Ciocirlan, B., & Campbell, C. (2013). Optimum selection of variable punch-die clearance to improve tool life in blanking non-symmetric shapes. *International Journal of Machine Tools and Manufacture*, *75*, 63–71.

Suwanpreecha, C., & Manonukul, A. (2022). A review on material extrusion additive manufacturing of metal and how it compares with metal injection moulding. *Metals*, *12*(3), 429.

Syrcos, G. P. (2003). Die casting process optimization using Taguchi methods. *Journal of Materials Processing Technology*, *135*(1), 68–74.

Taga, Y. (1997). Recent progress in coating technology for surface modification of automotive glass. *Journal of Non-crystalline Solids*, *218*, 335–341.

Tarng, Y. S., Tsai, H. L., & Yeh, S. S. (1999). Modeling, optimization and classification of weld quality in tungsten inert gas welding. *International Journal of Machine Tools and Manufacture*, *39*(9), 1427–1438.

Taub, A. I. (2006). Automotive materials: technology trends and challenges in the 21st century. *MRS bulletin*, *31*(4), 336–343..

Thirugnanam, M. (2013). Modern high pressure die-casting processes for aluminium castings. In *Transaction of Indian Foundry Congress* (Vol. 61, pp. 1–7).

Thomas, W. M., Nicholas, E. D., Watts, E. R., & Staines, D. G. (2002). Friction based welding technology for aluminium. In *Materials Science Forum* (Vol. 396, pp. 1543–1548). Trans Tech Publications Ltd.

Tillmann, W., & Vogli, E. (2006). Selecting surface-treatment technologies. *Modern Surface Technology*, 1–10.

Timings, R. L. (2009). *Manufacturing technology volume Ii, 2/E.* Pearson Education India.

Todor, M. P., & Kiss, I. (2016). Systematic approach on materials selection in the automotive industry for making vehicles lighter, safer and more fuel-efficient. *Applied Engineering Letters*, *1*(4), 91–97.

Totten, G. E. (2006). *Steel heat treatment handbook-2 volume set.* CRC Press.

Totten, G. E. (Ed.). (2006). *Steel heat treatment: equipment and process design.* CRC Press.

Totten, G. E., & MacKenzie, D. S. (Eds.). (2003). *Handbook of aluminum: vol. 1: physical metallurgy and processes.* CRC Press.

Tounsi, N., Vincenti, J., Otho, A., & Elbestawi, M. A. (2002). From the basic mechanics of orthogonal metal cutting toward the identification of the constitutive equation. *International Journal of Machine Tools and Manufacture*, *42*(12), 1373–1383.

Trzepieciński, T. (2020). Recent developments and trends in sheet metal forming. *Metals*, *10*(6), 779.

Tschätsch, H. (2010). *Applied machining technology.* Springer Science & Business Media.

Tschätsch, H. (2010). *Applied machining technology.* Springer Science & Business Media.

Tsukerman, S. A. (2013). *Powder metallurgy.* Elsevier.

Tsutsui, T. (2012). Recent technology of powder metallurgy and applications. *Hitachi Powdered Metals Technical Report*, *54*, 12–20.

Tzafestas, S. G. (Ed.). (2012). *Advances in manufacturing: Decision, control and information technology.* Springer Science & Business Media.

Upadhyaya, G. S. (1997). *Powder metallurgy technology.* Cambridge Int Science Publishing.

Urbański, T., & Taczała, M. (2022). Prediction of the welding distortions of butt welded joints using total moments method based on equivalent loads. *Journal of Manufacturing Processes*, *75*, 1039–1057.

Vetter, J. (2015). Surface Treatments for Automotive Applications. In *Coating Technology for Vehicle Applications* (pp. 91–132). Springer, Cham.

Vetter, J., Barbezat, G., Crummenauer, J., & Avissar, J. (2005). Surface treatment selections for automotive applications. *Surface and Coatings Technology*, *200*(5–6), 1962–1968.

Vignesh, S. (2017). A review of advanced casting techniques. *Research Journal of Engineering and Technology*, *8*(4), 440–446.

Vingas, G. J., & Zrimsek, A. H. (1964). Thermal stability of bentonites in foundry molding sand. *Clays and Clay Minerals*, *13*(1), 367.

Vyshakh, M., Krishnaraj, R., Sayooj, A. P., & Afzal, M. (2014). Experimental investigation on aluminium gravity die casting. *International Journal of Applied Environmental Sciences*, *9*(2), 213–222.

Wagoner, R. H., & Chenot, J. L. (2001). *Metal forming analysis.* Cambridge University Press.

Wang, H., Fang, Z. Z., & Sun, P. (2010). A critical review of mechanical properties of powder metallurgy titanium. *International Journal of Powder Metallurgy*, *46*(5), 45–57.

Wang, J., Chen, X., Yang, L., & Zhang, G. (2022). A strength estimation method for aluminum alloy butt welded joints with mismatched mechanical properties. *Thin-Walled Structures*, *179*, 109585.

Wang, L., & Speer, J. G. (2013). Quenching and partitioning steel heat treatment. *Metallography, Microstructure, and Analysis*, *2*(4), 268–281.

Wang, P. Y., Wang, Z. J., Xiang, N., & Li, Z. X. (2020). Investigation on changing loading path in sheet metal forming by applying a property-adjustable flexible-die. *Journal of Manufacturing Processes*, *53*, 364–375.

Wang, R. R., & Welsch, G. E. (1995). Joining titanium materials with tungsten inert gas welding, laser welding, and infrared brazing. *The Journal of Prosthetic Dentistry*, *74*(5), 521–530.

Weiler, J. P. (2021). Exploring the concept of castability in magnesium die-casting alloys. *Journal of Magnesium and Alloys*, *9*(1), 102–111.

Welbourn, D. B. (1984). Computer aided engineering in the foundry industry. *Materials & design*, *5*(1), 7–14.

Wickramarachchi, C. (2019). *Automated Testing of Advanced Cutting Tool Materials* (Doctoral dissertation, University of Sheffield).

Wilhelm, M. (1993). Materials used in automobile manufacture-current state and perspectives. *Le Journal de Physique IV*, *3*(C7), C7–31.

Woon, Y. K., & Lee, K. S. (2004). Development of a die design system for die casting. *The International Journal of Advanced Manufacturing Technology*, *23*(5), 399–411.

Wright, P. K., & Trent, E. M. (1974). Metallurgical appraisal of wear mechanisms and processes on high-speed-steel cutting tools. *Metals Technology*, *1*(1), 13–23.

Wright, P. K., & Trent, E. M. (2000). *Metal cutting*. Elsevier.

Wu, C. S., Wang, L., Ren, W. J., & Zhang, X. Y. (2014). Plasma arc welding: Process, sensing, control and modeling. *Journal of Manufacturing Processes*, *16*(1), 74–85.

Wu, S. H., Fuh, J. Y., & Lee, K. S. (2007). Semi-automated parametric design of gating systems for die-casting die. *Computers & Industrial Engineering*, *53*(2), 222–232.

Wu, S. H., Lee, K. S., & Fuh, J. Y. H. (2002). Feature-based parametric design of a gating system for a die-casting die. *The International Journal of Advanced Manufacturing Technology*, *19*(11), 821–829.

www.forging.org/forging/why-are-forgings-so-prevalent.html

www.ispatguru.com/forging-quality-steels/

Xia, J., & Jin, H. (2017). Numerical study of welding simulation and residual stress on butt welding of dissimilar thickness of austenitic stainless steel. *The International Journal of Advanced Manufacturing Technology*, *91*(1), 227–235.

Xiao, G., Ren, M., & To, S. (2018). A study of mechanics in brittle–ductile cutting mode transition. *Micromachines*, *9*(2), 49.

Yadav, P., & Saxena, K. K. (2020). Effect of heat-treatment on microstructure and mechanical properties of Ti alloys: An overview. *Materials Today: Proceedings*, *26*, 2546–2557.

Yaghoubi, M., & Tavakoli, H. (2022). Welded Joints. In *Mechanical Design of Machine Elements by Graphical Methods* (pp. 21–34). Springer, Cham.

Yan, S., Menary, G., & Nixon, J. (2017). A novel methodology to characterize the constitutive behaviour of polyethylene terephthalate for the stretch blow moulding process. *Mechanics of Materials*, *104*, 93–106.

Yang, X., Wang, B., Zhou, J., Dang, L., Xiao, W., & Wang, Y. (2020). Influence of process parameters on deep drawing of 2060 Al–Li alloy under hot stamping process. *International Journal of Lightweight Materials and Manufacture*, *3*(1), 36–42.

Youssef, H. A., & El-Hofy, H. (2008). *Machining technology: Machine tools and operations*. CRC Press.

Youssef, H., & El-Hofy, H. (2020). *Non-Traditional and Advanced Machining Technologies*. CRC Press.

Yu, C. J., Eifert, H. H., Banhart, J., & Baumeister, J. (1998). Metal foaming by a powder metallurgy method: Production, properties and applications. *Materials Research Innovations*, *2*(3), 181–188.

Żenkiewicz, M., Richert, J., & Różański, A. (2010). Effect of blow moulding ratio on barrier properties of polylactidenanocomposite films. *Polymer Testing*, *29*(2), 251–257.

Zhang, C., & Choi, H. (2021). Study of segmented chip formation in cutting of high-strength lightweight alloys. *The International Journal of Advanced Manufacturing Technology*, *112*(9), 2683–2703.

Zhang, S. H., Wang, Z. R., Wang, Z. T., Xu, Y., & Chen, K. B. (2004). Some new features in the development of metal forming technology. *Journal of Materials Processing Technology*, *151*(1-3), 39–47.

Zhangiang, L., & Xing, A. (2005). Cutting tool materials for high speed machining. *Progress in Natural Science*, *15*(9), 777–783.

Zhao, Q., Yang, F., Torrens, R., & Bolzoni, L. (2020). Allotropic phase transformation and high-temperature tensile deformation behaviour of powder metallurgy Ti-5553 alloy. *International Journal of Refractory Metals and Hard Materials*, *90*, 105235.

Zhu, Z., Zhao, X., & Liu, Y. (2022). A Novel Method for Online Control in Equipment Parameters of Blow Molding Process. In *Smart Communications, Intelligent Algorithms and Interactive Methods* (pp. 53–62). Springer, Singapore.

Zhuang, K., Fu, C., Weng, J., & Hu, C. (2021). Cutting edge microgeometries in metal cutting: A review. *The International Journal of Advanced Manufacturing Technology*, *116*(7), 2045–2092.

Zolotorevsky, V. S., Belov, N. A., & Glazoff, M. V. (2007). *Casting aluminum alloys* (Vol. 12). Amsterdam: Elsevier.

Zong, R., Chen, J., Wu, C. S., & Chen, M. A. (2016). Undercutting formation mechanism in gas metal arc welding. *Welding Journal*, *95*(5), 174–184.

Index

E

F

G

H

I

L

M

N

O

P

Q

R

S

T